Troubleshooting and Repairing
Solid-State TVs

Troubleshooting and Repairing Solid-State TVs

3rd edition

Homer L. Davidson

TAB Books

Imprint of McGraw-Hill

New York   San Francisco   Washington, D.C.   Auckland   Bogotá   Caracas   Lisbon   London
Madrid   Mexico City   Milan   Montreal   New Delhi   San Juan   Singapore   Sydney   Tokyo   Toronto

McGraw-Hill

*A Division of The **McGraw·Hill** Companies*

©1996 by The McGraw-Hill Companies, Inc.
Published by TAB Books, an imprint of McGraw-Hill

pbk   1 2 3 4 5 6 7 8 9   FGR/FGR   9 0 0 9 8 7 6
hc   1 2 3 4 5 6 7 8 9   FGR/FGR   9 0 0 9 8 7 6

Library of Congress Cataloging-in-Publication Data

ISBN 0-07-015754-5 pbk.
ISBN 0-07-015753-7 hb.

Acquisitions editor: Roland S. Phelps
Editorial team: Robert E. Ostrander, Executive Editor
          Aaron Bittner, Book Editor
          Jodi L. Tyler, Indexer
Production team: Katherine G. Brown, Director
          Lisa M. Mellott, Coding
          Rhonda E. Baker, Coding
          Rose McFarland, Desktop Operator
          Brenda S. Wilhide, Computer Artist
          Janice Ridenour, Computer Artist
Design team: Jaclyn J. Boone, Designer
          Katherine Lukaszewicz, Associate Designer

EL3
0157545

Contents

3 Servicing the horizontal sweep circuits *103*

8 Tuner problems and cures *355*

9 Troubleshooting the color circuits *401*

13 Troubleshooting picture tube circuits *535*

14 Servicing the black-and-white and small-screen chassis *571*

Practical
commonsense servicing

TELEVISION SERVICING HAS BEEN WITH US SINCE ABOUT 1950. At first, the vacuum-tube chassis developed many service problems, many of which were new for the electronics technician. Each problem was met, challenged, and conquered.

Then a new era of electronics servicing developed to deal with transistors in the hybrid chassis. It seems like only yesterday that tubes disappeared from the service scene. Besides the transistor, IC components have been placed in the TV set. Today, many ICs are found (along with transistors) in the solid-state chassis (Fig. 1-1). Next, perhaps, the solid-state chassis will be taken over by only a few IC components.

Many new (and some variations of old) circuits have crept into the present TV receivers. For example, many remote-control receivers are being sold today (Fig. 1-2). Quiet, foolproof, portable transmitters have appeared with the infrared remote chassis. New circuits include the surface acoustic wave (SAW) filter device, which establishes the proper IF response in the IF stages. The comb filter circuit separates luminance and chroma video information so that high-frequency luminance detail can be seen without cross-color information. The integrated horizontal output transformer has replaced the old output tube flyback and high-voltage rectifier tube. The color picture tube is getting squarer, bigger, and flatter. Now there is even projection TV.

Today, picture tube sizes have greatly increased. Panasonic has a 27" direct view (CT27SF11), JVC has a 31" size (AV31BM5), Sony has a 32" (KV32S10), RCA has a 35" screen (F35672M8), and Mitsubishi has a unit with a 40" direct-view screen (CS40503). Projection TVs have increased in size, as well as in brightness and picture quality. An example can be found in Pioneer's 45" screen (SDP4573), Zenith's 46" size (PV4660), Mitsubishi's 50" (VS5051), Magnavox's 52" screen

■ **1-1** *Today, all the devices in a TV chassis are solid-state except the picture tube.*

■ **1-2** *Most modern color TV sets, including small portable sets, are operated with a remote control unit.*

(F15330W), Sony's 53" (KP53S55), and RCA's 60" (P60752) projection TVs.

Solid-state TV servicing is here to stay. This is not a heavy-theory electronics book. It is a book filled with practical methods of servicing. You will find in the following pages a collection of 38 years of electronics servicing information about the TV chassis. Servicing the solid-state TV chassis can be quite rewarding, profitable, and just plain fun.

Time is money

Time is one of the electronics technician's greatest tools. You can waste it, or make money with it. The time spent servicing the "tough dog" or intermittent problem will determine if the repair job is profitable. Anytime the technician spends over 1 hour on a given electronics problem without locating the defective component is lost time. Lost time can also occur when a coffee break extends from 15 minutes to 30 minutes, or even longer.

Callbacks or repeated repairs cost the electronics technician extra money. Doing a thorough repair job at the beginning eliminates repeated calls. Of course, I must admit the TV chassis does produce a lot of service problems that can happen after a repair. Always charge for repeated service when the original repair has nothing to do with the present problem. Remember, the doctor or auto mechanic charges for the additional call.

The best things that can be said to a young person entering the electronics business are:

☐ Do the best repair job you possibly can.

☐ Do not work for nothing—make a fair charge.

☐ Always collect the bill before the repaired device leaves your shop.

By doing these three things, you will become a successful businessperson.

Basic test equipment

Back in the vacuum tube days, many TV sets were repaired with a tube tester, a VTVM or VOM, and an oscilloscope. Today a digital multimeter (DMM), oscilloscope, CRT checker, color-dot-bar generator, and tube test jig are the basic test equipment. Critical voltage and resistance measurements can be made with the DMM. The DMM can even check diodes and transistors in or out of the circuit. The oscilloscope is required to look at waveforms in the various circuits. Most waveforms are checked with the color-dot-bar generator connected to the antenna terminals to provide a signal. A tube test jig is the ideal monitor when servicing the pulled chassis (Fig. 1-3). A good CRT tester eliminates guessing about the status of the picture tube.

■ **1-3** *All portable TVs can be serviced by pulling the chassis back with cables extended.*

As new circuits arrive and old test equipment is replaced, an older oscilloscope should be replaced with a dual-trace 35- or 60-MHz unit. An isolation variable line transformer is a *must* item when servicing the new integrated flyback chassis. A capacitance meter and CRT restorer-analyzer are handy test instruments to have around. A frequency counter and sweep-marker generator are required for TV alignment purposes. To check high voltage at the CRT, select a good high-voltage probe.

With the latest portables and console TVs, the chassis can be loosened up, slid backwards, or tilted on edge or side for easy servicing. Often, the connecting lead wires and cables from the TV chassis to other components mounted within the plastic portable case can be untied and extended so the chassis can be slid backwards or actually removed from the cabinet. Likewise, remove chassis bolts and screws so the small TV chassis inside the large TV console can be moved to test most parts within the wooden cabinet. Check the list that follows for required basic TV test instruments.

You might find the busy electronics technician uses only three or four test instruments in his daily service routine. There are many test instruments collecting dust on the service bench and not being used. Knowing how the basic test equipment operates and how to use them is the secret to quick, efficient, and practical electronics servicing. One quick way to learn how the test instruments operate is to attend the classes sponsored by many test instrument distributors and manufacturers.

Basic test equipment

1. Digital multimeter (DMM)
2. VOM, VTVM, or FET voltmeter
3. Oscilloscope (at least 40 MHZ)
4. Variable isolation transformer
5. CRT tester and rejuvenator
6. Diode, SCR, semiconductor tester
7. Capacitance tester
8. High-voltage probe
9. Color-dot-bar generator

Schematics and service literature

Although many TV technicians might service a TV chassis without a schematic, it is not practical over the long run. You can lose a lot of valuable service time without a good TV schematic (Fig. 1-4). Several years back, the Japanese TV chassis was difficult to service because wiring diagrams and parts were difficult to obtain.

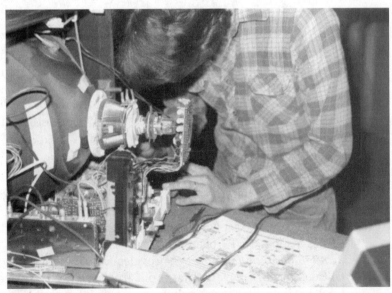

■ **1-4** *The TV schematic is required when servicing a difficult or "tough dog" chassis.*

This is not so today, as most foreign imports are covered by schematics and service literature. *Howard Sams Photofacts* covers most manufactured TV chassis. Manufacturers Profax schematics are found in the pages of *Electronic Servicing & Technology* magazine.

Service literature, including detailed diagrams, is published by each electronics manufacturer. Here, additional data as to how each stage operates might clear up the tough job. This service literature can be purchased yearly or for each individual TV chassis. You will also receive important production and modification information for each chassis. Many TV set manufacturers hold service clinics several times a year. Take a day off and attend these meetings, because they always provide crucial servicing data. Besides, you might talk to a fellow electronics technician who has just licked the same service problem that has you stumped. A day away from the shop might bring future rewards.

Howard Sams Photofacts

Although *Sams Photofacts* do not cover every TV chassis manufactured, they do carry the major models. When possible, secure the exact schematic from the manufacturer. *Howard Sams Photofacts* provide more data than just a wiring diagram (Fig. 1-5). Critical voltage measurements and exact waveforms are found for each transistor, IC, or microprocessor. Some manufacturers' schematics do not have critical voltage measurements.

Besides these features, the *Photofacts* have photos and charts to locate the exact part you diagnosed as causing the service problem. Most chassis photos will have a list of capacitors, resistors, semiconductors, and inductances listed on separate illustrations. Complete part lists with universal semiconductors are listed beside the original component. For quicker, more efficient TV servicing, choose *Howard Sams Photofacts*.

Electronic Servicing and Technology

You will find a different schematic of TV chassis and VCRs in each monthly issue of the *Electronic Servicing* magazine. Usually, these are exact replicas of recent manufacturer schematics. Besides a separate TV chassis in every issue, *Electronic Servicing and Technology Magazine* provides a 13th month issue of TVs, VCRs, and CD players. The profit from one TV repair can easily pay for the whole year's schematics, plus information on the critical servicing problems of other electronic technicians.

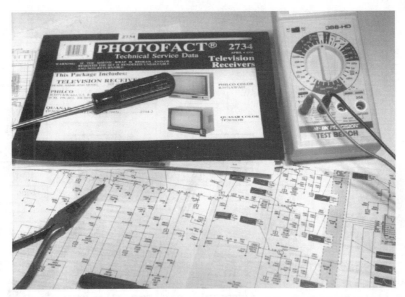

■ **1-5** *Most TV chassis are covered by* Howard Sams Photofacts.

The block diagram

Along with the TV receiver schematic, the block diagram helps to locate the stage or section in which the trouble might occur. The functional block diagram ties the various circuits together. Often, the block diagram is located in the front of the service manual. Always have the TV schematic handy for easy troubleshooting. Doing any servicing without the schematic diagram is quite difficult in today's TV chassis.

RCA CTC130 block diagram

In the RCA CTC130 chassis, the tuner control section controls the frequency synthesis (FS) tuning system. The IF signal is fed to a SAW filter network with the video signal going to the chroma/luminance IC. The color output feeds the color drivers and color picture tube (Fig. 1-6).

The low-voltage power supply consists of a variable-frequency switching power supply (VIPUR). A hot and cold ground is required with this type of operation. The prime reason for using the variable frequency switching power supply is to provide "hot-to-cold" ground insulation between the primary TV power supply ("hot" ground) and the video/audio input/output circuit (with a "cold" ground).

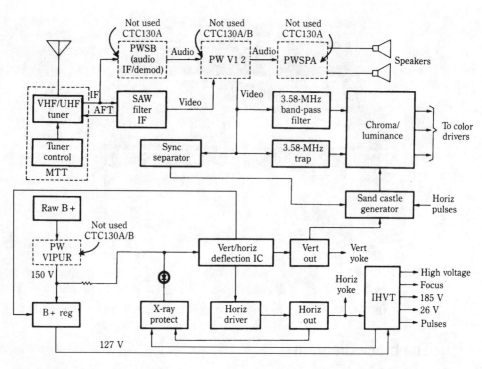

■ 1-6 *Check the block diagram to locate the possible troubled section with screen symptoms. This RCA CTC130 chassis contains a frequency synthesis (FS) tuning system.*

The 150-V source feeds to the vertical/horizontal deflection IC. The output from the B+ regulator (127 V) feeds the horizontal driver and output transistors. The flyback transformer (IHVT) supplies high voltage and focus voltage to the CRT. Besides two low-voltage sources, flyback pulses are provided at the horizontal output transformer. Both vertical and horizontal drive output sources are fed to the respective windings of the deflection yoke.

A vertical and horizontal output pulse and sync pulse are applied to the sand castle generator. The sand castle waveform is developed by combining the horizontal blanking, vertical blanking, and burst keying pulses and feeding them to the chroma/luminance IC. The sand castle generator eliminates cross or color patterns in the picture.

Sylvania C9 chassis block diagram

The incoming TV RF signal is applied to the RF switches, which are used in selecting the RF input desired. The RF signal is then sent to the tuner, where it is converted to a 45.75-MHz IF signal. The IF signal is coupled to the IF preamp (Q240) where it is amplified and applied to SAW filters Y200 and Y201. SAW filter Y200 feeds the IF signal to the video/IF sync IC (IC201) for video IF processing. The automatic gain control (AGC) and automatic fine tuning (AFT) signals are developed within IC201 and sent to the tuning system for RF amp gain control and oscillator frequency correction (Fig. 1-7).

The IF sound signal processing is performed by coupling the IF signal via SAW filter Y201 to the sound IF (IC202). The 4.5-MHz IF sound signal is detected within IC202. Base band audio (50 Hz to 100 kHz) is output from pin 5 and applied to audio buffer Q247. The audio signal is sent to the stereo decoder module for processing. The stereo decoder module has its own audio output stages.

IC201 detects the incoming 45.75-MHz IF video signal producing the composite video that is sent from P17 and passed through a 4.5-MHz trap (Y202) to remove any 45.75-MHz sound products. This signal is coupled by the video buffer (Q244) to the chroma/luminance IC (IC640) through a comb filter circuit.

The RGB signals are developed within IC640 to reproduce the original color signals. The chroma/luminance IC receives tint, brightness, picture, and color control via IC312, a digital-to-analog converter (DAC). The DAC receives its data in serial form from the tuning system microcomputer. The DAC also sends control signals to the stereo decoder for volume, treble, bass, and balance.

The CRT board receives the RGB signals from the chroma/luminance IC via RGB buffer transistors Q643, Q641, and Q642. The RGB buffer transistors also allow input of the on-screen graphics from the tuning system to the CRT. When the tuning system sends on-screen information to the CRT, the character blanking transistor (Q640) is turned on by the fast blanking input to eliminate the video drive from the chroma/luminance IC. The on-screen graphics appear with a black background. This background can appear in other colors, depending on the output from the microcomputer. The CRT board amplifies the signals to drive the picture tube. Also, IC201 develops the horizontal and vertical drive signals and outputs these signals via pins 26 and 3. Synchronization is obtained by taking some of the composite video signal from the output of the video buffer (Q244) and returning it to the sync input at pin 25.

The vertical signal is applied to pin 6 of the vertical output (IC580). The vertical signal is amplified to drive the vertical windings of the deflection yoke. The return path is applied to pin 4 for feedback and linear control.

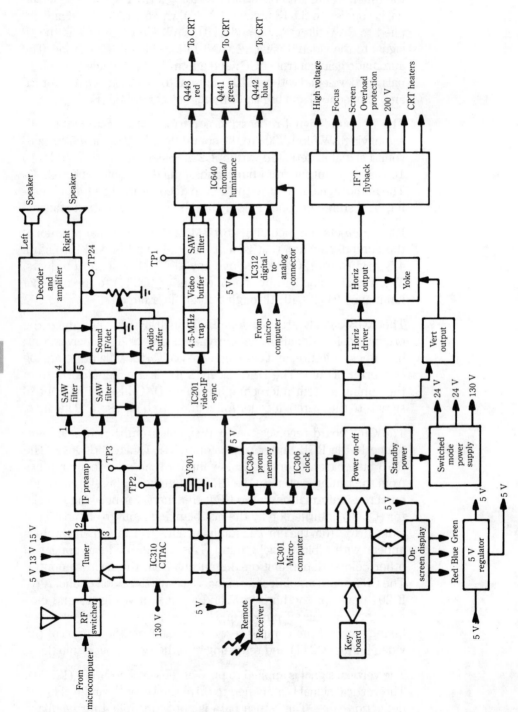

■ 1-7 A block diagram of a Sylvania C9 TV chassis.

The horizontal drive signal is sent to the horizontal driver (Q500) where it is amplified to drive the horizontal output transistor (Q501). The horizontal signal drives the horizontal windings of the deflection yoke and the integrated flyback transformer (IFT). The IFT provides the high voltage, focus voltage, screen voltage (G2), filament voltage, and 200 V. The focus and screen voltage are adjusted by manual controls mounted on the IFT.

The switched-mode power supply develops the remaining dc voltages for the chassis. These sources are the 13-V, 24-V, 24/27-V (in 20-, 26-, and 27-inch chassis), and 130-V supplies. Other voltages, such as 5 V, 10.5 V, and 11 V, are derived from the basic voltages.

Read, read, and study some more

With the electronics service business changing from year to year, TV technicians must keep abreast of each new development. Besides attending the manufacturers' service meetings, they must read books and magazines on TV servicing. Howard W. Sams, Parker, Hayden, and TAB Books, (an imprint of McGraw-Hill), cover many subjects related to solid-state TV servicing. Subscribe to *Electronic Servicing & Technology* and *Electronics Now* magazines for additional electronics servicing information.

Set aside one hour a day for study. This hour a day can keep you abreast of new developments and provide more efficient use of your time in TV servicing. Fifty years ago, an old-time motor rewinder would study one hour each day. He would go to work at 7 a.m. and study from 7 to 8 a.m. Motors were brought to him from 100 miles around, he turned down more work than he could handle, and he was successful. He was efficient, furthered his knowledge about motors, and made money.

Sight, sound, and smell

The three *S*s might be your most important tool. Sight, sound, and smell solve a lot of TV repair problems. You can *see* a burned resistor, fried flyback transformer, or lightning damage. A leaky electrolytic capacitor might have a black or white substance oozing out at the bottom connections (Fig. 1-8). Poor board connections might indicate an overheated solder connection. Cracked or overheated connections of a 15-W or 20-W resistor might indicate possible trouble. Definitely, you can see a spark gap or tripler unit arc over. Above all, your eyes identify the trouble symptoms from the front of the picture tube.

■ **1-8** *Locate the possible problem with your eyes (sight), ears (sound), and nose (smell).*

Insufficient or distorted audio *sound* might be traced to the sound stages. Arcover at the picture tube or high-voltage transformer might be heard. The ear might pick up the tic-tic sound of the flyback transformer with high-voltage or chassis shutdown. Some TV technicians can hear the 15,750-Hz sound of the horizontal output transformer indicating the horizontal oscillator stages are performing. Noisy or improper horizontal sweep might be heard from the flyback transformer.

You might *smell* an overheated voltage-dropping resistor or degaussing thermistor. Ozone smell might be traced to the flyback transformer, tripler unit, screen-focus assembly, focus controls, or the anode connection of the picture tube. Not only can you smell an overheated transistor or diode, but you can feel them. Overheated components in the TV chassis are always a source of trouble.

Chassis comparison

Many times the same TV chassis might be found in more than one TV set awaiting repair. This is especially true if you are servicing chain store merchandise. It's possible to have trouble with a new TV right out of the box. The new TV chassis might also be compared with those in the showroom if you carry that certain brand of TV.

If, by chance, you are stumped on a new circuit or do not have a schematic handy, comparison voltage and resistance measurements of the working chassis can help solve the problem (Fig. 1-9). Many chain store TVs are manufactured for them by another source, and many different TV brands are actually the same. Of course, comparison checks take a lot of time and should only be made when you have no other choice.

■ **1-9** *Compare voltage and scope waveforms by comparing them with those from another TV of the same make.*

"Tough dog"

Any TV malfunction that takes over 3 hours to locate might be called a "tough dog" by the professional TV technician. Sometimes a repair might be called tough by one technician while the next technician might locate it at once. Always set the chassis aside after working on it for an hour or so. Time is lost if you keep working at it. Confusion and the "I don't know" syndrome sets in. Try tackling it the first thing in the morning when your mind is clear.

Ask for help if you cannot locate the defective component after several attempts. Go to the TV distributor for help. Usually, they have service personnel available. If not, contact the factory service representative or national service manager. A simple telephone call and 5 minutes of your time might save several hours of frustration. Check with a local technician who services that brand of

TV. That's why it pays to be friendly with your competitor. Most people connected with electronics servicing are a bunch of nice people just waiting to be helpful.

The intermittent

The intermittent chassis, in which the malfunction comes and goes, is the most difficult service problem within the TV chassis. They seem to come in bunches. In fact, in one week you might see all of them for the whole month. Sometimes the intermittent problem can be caused in the home. A defective antenna system or ac power problem can cause the intermittent. Intermittents take a lot of time and hours of thinking power. The intermittent can be compounded by more than one defective component or problem in the chassis.

The intermittent transistor or IC component might be located with a can of freeze spray or heat application (Fig. 1-10). Approach the intermittent slowly. Be careful in handling the chassis. Do not hammer on the TV chassis with a screwdriver handle. You might disturb the intermittent and it might not act up again for weeks. Sometimes a loose or cracked board connection can be located with coolant or by moving and prying on sections of the PC board. First try to isolate the intermittent to a certain section of the chassis.

■ **1-10** *A can of coolant and heat spray can locate that tough intermittent TV problem.*

The touchy intermittent might be caused by a cracked board or poorly soldered connection. After isolating the intermittent to one section of the board, carefully check the etched wiring with a light and a magnifying glass. Sometimes you can find large cracks in the PC wiring or cracked component connections. The intermittent connection might be under a large blob of solder. A poorly tinned component lead might cause an intermittent soldered connection. Touch up all soldered connections and feed-through eyelets in that area of the board.

Sometimes intermittents might take a couple of hours or days to act up. Monitor the chassis with critical voltage tests and scope waveforms. You might have to return the TV to the customer's home if the chassis will not act up after a week or so. Have them call the next time it does. Call the manufacturer's service manager or distributor for help. The same intermittent might have occurred with some other TV technician and they might have the same trouble symptom in their files. Often, intermittent problems are not easy, they eat up valuable service time, and they cost you in the end.

The filter capacitor

Filter capacitors have caused many strange symptoms in the TV chassis (Fig. 1-11). A leaky electrolytic capacitor might have a white or black substance oozing out of the connection terminals. The defective electrolytic capacitor might dry up or become leaky. Some capacitor terminals will break just inside the connecting ter-

Filter capacitor

■ **1-11** *Bad filter capacitors can produce funny and universal problems within the TV chassis.*

minals, resulting in an open capacitor. The outside metal case of a filter capacitor should never run warm. You might save a lot of valuable service time by going directly to the suspected filter capacitor.

The open or dried-up filter capacitor can cause a hum in the sound and hum bars in the raster. These bars can slowly creep up the raster. You might hear the on-off relay chatter with a defective filter capacitor. Low-voltage sources can result from a bad capacitor.

Poor AGC and sync problems might relate to a defective electrolytic capacitor. Small electrolytic capacitors in the AGC circuits can produce vertical bars in the raster. Vertically oscillating white lines in the raster can be caused by filter capacitors in the horizontal circuits. Some TV technicians automatically shunt each electrolytic capacitor when they have an unusual TV symptom.

The defective filter capacitor can be located by shunting a new one across the suspected one. Always clip the capacitor in place, or tack it across the PC board with the power *off*. Shunting a capacitor with the power *on* might damage transistors and IC components in the solid-state chassis. Replace the entire multipart capacitor when one part is found to be defective. Often, each capacitor section will rapidly fail if it is left in the circuit. Never tie single electrolytic capacitors across individual sections; replace the whole unit. A low-priced digital capacitance meter can help locate the defective electrolytic capacitor (Fig. 1-12).

■ **1-12** *Check those suspected filter or electrolytic capacitors with a capacity meter.*

Coke, Pepsi, or water

Besides being a big cleaning mess, the chassis might not be repairable when Coke, Pepsi, or any other liquid has been spilled inside the back of the TV. It is very difficult to clean up small resistors and capacitors laying flat against the PC board. Often, the chassis is turned on, causing excessive arcing in the horizontal and high-voltage circuits.

Sometimes when a voltage arc between components occurs, the arc breaks down the PC board and can cause burning of the board. The board can be saved by cutting out the burned area. Use a pocket knife or drill out the burned area, then replace burned components across the area. Place hookup wire around burned-out PC wiring connections. The wiring schematic and component replacement charts can come in handy.

When a soft drink or liquor is spilled in the chassis, the PC board and components become sticky and can be difficult to repair. Place a fan near the chassis to dry out areas when water has been spilled inside the TV chassis. After cleaning and repairing the board, spray the entire area with clear high-voltage insulation spray after the chassis operates for several hours. The chassis should then operate for several days to make sure no further breakdowns occur. Sometimes you will find that the TV cannot be repaired and must be replaced.

Checking the solid-state devices

Solid-state devices such as transistors, diodes, and linear ICs require the following diagnostic procedures.

Transistors

Besides using the voltage and resistance methods, the transistor can be checked in or out of the circuit with any of the many different transistor testers on the market. The suspected transistor can be checked out of the circuit with resistance measurements of the VOM or DMM (Fig. 1-13). Leaky or open transistors can be located with the ohmmeter scale.

A quick method to check a transistor in or out of the circuit is with the diode or transistor test of the digital multimeter (DMM) (Fig. 1-14). Comparable resistance measurements of the diode junction from base to collector and base to emitter will identify an npn or pnp transistor, and will indicate if a leakage or open junction oc-

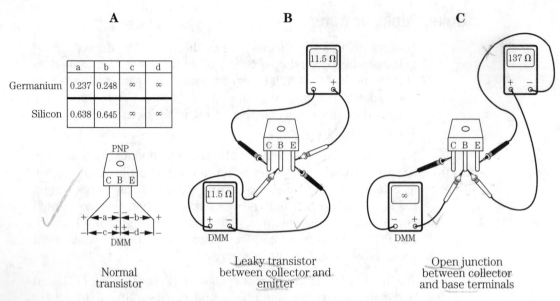

	a	b	c	d
Germanium	0.237	0.248	∞	∞
Silicon	0.638	0.645	∞	∞

Normal
transistor

Leaky transistor
between collector and
emitter

Open junction
between collector
and base terminals

■ **1-13** *Transistors can be checked with the ohmmeter or diode scale of a digital multimeter (DMM).*

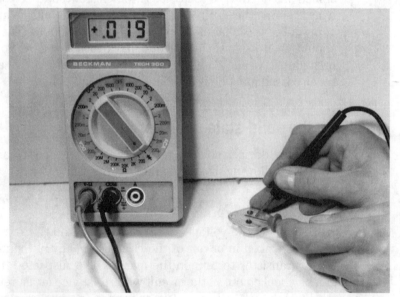

■ **1-14** *Check those transistors with the diode-transistor scale of the DMM.*

curs between the elements. The leaky transistor resistance might be low between collector and emitter or base and emitter in both directions (Fig. 1-15). Most transistors become leaky between the emitter and collector terminals.

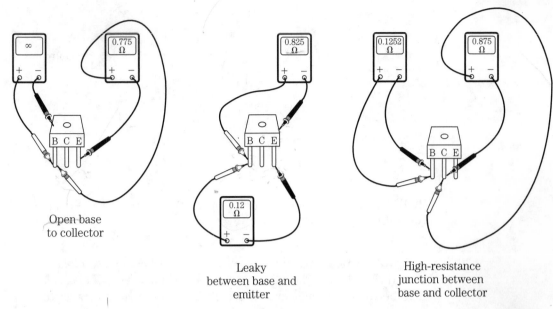

1-15 *The base terminal is common to both the collector and emitter when testing a transistor. Check transistors in or out of the circuit.*

The transistor might be open between the base and collector or base and emitter terminals, with no measurement on the DMM. A transistor with a high-resistance joint can cause a weak or dead response when the leakage is greater with one set of transistor elements than the other set. A high-resistance joint exists when the measurement is several hundred ohms different between two elements and not the other two. Transistors can be quickly checked with the DMM diode or transistor tests.

Transistor voltage tests

Taking crucial voltage measurements on the transistor elements can determine if the transistor is normal. Suspect an open transistor when the collector voltage is much higher than normal, with no voltage on the emitter terminal. An open-emitter resistor can give the same voltage readings (Fig. 1-16).

Practically the same voltage measurements on all three terminals can indicate a leaky or shorted transistor. The collector and emitter terminals become leaky in most transistors. The voltage measurement on both terminals can be quite close with a direct collector-to-emitter short. Discard transistors with any signs of leakage between any of the terminals.

Checking the solid-state devices

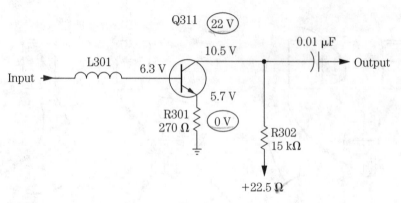

■ 1-16 *An open emitter resistor or emitter element can show a higher collector voltage than normal.*

The transistor is usually good when normal base-emitter bias is found (Fig. 1-17). The silicon transistor will have a 0.6-V bias voltage between the emitter and base, while a germanium transistor will have a bias voltage of 0.3 V between these same elements. Of course, with an improper or no voltage source, very little voltage is found on any transistor terminals.

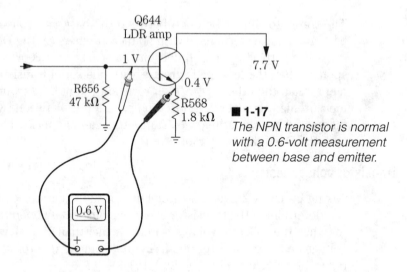

■ 1-17
The NPN transistor is normal with a 0.6-volt measurement between base and emitter.

First, measure the voltage on all elements to ground. Then check the base-emitter bias voltage. If you're not convinced of the legitimacy of the results, remove the transistor from the circuit and check it out of the circuit with a DMM or transistor tester. Sometimes when the transistor is removed, or when the in-circuit tester is applied to the transistor, the defective or intermittent transistor can test normal.

Diodes

Low-voltage diodes can be checked for open or leaky conditions with the diode test of a DMM. Make sure a power transformer winding, low-ohm resistor, or transistor is not across the path of the diode when making in-circuit tests. It's best to remove one end of the diode for accurate leakage tests (Fig. 1-18).

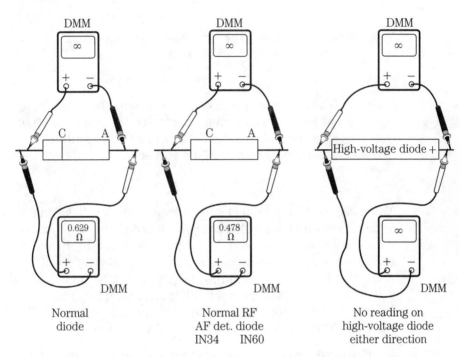

■ 1-18 *Diodes can be checked for open or leaky conditions with a VOM or DMM.*

Most low-voltage and damper diodes can be checked for open or leaky conditions with an ohmmeter or DMM. The RF, video, and audio detector diodes can be checked with an ohmmeter and DMM, but will have a higher resistance measurement. Some older boost rectifiers can only be checked for heavy leakage. The high-voltage rectifier found in the black-and-white TV chassis cannot successfully be measured with the VOM or DMM tests.

SCRs

The SCR found in low-voltage power supply and high-voltage rectifier circuits can be checked with resistance measurements (Fig. 1-19). A low resistance measurement between the gate (G) and cathode (K) is normal. Replace the SCR if the measurement is be-

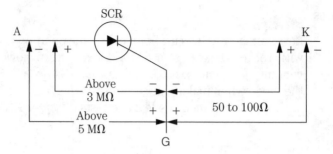

1-19 *Check the SCR with a tester, or take resistance measurements with the SCR out of the circuit.*

low 50 Ω. If any resistance measurement is found between the anode (A) and cathode (K) terminals, replace the leaky SCR. Like the transistor, the SCR can break down or become intermittent under load and should be replaced if suspected.

ICs

The defective IC component can be located with voltage and resistance measurements. Check the signal in and out of the IC component with the scope. Very low voltage on several terminals of the IC can indicate a leaky component. Low voltage at the IC terminal supplied directly from the low-power supply can indicate a leaky IC. Remove the terminal pin from the circuit and take a resistance measurement between the pin and chassis ground. Replace any IC that has a low resistance measurement.

Internal open connections inside the IC can cause a dead or intermittent symptom. If the signal is applied to the input and there is no signal or a low signal at the output terminal, suspect a defective IC. Take accurate voltage measurements and check them against the schematic to determine if the IC is defective. Sometimes voltage and resistance measurements are normal compared to the schematic, and replacing the IC is the only way to solve the problem. Replace the suspected IC at once if it is a plug-in socket type.

Surface-mounted components

Besides being used in CD players and camcorders, surface-mounted components now are found in TV sets. Within RCA's CTC140 and CTC145E (1990), the main heavy components are mounted on top of the PC chassis with surface-mounted components underneath (Fig. 1-20). The surface-mounted components, such as resistors, ca-

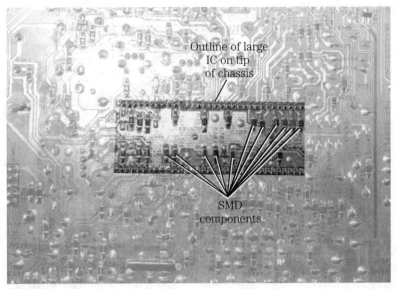

Outline of large
IC on tip
of chassis

SMD
components

■ **1-20** *The RCA CTC140 chassis has SMD components on the bottom side, and regular components on the top side.*

pacitors, and diodes, might have two or three ends to connect to the circuit. The microprocessor can have gull-type wings or elements that solder directly on the PC wiring (Fig. 1-21). These processors or ICs might have over 80 different soldered elements. Naturally, the surface-mounted transistor has three leads, with a small body that mounts flat against the wiring.

The resistors, capacitors, and transistors are very small components that are either black or brown, and which are soldered into the circuit (Fig. 1-22). It's wise to acquire the exact part from the manufacturer for replacement, although cards of different value capacitors and resistors are being manufactured for universal replacement (Fig. 1-23). You might find a part that looks like a resistor, but is actually a solid-tie feed-through component. A double resistor chip contains two resistance elements of different value in one chip. The small transistor might have all three elements out of the bottom, or the collector out of the top. You might find more than one diode in the surface-mounted component. A different hookup connection can be made with positive and anode connections.

The digital transistor might have the base and bias resistors mounted inside the surface-mounted component, while the standard resistor has leads directly to the internal elements. Note that the pnp digital transistor has the base and bias resistors tied

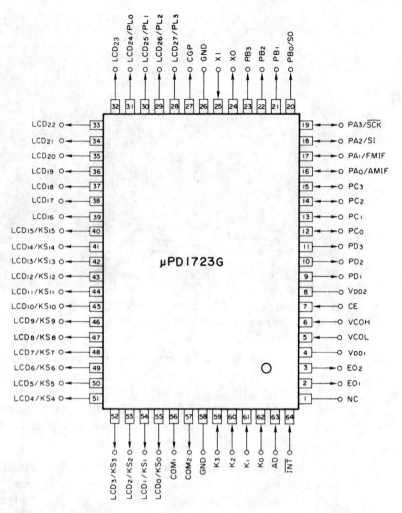

■ 1-21 *Surface-mounted IC processors and ICs are found in the latest TV chassis.*

to the emitter terminal. Likewise, the npn transistor is the same. When making transistor tests on digital transistors, allow for the resistor in series with the base terminal and a resistance leakage between the base and emitter terminals. Determine whether the transistor is a standard or a digital type before testing (Fig. 1-24).

■ **1-22** *SMD resistors and capacitors can be very small, and are mounted directly on the PC wiring.*

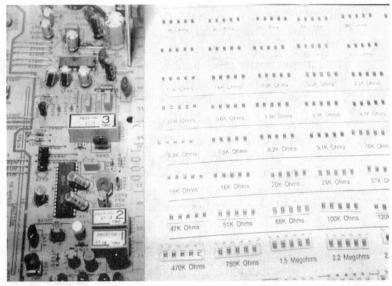

■ **1-23** *The top side of an RCA CTC146 chassis contains regular-sized components.*

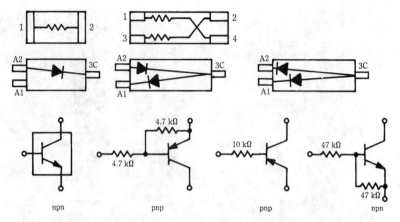

1-24 *The various surface-mounted components with different end connections. Digital transistors can have resistors in the base and emitter terminals.*

How to read SMD values

Surface-mounted parts such as resistors, capacitors and transistors may have numbers and letters stamped upon the top side of the component. Fixed resistors can be marked with white stamped numbers. A resistor with the numbers 123 has a value of 12 kΩ. The first two numbers indicate the resistance value, with the last number indicating the number of zeros that follow. In the last position, the number three equals 000, while a number 4 equals 0000 (Fig. 1-25). For instance, the numbers 692 equals 6.9 kΩ.

Fixed capacitors and transistors may contain stamped letters and numbers on the top side of the component. Note that a feedthrough SMD part can be easily mistaken for a fixed resistor. When a resistance measurement is taken, a zero or shorted reading is obtained. These chip components take the place of wire jumpers to tie two different circuits together on the regular PCB. You may find round capacitors and resistors instead of flat-chip components in some TV chassis.

Replacing surface-mounted components

Do not apply heat for more than 4 seconds; you might destroy the new chip or PC wiring. Avoid using a rubbing stroke when soldering. Do not bend or apply pressure to the transistor or IC terminals in the replacement. Use extreme care not to damage the new chip.

Replace all chip components except semiconductors and IC processors by preheating with a hair dryer for approximately

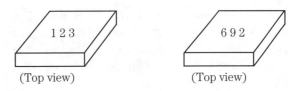

(Top view)   (Top view)

■ **1-25** *Check the top numbers on the SMD part for correct resistance or capacity.*

2 minutes. Tin or presolder the contact points on the circuit pattern or wiring. Press down the component with tweezers, and apply the soldering iron to the terminal connections. Apply more solder over the terminal if it is needed for a good bond. Do not leave the soldering iron on the transistor or IC terminals for too long; you might damage the internal junctions of the semiconductor. The large IC or processor should be flat and should mount right over each correct terminal connection.

Removing surface-mounted components

To remove a defective resistor or capacitor, remove the solder at one end. Grasp the end of the component or lead and melt the solder from the other end. Hold the part with a pair of tweezers and twist as the component is removed. Throw away all surface-mounted parts removed from the chassis. They cannot be used after removal.

To remove a surface-mounted transistor or diode, melt the solder at one end and lift the lead upward with a pair of tweezers. Do this to each terminal until all are removed. Some larger components might be glued underneath. Cementing the new component is not necessary. Remove jagged or overlapped solder from the PC wiring. Clean off lumps of solder from the wiring with the soldering iron.

Do not use a large soldering gun and solder wick to remove small surface-mounted components. Use a 30-W pencil soldering iron. Use extreme care so as not to damage the PC wiring. Make sure all solder is removed before lifting small components.

Preventive maintenance

Repeated callbacks can be eliminated if preventive maintenance procedures are followed in servicing the TV chassis. Often, it takes just a few minutes to check various items on the chassis after the original repair is made. Check for poor or burned connections of

27

components on the PC board or chassis (Fig. 1-26). Check for good soldered connections on high-wattage resistors. Check for poorly soldered terminals and hookup wires around the large resistors. Repair or replace burned terminal strips connected to various heated components.

Large white high-wattage resistors

■ **1-26** *Check for poor or melted solder joints where large power resistors are connected to the foil pattern.*

Always check each dial light and see if it is lit. Replace any dial light that comes on, but the glass remains black. This indicates the bulb will not last very long. The same applies to neon dial lights or neon bulbs found in the various circuits. Check for a loose dial bulb assembly or poor soldered connection if the lamp is intermittent.

Do not forget to clean up the tuner assembly after each repair job. Most rotary tuners need cleaning once a year. Squirt an approved cleaning fluid inside volume, color, tint, and contrast controls. Lubricate tuner bearing areas. If a small control is frozen, squirt silicone lubricant inside the shaft area and work the control back and forth to free it. Check each knob. Make sure it fits properly and is not loose. Replace the spring, or the entire knob if necessary.

Take a peek at each large electrolytic filter capacitor. If a white or black substance is oozing out around the terminal connections, replace it. Enlarged areas on the body of the capacitor might indicate the need for replacement. Check the ends of small filter or bypass capacitors for signs of deformed or cracked ends where each terminal is inserted. Sometimes the capacitor can explode, leaving blown pieces laying around. Usually, the ends come loose and make an intermittent connection.

Make sure the picture tube is good. Always test the picture tube with each repair job. A weak tube might last only a few months. Of course, you cannot tell how long the tube will last, but you can warn the customer he has a weak or blotchy picture tube. Often, picture tubes are not recommended to be replaced after 8 years, because other components like transformers and filter capacitors might also go bad, resulting in additional repairs in a few months. Finally, make all manufacturer modifications for that particular chassis.

The estimate

An estimate should be given on each repair job so the customer knows about how much it will cost. Some TV shops give two estimates. One is to just fix the present trouble; the second is for a complete overhaul. It is surprising how many customers go for the overhaul because they do not want the TV to return to the shop.

The overhaul should include the following:

☐ Repair the original trouble.

☐ Check and clean the tuner.

☐ Test and replace the picture tube, if needed.

☐ Inspect all filter capacitors.

☐ Inspect the flyback transformer.

☐ Inspect and replace all overheated components.

☐ Check all small controls—replace the ones that are worn and erratic.

☐ Make sure all dial lights are working.

☐ Check the cabinet for dents or scratches and loose grilles.

☐ Polish and clean the entire cabinet, picture tube, and all control knobs.

☐ On sets hit by lightning, determine if the chassis is worth servicing (Fig. 1-27).

☐ Check all tubes in hybrid chassis.

A good book on TV pricing and estimates is John Sperry's pricing book. This book can be used on house calls or with bench service charges. Each repair item is listed with several different charging methods. You can pick out your hourly rate for the different repaired items. Order the book by calling 1-(800)-228-4338.

■ **1-27** *This TV chassis was damaged by a heavy power-line voltage charge.*

Correct part replacement

Most transistors and IC components can be replaced with originals or universal replacements. Use original transistors in tuners and IF stages. Large, crucial IC components should be replaced with the original part numbers (Fig. 1-28). Foreign and Japanese transistors and IC components are now found at most wholesale or mail-order TV parts establishments.

The volume control with tone taps and special switches attached to the back should be replaced with originals. An ordinary volume control can be replaced with a universal type by cutting the shaft to the correct length. Small color, tint, brightness, and contrast controls can easily be replaced with universal types. When two or three controls are found mounted together (such as height, linearity, and AGC) in one component, replace the one control that is defective. If the complete assembly cannot be located, mount a separate control on a bracket at the chassis and connect hookup wire in place of the original control terminal connection.

Some technicians replace low-voltage diodes with the 2.5-A and 3-A variety even if the original was a 1-A diode. Make sure the operating voltage is the same or higher (1 kV). When replacing damper diodes, make sure the peak voltage is correct in universal replacements.

Large IC with
numbers on top

■ **1-28** *Large ICs may have the part number stamped on the top side. Replace them, when necessary, with an exact replacement part or a known universal replacement.*

Universal electrolytic capacitors can be used if the capacitor has the same (or higher) working voltage. Original filter capacitors fit nicely and correctly when replacing parts in the TV chassis. If the original cannot be obtained, substitute a universal replacement (Fig. 1-29). Always replace the entire can even if only one is found defective.

Filter capacitors can be added or paralleled when the original is not available. For example, if you need a can-type filter of 250 μF, 200 μF, and 10 μF at 180 V and you have a universal replacement 200–300 μF, 200–400 μF, 20–50 μF at 250 V, use it. Often in filter circuits, the added capacitance will improve filtering action. Be sure the operating voltage is the same or higher with universal replacements. The TV chassis should not be tied up when electrolytic capacitors can be connected together to safely place the chassis back into service. Test all new parts before installing them in the TV chassis.

When a special bridge rectifier is not available, use single diodes in the circuit. Because 2.5-A silicone diodes will serve most current requirements in the TV chassis, connect four single diodes in a bridge circuit (Fig. 1-30). Make up the bridge unit with diodes mounted and soldered close together. Then place the correct polarity leads into the bridge rectifier mounting holes.

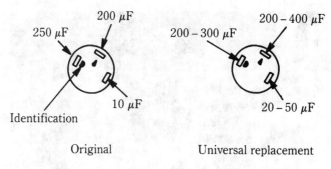

Original   Universal replacement

Bottom view of filter capacitors

■ **1-29** *You can replace most electrolytic filter capacitors in the TV chassis with universal filter capacitors.*

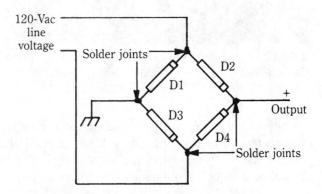

■ **1-30** *Connect four individual 2.5-A diodes in a bridge circuit when a bridge rectifier component is not available.*

Always mark each terminal lead of a power or flyback transformer before removing the component. This also applies to any component with several color-coded leads. You might be called away from the bench and not return until the next morning. It's difficult to remember where each lead connects and this means tracing out each wire on the schematic. Of course, a lot of time will be lost if the schematic is not available.

Case histories

The tough or unusual repairs in TV servicing will not be forgotten if they are recorded. Some technicians place the repairs in a 3-x-5 file card system. One quick method is to list the problem along the side of the schematic diagram. Each part can be identified on the schematic for easy reference. Long troubleshooting techniques can also be stapled to the service literature.

One case history might save hours of time in the busy world of TV servicing. You cannot possibly remember all of the various problems related to a special TV chassis, so it's wise to keep a file or mark the case history on the schematic. Additional case histories can be found in *Radio-Electronics* and *Electronic Servicing & Technology* magazines (Fig. 1-31).

■ **1-31** *Check the electronics magazines, including* Electronic Servicing & Technology, *for TV schematics.*

Safety precautions

Besides having each service bench fused with grounded polarity outlets, the service technician must follow safety procedures when repairing the TV chassis. Use isolation power line transformers with all TV chassis. Today, most TV chassis are ac-grounded without a power transformer. Keep the chassis away from grounded pipes or metal bench posts. Work on a rubber mat to protect yourself not only from electrical shock, but also to save those tired feet.

Think before you leap when working in the ac power line or high-voltage circuits. Make sure voltage test instrument cables are good, with no bare wires or exposed meter terminals. Discharge the picture tube before attempting to remove it. Wear safety goggles when replacing a picture tube. It takes only one picture tube explosion to damage the eyes for life. Discharge filter capacitors before checking voltages in the power circuits.

Be careful in parts replacement. An electrolytic capacitor replacement with lower voltage than required might blow up in your face. The same can occur if the capacitor is installed backwards. Make sure the power cord is disconnected when touching a component such as the flyback transformer, tripler, or power output transistor. Sometimes a leaky horizontal output or power regulator transistor can be red hot, and you will come away with burned fingers. Look and think before placing a meter probe or finger into the TV chassis.

Before buttoning up the rear cover, make sure no leads or cables are laying on parts of TV chassis. Tie cables up by folding long wires, and wrap them up with plastic ties and rubber bands. Make sure the speaker wires or cables are not pinched between chassis and bottom slide areas. Keep the high-voltage lead tied up away from components on the TV chassis.

Take the antenna cold check and leakage current hot check. For the antenna cold check, remove the ac power cord, turn the ac switch on, connect one lead of ohmmeter to the ac plug prongs, and touch the metal tuner with the other probe. If the resistance measurement is less than 1.0 megohms or greater than 5.2 megohms, something is wrong, and it should be corrected before the TV set is returned to customer.

Make the hot current test as follows: Plug the TV set into the ac outlet, turn the power on, connect the leakage current meter probe to the tuner, screw leads, or metal control shafts, and then connect the other probe to an earth ground or water pipe (Fig 1-32). This current measurement check should be less than 0.5 milliamperes. Now reverse the power cord in the outlet and repeat the test. If the current reading is higher, check the TV set component leakage to power line components.

Safety capacitors

Bypass and voltage holddown capacitors found in the horizontal output transistor, yoke, and flyback circuits should be replaced with factory-spec original parts (Fig. 1-33). Always replace with the exact capacity and working voltage. If the manufacturer's exact part number is not available, replace with an exact-capacity, higher-operating-voltage universal replacement. Remember that the safety capacitors in the horizontal output transistor collector

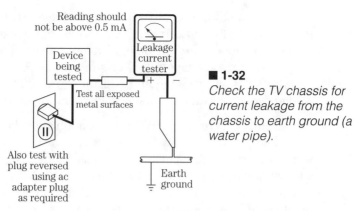

Reading should
not be above 0.5 mA

Device
being
tested

Leakage
current
tester

Test all exposed
metal surfaces

Also test with
plug reversed
using ac
adapter plug
as required

Earth
ground

■ **1-32**
*Check the TV chassis for
current leakage from the
chassis to earth ground (a
water pipe).*

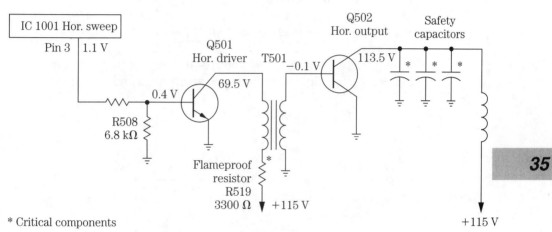

IC 1001 Hor. sweep

Pin 3 | 1.1 V

Q501
Hor. driver

R508
6.8 kΩ

0.4 V

69.5 V

T501

Flameproof
resistor
R519
3300 Ω

−0.1 V

+115 V

Q502
Hor. output

Safety
capacitors

113.5 V

35

* Critical components

+115 V

■ **1-33** *Replace safety or holddown capacitors with ones having exactly the same capacity and working voltage as the originals.*

circuit hold down the high voltage at the flyback and anode terminal of picture tube. If the exact capacity is not used, the high voltage can be raised extremely high, or lowered unacceptably.

Preventing fires

Always replace critical components with the exact part numbers
that have shaded or star indicators. Some critical parts are identified by a black triangle symbol found in some chassis. These critical
components are found in the low voltage power supply, regulator,
high voltage, sweep, and horizontal output circuits.

Safety components found in the low voltage power supply consists
of capacitors, fuses, silicon diodes, and power resistors. You may

find oxide metal-film type resistors in voltage divider networks. Replace critical capacitors and resistors in the horizontal output circuits with exact part, or with equivalent universal replacements (Fig. 1-34). High-voltage fixed resistors should be replaced with parts of the same wattage and resistance ratings. Replace wire-wound resistors with the same or higher-wattage-rated parts.

Route wires covered with PVC tubing, double insulated wires and high voltage leads away from components that operate fairly warm. Use specified insulated materials for hazardous live parts; such materials include insulation tape, PVC tubing, spacers, and insulators for transistors. Keep all loose wire and cables away from overheated components such as power transistors, heatsinks, oxide metal-film resistors, fusible resistors and power resistors. Replace the new component in the exact space as the original. Do not leave real long leads on capacitors and resistors, that can flop around. Make clean soldered joints or bonds. Sloppy part replacement can cause fires resulting in extensive fires and even death.

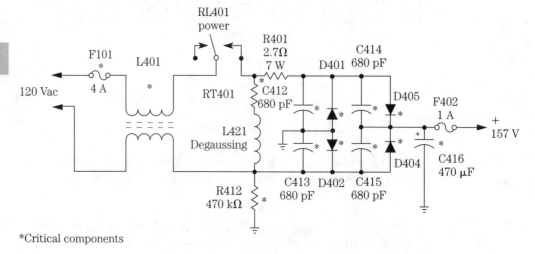

*Critical components

■ **1-34** *Replace oxide film resistors, metal oxide resistors, and capacitors with exact replacements in horizontal, high voltage, and low-voltage circuits; it will help to prevent fires.*

RCA DSS satellite antenna

The RCA DSS System (Digital-Satellite-System) is a satellite system that uses direct broadcast from a satellite. It enables millions of people to receive over 100 channels of high-quality digital programs from anywhere in the USA. The system provides digital data, video, and audio to the various homes via high-powered Ku-band satellites. The broadcast signal is transmitted to the DBS satellites (orbiting 22,000 miles above the earth) from a site in Colorado Springs, Colorado. The signal is relayed back to earth and decoded with a receiver unit that rests on top of the TV set.

There are two different packages for the homeowner. The model DS1120RW is a basic package with an antenna (dish) and a single-output, low-noise block converter (LNB) for $699.95. This satellite receiver has a DSS/TV universal remote.

The deluxe package contains a sheet molded compound (SMC) antenna with a dual polarity LNB (Fig. 1-35). The satellite receiver has an output so that two different TVs can be plugged in with additional audio/video jacks and sells for $899.95 (DRD203RW). A fully universal remote is included that can also operate the satellite receiver and VCR (Fig. 1-36).

The dish is an 18-inch-diameter job with a slight oval shape for the KU-band antenna. The LNR is offset so it is out of the way and does not block any surface area of the dish (Fig. 1-37). The low-noise block converter converts the 12.3 GHz to 12.7 GHz down signal from the satellites from 950 MHz to 1450 MHz signal picked up by the receiver. The dual output LNR is found in the deluxe package. The dual output LNR can be used to feed a second receiver or other form of distribution system (Fig. 1-38).

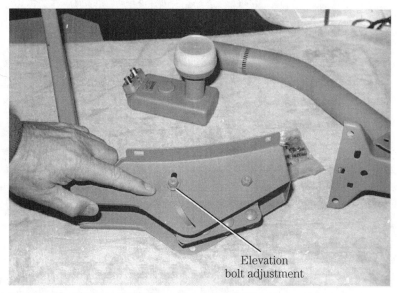

Elevation
bolt adjustment

■ **1-35** *The deluxe DSS antenna has a dual LNR assembly to connect two different receivers. The finger points at the elevation bolt assembly.*

■ **1-36** *The deluxe remote can operate the satellite receiver and VCR.*

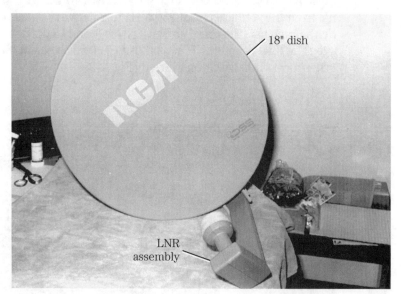

18" dish

LNR
assembly

■ **1-37** *The LNR is offset out of the way of the antenna dish, and picks up the reflected signal.*

The price of the DSS system is quite low compared to the early large-antenna satellite systems that were introduced for above $10,000. This DSS system has a regular satellite and cable system beat at every turn, in cost as well as excellent digital video and sound quality. There are many good TV programs to select

in several different combinations. The DSS system can be used by the those who live in homes in locations that cannot hook up to cable TV.

Installing a DSS antenna system

Select a mounting site for the dish that is clear of all buildings and trees between the antenna and satellites in the sky. The DSS satellite is located over the equator at 101 degrees west longitude. With the azimuth and elevation known, find the correct angle with a compass, level, and protractor for the approximate location of the satellite and the angle of the antenna. Here in the Midwest, the 18-inch dish would be pointing south and a little west, at a 36-degree angle upwards towards the sky (Fig. 1-39).

The DSS dish has two different positioning adjustments, azimuth and elevation. The azimuth adjustment is the side movement of the dish. You can make this adjustment by rotating the dish on the mounting or post. The elevation adjustment is done with the LNB support arm. Notice the side of this arm; the dish and arm can be adjusted up and down. A reference scale on the side of the dish is calibrated in degrees. Adjust the LNB arm and dish for the correct degree of elevation.

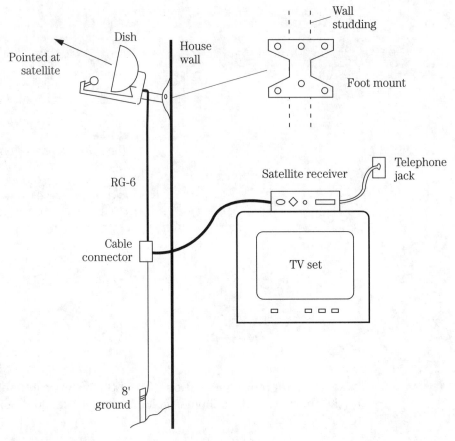

■ 1-39 *Point the dish towards the equator and adjust the elevation for maximum signal.*

To adjust the screw for elevation, loosen two nuts securing the dish. Now the dish will move up and down. Line up the elevation indicator with the elevation degree angle. Tighten the two nuts securing the dish elevation movement.

Point the dish in the general direction of the signal satellites. Follow the instruction with the service literature on how to make the correct azimuth adjustments with a compass and azimuth coordinates. Fine tuning of the dish position can be made with a signal-strength meter (and receiver audio-tone indicator on the deluxe satellite receiver).

The DSS receiver signal-strength screen uses two systems to help fine-tune the location of the dish. Listen to the audio tone by connecting a TV, headphones, or amplifier to the right jacks on the rear of the receiver. When the dish is pointed at the satellites, a continuous tone is heard. When not pointed at the signal, short bursts of tone can be heard.

The second tuning system uses the television on-screen display. This display includes both a bar and numeric display. The numeric display range is from 0 to 100. The higher the number, the greater the signal. The stronger the signal with bar display, the farther the bar stretches across the screen. By listening to the audio tone and looking at the display, you can fine-tune the position of the dish antenna with a satellite broadcast signal. An excellent signal is from 75 to 85 on the display.

The antenna dish lead-in wire should never be more than 100 feet from receiver with RG-6 coaxial cable. Twenty feet of telephone wire is supplied to connect the telephone cable to the satellite receiver. The shorter the lead-in, the greater the reception. Route the cable through the shortest distance. A telephone jack connection should be installed next to the DSS receiver, for communication.

Mounting the dish

The DSS dish can be mounted just about anywhere, as long as it is free of all objects between the satellites and dish. Level the mast holding the antenna with a bubble level or plumb bob. A metal foot mount is found at the base of the mast (Fig. 1-40). Do not mount

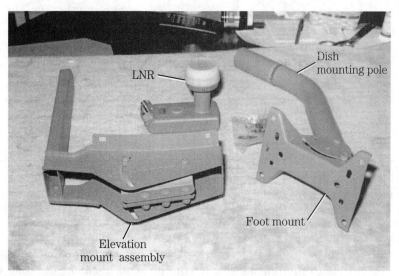

■ **1-40** *The foot mount is bolted to 2 × 4s and leveled vertically in all directions.*

the foot mount over vinyl or metal siding. When mounting the unit on lapped wood siding, make sure the center part of the foot mount is over a stud. Do not place the foot mount on chipboard, fiberboard, or particleboard.

When installing the foot mount on a flat or slanted roof, make sure the mount straddles a rafter. The dish can be mounted into the ground, by digging a 2½-foot-deep hole, 10 inches in diameter. Use a 1¼-inch galvanized steel pipe and pour concrete into the hole. Make sure the mast is plumb vertically, in all directions. Pound an 8-foot ground rod into the ground nearby and clamp a ground wire to the antenna mast (Fig. 1-41). The antenna dish can be mounted on a chimney with a chimney mounting kit. It is best to use a chimney that is not active with a lot of smoke. Again, the foot mount must be mounted solidly at the chimney corner with mounting straps. Place the chimney mount so the antenna has no objects between it and the satellites.

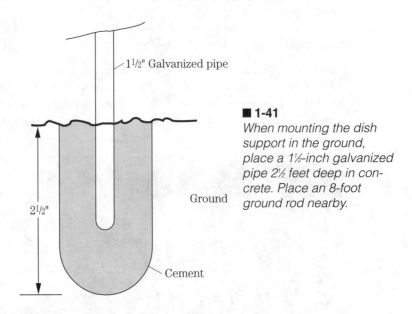

1½" Galvanized pipe

Ground

2½"

Cement

■ 1-41
When mounting the dish support in the ground, place a 1½-inch galvanized pipe 2½ feet deep in concrete. Place an 8-foot ground rod nearby.

Installing the LNB cable

If you have installed the deluxe RCA DSS system, three different cables must be installed. Two RG-6 (75-ohm) cables must be attached to the LNB at the dish and run to each receiver on the two different TV sets. The third cable is a telephone wire for each receiver. Only two different cables are required when installing the basic DSS system: one cable from the dish carries the TV signal from the LNB to the receiver, and a separate telephone cable (Fig. 1-42). The tele-

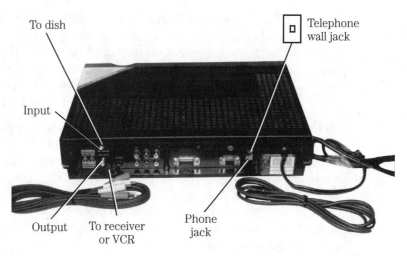

To dish

Telephone
wall jack

Input

Output To receiver
or VCR

Phone
jack

■ **1-42** *Connect the dish's RG-6 cable into the input terminal of the satellite receiver, and connect the receiver's output to the VCR antenna or TV connection.*

phone jack on the satellite receiver is connected to the telephone line. Now sit back and enjoy the programs and pictures.

High-definition TV

High-definition TV (HDTV) is a new system that improves picture resolution, giving twice the horizontal and twice the vertical resolution. Besides improved picture definition, improved audio is also part of the package; it is somewhat like the audio from a compact disc (CD). The new system must not interfere with existing transmission of the present-day television channels set by the National Television Standards Committee (NTSC).

The Federal Communications Commission (FCC) provides standards that require transmission systems to operate within specific widths and frequencies. Within these parameters, there are gaps where nonactive channels exist. HDTV will use these nonactive channels and at the same time broadcast over existing channels.

To enjoy the benefits of HDTV, you must purchase a new TV with HDTV features. Several American and Japanese firms planned to manufacture HDTV receivers by 1995. Although the official standards have not been set by the FCC, several different systems are now competing to set the HDTV standard. The new standard must be compatible with present TVs operating in the United States, so your set will not be obsolete.

43

The present-day standard TV has 262.5 scanning lines in the odd and even field (Fig. 1-43). The interpolated-definition TV (IDTV) introduced by foreign manufacturers has 525 scanning lines in the odd and even field. The interpolating digital memory circuits provide additional scanning lines between the broadcast signal lines, thus showing the entire image on the TV screen (Fig. 1-44).

On December 7, 1989, the Duran-Leonard prizefight took place and was shown in four major cities, including Los Angeles and Chicago. The fight was broadcast by the Japanese via satellite to demonstrate the transmission system known as HDTV. Zenith Electronics, in cooperation with the American Telephone and Telegraph Company (AT&T), are working on a $24-million research and development program to develop an HDTV system, which, unlike the Japanese system, is full-spectrum compatible.

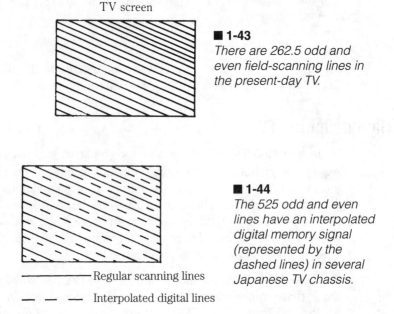

TV screen

■ **1-43**
There are 262.5 odd and even field-scanning lines in the present-day TV.

——————— Regular scanning lines

— — — Interpolated digital lines

■ **1-44**
The 525 odd and even lines have an interpolated digital memory signal (represented by the dashed lines) in several Japanese TV chassis.

Troubleshooting the low-voltage power supply

THE LOW-VOLTAGE POWER SUPPLY OF THE SOLID-STATE TV receiver supplies voltage to the transistors, ICs, and in some cases, filament voltage to the picture tube. In the latest solid-state TV chassis, the filament of the picture tube and other circuits might be powered with voltage developed from the flyback or horizontal output transformer. A power transformer was found in earlier chassis, while most of today's low-voltage power supplies operate directly from the power line (Fig. 2-1).

Most line-connected low-voltage power supply circuits consist of either a half-wave silicon rectifier or a full-wave bridge circuit using four different silicon diodes. Either circuit can have a zener diode, transistor, and SCRs in a filter regulator circuit. You might find very large filter capacitors in line-connected power supplies.

The integrated or flyback low-voltage power supplies are rectified and filtered from a separate winding of the horizontal output transformer. A single silicon diode rectifies the high ripple voltage from the flyback winding. The horizontal circuits must be operating before any voltage is developed by the output transformer.

The following test instruments are used for servicing the low-voltage power supply:

☐ VTVM, either VOM or DMM.
☐ Oscilloscope.
☐ Rectifier tester (could be included in DMM).
☐ Capacity tester.
☐ Isolation transformer.

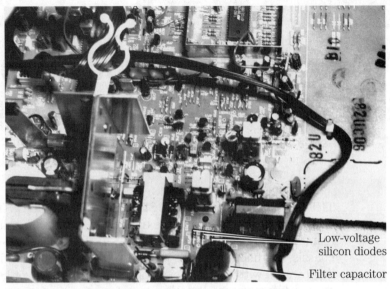

Low-voltage
silicon diodes

Filter capacitor

■ **2-1** *Locate the low-voltage circuits by finding the large filter capac-itor and parts nearby.*

Low-voltage power supply symptoms

The low-voltage power supply is the heart of any consumer elec-tronic product. Nothing operates without power or voltage furnished from the low-voltage source. When trouble occurs in the TV chassis, most electronic technicians check the low-voltage sources the first thing. A defective power supply can cause a dead, intermittent, hum and improper TV symptom.

The low-voltage power source can cause chassis shutdown due to excessively high or low output voltage. Poor or narrow width can be caused with low power-supply voltage, while high-voltage out-put can cause excessive width and high voltage at the CRT. De-fective or dried-up filter capacitors can produce hum in the sound and crawling pictures in vertical circuits. Chassis shut-down can result from a defective component in the low-voltage power supply.

Isolation transformer

The isolation transformer is a "must" item when servicing the transformerless ac-dc chassis. Besides providing direct ac voltage on the TV chassis, you will prevent fuse and low-voltage circuit damage when attaching the test instrument (Fig. 2-2). If you do

Power
isolation
transformer

■ **2-2** *Always use a variable isolation transformer when servicing today's TV chassis.*

not use the isolation transformer, ac-operated test equipment cannot be used directly on the live chassis.

Choose an isolation transformer with variable voltage taps or continuous voltage control. The isolation transformer can vary from 0 to 150 Vac with an isolated outlet. Some of the transformers are fused with a 10-A fuse. Select the isolation transformer that has a 1.5- to 5-A outlet. You might find the isolated transformer is contained with the low-voltage power supply.

You can make your own isolation transformer by ordering a regular line isolation transformer (115 Vac to 115 Vac). These transformers might have to be ordered through an electronics wholesale distributor. Of course, you might have to wait 30 days for delivery, but they are worth it. Stancor and Thordorson have line isolation transformers. Remember, these transformers weigh between 4 and 10 lbs.

Choose a line transformer from 115- to 125-Vac output for ordinary ac line operations. Although this type of isolation transformer is not a variable type, it will isolate the ac-dc TV chassis away from the power line. Isolation is required when connecting test equipment to the ac-dc TV chassis. If not, you might damage components in the TV chassis.

The isolation transformer might be included in the same cabinet as the other power supplies. A separate ac outlet is placed on the outside of the cabinet to plug in the TV chassis or other equipment. With this method, the isolation transformer can be used without cranking up the other power supplies.

The half-wave rectifier

The half-wave rectifier circuit was once found only in the portable TV chassis. Today, the half-wave circuit, such as the one in Fig. 2-3, can be found in both console and portable TV chassis. In fact, a modern TV chassis is rather small compared to the early solid-state chassis. The power-line operated power supply does not contain a heavy power transformer, large choke filter, or several large filter cans (electrolytic capacitors).

The circuits of the entire receiver are usually protected by a 5-A fuse located on the main chassis board. Most ac and dc circuits are fused instead of using circuit breakers. The power plug is polarized to prevent a "hot" metal chassis. Always use an isolation power transformer when servicing a line-connected TV chassis to prevent injury from the power line.

L900 and C900 (Fig. 2-3) prevent spurious signals from entering the power supply from the power line. Power is applied to the TV chassis when the on/off switch (SW1) is closed. The silicon rectifier (D900) rectifies one-half of the full cycle, and the ripple dc voltage is filtered by L101, C901, and C902. Because silicon diodes

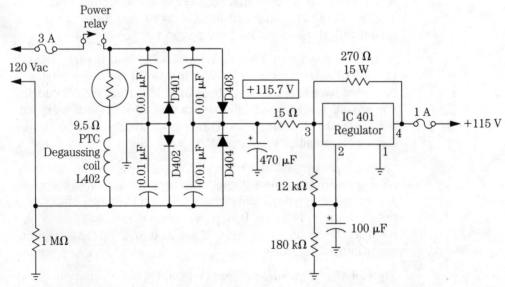

■ **2-3** *The low-voltage power supply may have one large power-line IC regulator.*

have a tendency to radiate spurious signals, C903 keeps the diode from radiating noise lines into the picture. The half-wave circuits have RC (resistance-capacitance) or LC (inductance-capacitance) filter networks.

The degaussing coil is energized through the R900 (3-Ω cold thermistor). After a few seconds, R900 heats up and increases in value to prevent current from entering the degaussing coils. Thus, the face plate inside the picture tube is degaussed for a few seconds only when the receiver is turned on. This prevents spotted or magnetically contaminated areas on the face of the picture tube.

F800 protects the power supply circuit if the horizontal output, damper, and flyback circuits become leaky. Besides feeding the horizontal circuits, the 140-Vdc supply can be tied into transistorized or SCR-regulated low-voltage circuits. The low-voltage regulated circuits can furnish voltage to sound, IF, video, sync, horizontal, and vertical oscillator countdown stages. The power supply circuits are easily serviced by making voltage and resistance measurements.

Fuse and silicon diode replacement are the most commonly needed fixes within low-voltage power supplies. Always replace the blown fuse with another of the exact same current rating. D900 can be replaced with a 2.5-A silicon diode. Some half-wave rectifiers are replaced with 3-A types. The fuse and diode can open due to overloaded components in the power supply or in connecting circuits. The fuse and silicone diode can also be damaged by lightning or power-line spikes.

The defective silicon diode can become leaky, shorted, or open. A shorted diode can blow the line fuse and open the line isolation surge resistor. The leaky diode of high resistance may not blow the line fuse at once. Look for a shorted diode or electrolytic capacitor when the fuse is exploded, seared, or black. Besides a shorted silicon diode, the fuse may open with a leaky low-voltage dc line regulator.

Hum bars or audio hum can mean bad filter capacitors (Fig. 2-4). Defective filter capacitors can cause unstable vertical rolling and horizontal sync problems. Chattering relays and improper operation of remote-control systems can also be caused by defective filter capacitors.

The filter capacitors (C901 and C902) in Fig. 2-3 can be replaced with the same (or higher) capacitance and voltage. Notice how large the filters are in the half-wave circuit (700 μF/800 μF). If they are replaced with units of lower operating voltage, the capacitors could run warm and blow up in your face. Replacing a filter

Filter
electrolytic
capacitor

■ 2-4 *Hum bars in the sound and crawling pictures can result from de-fective filter capacitors.*

capacitor with one of a lower capacity rating can produce more hum and an unstable picture. Check the hum bar waveform with the scope when a larger capacitor is added to the filter circuit.

Filter capacitors can be paralleled to provide greater filtering action in most power supplies. For instance, if you have on hand a capacitor rated 400 μF, 400 μF, 700 μF at 200 V, the capacitor could be substituted for C901 and C902. Just connect the two 400-μF sections together as a 800-μF capacitor (C902) replacement and use the 700-μF section to replace C901 (Fig. 2-5). The original capacitors (C901 and C902) were rated at 700 μF and 800 μF with 185-V operating voltage. Sometimes you might have to add more capacity to the power filter circuits to eliminate hum.

The defective filter capacitor can be located by checking the waveform with a scope, or by shunting another capacitor across the suspected one. Choose a capacitor with the same or higher capacity and voltage. Most filter replacements are of a higher operating voltage than the original one.

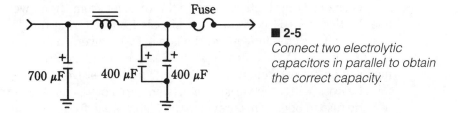

■ 2-5
Connect two electrolytic capacitors in parallel to obtain the correct capacity.

Use clip leads to clip a new filter capacitor across the suspected one. Observe correct capacitor polarity. Disconnect the power cord and discharge the shunted capacitor each time a different capacitor is shunted to prevent damage to transistor and IC components. Replace the defective capacitor when hum bars or an unstable picture disappear from the screen.

Full-wave rectifier circuits

Although the power transformer was found in the early solid-state chassis, very few were replaced unless damaged by lightning. In some chassis, a small power transformer might be used to supply low-voltage circuits and a power line full-wave rectifier circuit is then used to provide a higher voltage to the horizontal output transistor and the sound circuits. The full-wave rectifier can use single silicon diodes or a bridge-type rectifier. Figure 2-6 shows a typical full-wave rectifier circuit.

51

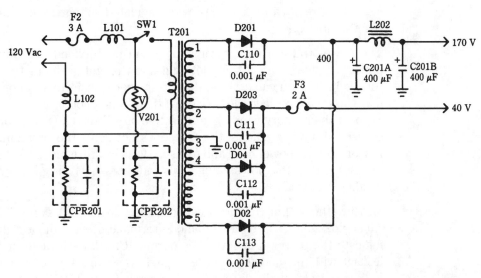

■ **2-6** *Full-wave rectification can use individual silicon diodes or one bridge rectifier component.*

A 3-A fuse (F2) protects the entire TV receiver from the power line. L101 and L102 prevent noise signals from entering the TV from the power line. In many of the latest TV receivers, a line-operated varistor (V201) protects the receiver from lightning or high-voltage surges from the power line. When excessive voltage is found across V201, the varistor shorts and arcs over, causing the 3-A line fuse to open. This prevents the high voltage from entering the primary winding of the power transformer. Capacitors CPR201 and CPR202 prevent static charges from nearby metal objects to come in via the ground when the receiver is turned off or on.

The power transformer (T201) has two separate full-wave power circuits supplied by the power transformer in this special circuit. Terminals 1 and 5 with center tap 3 supply high ac voltage to D201 and D202. Terminals 2 and 4 with center tap ground 3 provide another full-wave rectifier circuit to D203 and D204 with a lower voltage source and output. Each single silicon rectifier can be replaced with a 2.5-A diode. Capacitors C110 through C113 prevent stray voltages from entering the TV chassis via the low-voltage power supply. The four diodes can be checked within seconds with the diode tester of a DMM.

The full-wave 120-Hz ripple voltage is filtered out by capacitors C201A and C201B. The filter choke L202, C201A, and C201B provide an LC filter network. F3 provides protection to the low-voltage circuits. The 40-V output can have zener diode or transistor regulator circuits attached to provide regulated voltage to critical receiver stages.

The power transformer power supply can be serviced like any power supply circuit. If the pilot light and the end of the picture tube light up, you can assume that the power transformer is normal. Check and replace the 3-A fuse if there is no pilot or dial light. Suspect the primary winding, a defective on/off switch (SW1), or chassis wiring if replacing the fuse does not solve the problem. Disconnect the power cord and measure for a low resistance (50 Ω or less) across the power plug. This indicates that the primary winding is normal.

Measure the output voltage at capacitor terminals C201A and F3. An ac voltage measured at the anode terminal of each diode will indicate the power transformer is normal. Check each diode for leakage with the diode test of the DMM. Low voltage at C201A can indicate a leaky diode, or that the filter capacitor (C201A) is open.

If the 3-A fuse (F2) opens each time the receiver is turned on, suspect an overload in the secondary winding of the power transformer.

A 100-W bulb can be clipped in place of F2 to prevent replacing the blown 3-A fuse each time. Check each diode for leakage. Disconnect the power cord and remove the collector lead of the horizontal output transistor, or remove the transistor from the circuit. Suspect a leaky horizontal output transistor damper diode loading down the high-voltage source of the low-voltage power supply.

A shorted filter capacitor (C201A or C201B) will cause F2 to open. Most filter capacitors open or dry out, producing hum and a low-voltage source. A white or black substance leaking from the capacitor (or a bulging capacitor) indicates a defective filter capacitor. Leaky filter capacitors cause severe power supply hum, with 60-Hz or 120-Hz hum bars in the raster. An open filter capacitor can cause hum in the speaker even if the volume control is turned completely down. Replace the entire unit when any one capacitor is found to be defective.

Besides hum and low-voltage problems caused by a defective filter capacitor, rather odd symptoms can be found in the RCA CTC47 chassis (Fig. 2-7). You might hear relay K201 chatter, and the picture might appear very noisy with the brightness turned up or when the receiver is first turned on. A low whistle might be heard in the sound when the on/off switch is turned off and on. These symptoms can be caused by a defective filter capacitor (C206). Shunt another 400-μF electrolytic capacitor across each section to locate the defective capacitor. Remember, defective filter capacitors can produce odd symptoms and trouble in the TV chassis.

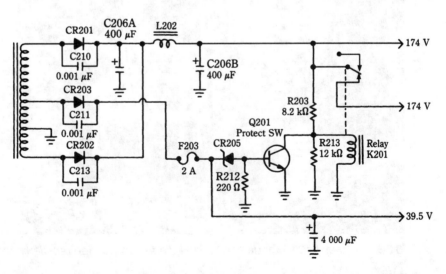

■ **2-7** *A defective 400-μF electrolytic capacitor caused the relay to chatter or shut down the TV chassis.*

Bridge rectifier circuits

The bridge rectifier circuit can be operated from a full-wave power transformer or directly from the power line (Fig. 2-8). The bridge rectifier can consist of four separate silicon diodes, or of four diodes molded into one component. Sometimes two silicon diodes are found in one envelope, and it takes two separate units to complete the bridge circuit (Fig. 2-9). Replace the whole unit even if it has only one leaky diode.

A 3-A fuse offers power supply protection of the full-wave line-operated power supply in Fig. 2-10. Line filter L101 prevents noise from entering the power supply of the power line. Thermistor T701 provides low resistance to the ac line voltage to energize the degaussing coil for a few seconds. This happens whenever SW1 is turned on; then T701 increases in resistance to shut off any current flow to the degaussing coils.

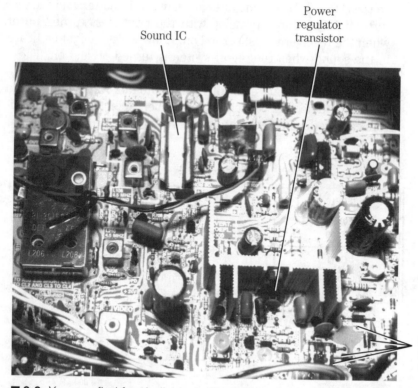

Sound IC

Power regulator transistor

Single diodes form a bridge circuit

■ **2-8** *You may find four individual silicon diodes or a bridge rectifier in the low-voltage power supply.*

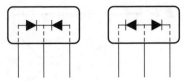

■ **2-9** *You might find two silicon diodes in one molded body in the bridge-type rectifier. It takes two such rectifiers for a full-wave bridge circuit.*

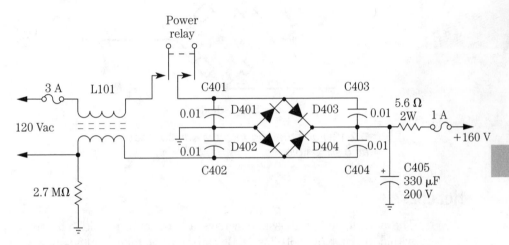

■ **2-10** *A typical remote low-voltage power supply with separate silicon diodes.*

The ac input voltage is rectified by a bridge rectifier circuit (D401, D402, D403, and D404). Capacitors C402, C403, and C404 bypass the silicon diodes to keep them from radiating noise into the dc voltage source. C405 is a capacitor filter at the circuit output.

A bridge-type symbol might be found in some of the Japanese models (Fig. 2-11). The diode symbol "points" to the rectified output voltage. All four silicon rectifiers can be replaced with single 2.5-A diodes if one whole bridge rectifier component is not readily available. Simply connect the diodes as shown in Fig. 2-12 and solder them into the respective holes of the PC board. Likewise, both rectifiers can be replaced in the half-molded bridge unit mentioned earlier with two separate 2.5-A diodes.

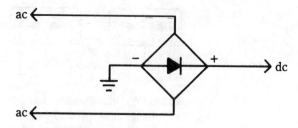

■ 2-11 *You may find a diode symbol on top of the bridge circuit in the Japanese chassis.*

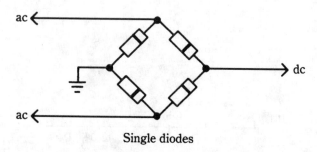

Single diodes

■ 2-12 *Form the bridge circuit with four separate 2.5 amp silicon diodes.*

56

Hot chassis

The new ac-dc TV chassis can have a "hot" and a "cold" chassis within the large overall chassis. Be careful when taking voltage measurements around these chassis. Often the "hot" chassis is above regular chassis ground and has a different ground marking (Fig. 2-13). If the "hot" ground is not used and you use common chassis ground, the voltage measurement is incorrect. Always use the "hot" ground when making voltage measurements on components in that area.

RCA CTC130 variable-frequency switching power supply

The variable-frequency switching power supply (PW VIPUR) used with the CTC130C chassis is similar to that used in previous chassis (Fig. 2-14). In this chassis, the components are mounted on a separate circuit board assembly (PW VIPUR). The prime reason for using the variable-frequency switching power supply is to provide hot-to-cold ground isolation between the primary TV power supply source (hot ground) and the video/audio input/output circuit PW V12 (cold ground).

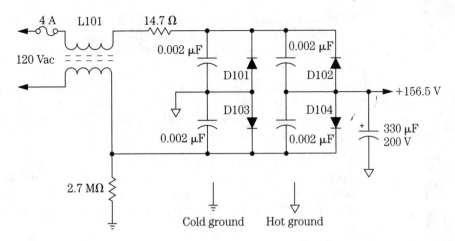

2-13 *Notice that the low-voltage power supply has both cold and hot grounds.*

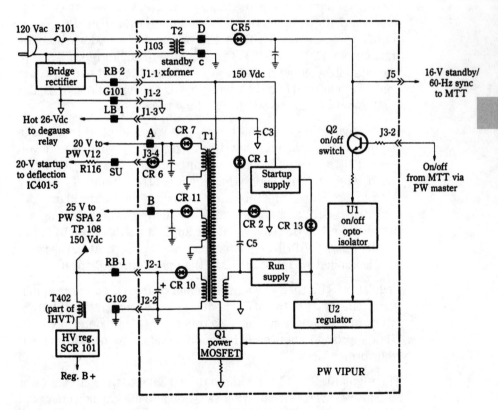

2-14 *A block diagram of the variable switching power supply in the RCA CTC130 chassis.*

This is accomplished by taking the hot 150 Vdc generated by the TV chassis primary power supply and applying it to the PW VIPUR circuit board. The PW VIPUR uses this (hot 150-V) dc supply to generate a (cold 150-V) dc supply that is routed back to the main chassis. Because the raw B+ that is supplied to the main chassis is cold, this makes the CTC130C chassis a cold chassis (isolated from earth ground) similar to a TV chassis with a separate power transformer.

The PW VIPUR module, when used, also develops the standby supply voltages for the MTT001 tuner/tuner control module. It develops dc supply voltage for the stereo broadcast circuit board (PW SBIB), the video/audio input/output circuit board, and the stereo power amp circuit board.

While the 150-V output of the PW VIPUR is regulated, it is also applied to the familiar SCR regulator circuit on the main chassis circuit board to generate the 127-V regulated B+. The 120 Vac is routed through fuse F101 to the primary of stepdown/standby transformer T2 (Fig. 2-15). The secondary transformer (T2) supplies standby voltage via half-wave rectifier CR5 to the tuner/tuner control module (MTT001A), when used. The output of CR2 is also used to power Q2, the on/off switching transistor.

On the main chassis circuit board, 120 Vac is applied to a bridge rectifier circuit when a (hot 150-V) dc supply voltage is developed. This dc supply voltage is routed to the PW VIPUR circuit board, where it is applied to the primary of transformer T1. During the initial startup cycle, this same 150 Vdc is routed through dropping resistor R2 and applied to zener diode CR3, generating approximately 11 Vdc. This voltage is applied via diode CR13 to pin 7 of IC U2, the VIPUR regulator integrated circuit. Also, the same 11 V is applied via resistor R12 to the on/off optoisolator IC (U1).

Regulator IC U2 receives on/off commands from the remote amplifier circuit (part of the MTT module) or the manual on/off button on the instrument front panel via transistor Q2 and optoisolator IC U1. U1 provides hot-to-cold ground isolation along with VIPUR transformer T2.

The regulator IC (U2) provides a drive signal to the gate of power MOSFET Q1. This signal switches Q1 on and off, causing energy to be stored in and discharged from the primary winding of switching regulator transformer T1. This energy is coupled to the secondary windings of power transformer T2, where it is rectified to provide regulated supply voltages of 150, 25, and 20 Vdc.

A hot supply voltage (12.5 V) is generated and used as a run supply voltage for the switching regulator IC (U2). The 12.5-V source is also applied through the voltage divider of R8 and R7, and trim resistors R4, R5, and R6, to pin 1 of regulator IC U2. The voltage at pin 1 is applied to an error amp internal to the IC. The output of the error amp is then used to control the duty cycle of the VIPUR output transistor (Q1). The hot supply secondary winding (used to generate the 12.5-V run supply for U2) is also used to generate a 26-V supply to power the instrument degaussing coil.

A slow-start circuit begins inside IC U2. The long time constant network consisting of resistors R1 and R3 and capacitor C7 controls the startup regulator IC by gradually allowing the pulse width modulator to reach its operating duty cycle. This prevents overheating of the MOSFET transistor (Q1) during startup.

Inside IC U2 is an overcurrent protection circuit. Should an overcurrent fault occur in the VIPUR output stage, it will be sensed by drain resistor R16. It is then applied to pin 11 of U2, shutting down the latch circuit (internal to IC U2). This will shut down the regulator circuit, turning off all power to the instrument. Remember to use an isolation transformer while working on this chassis. If not, damage can occur to the chassis or the test equipment, or the technician might be injured.

Sylvania C9 switched-mode power supply

The switched mode power supply used in the C9 series chassis is a self-oscillating, transformer-coupled voltage regulator (Fig. 2-16). The switched mode transformer (T401) is driven from the ac-line-derived 155 Vdc. Line insolation for the chassis is provided by the switched-mode transformer and the optoisolator (IC404). The 155 Vdc is applied to the transformer primary and controlled via the switched-mode regulator transistor (Q400).

Magnetic energy is stored in the transformer during the on time of the transistor, and is transferred to the secondary during the off time of Q400. This collapsing magnetic energy replenishes the voltage lost by the storage capacitors to the load during the on time of Q400. The amount of energy transferred is controlled by the self-oscillating frequency and the on/off time of transistor Q400. The self-oscillating frequency of the power supply under normal load conditions ranges from 20 kHz to 40 kHz. Depending on the load, the frequency can range from 10 kHz to 60 kHz.

VARIABLE SWITCHING POWER SUPPLY

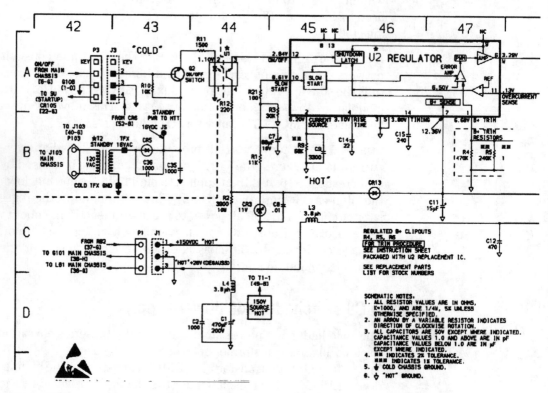

■ **2-15** *The complete variable switching power supply in the RCA CTC130 chassis.*
Thomson Consumer Electronics

60

(PW VIPUR) SCHEMATIC

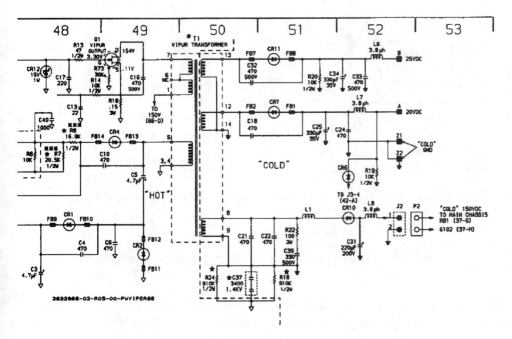

61

Sylvania C9 switched-mode power supply

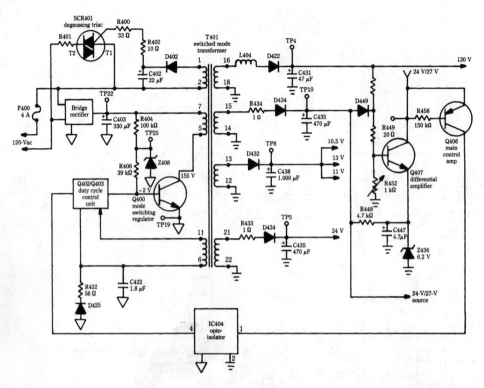

■ 2-16 *Block diagram of Sylvania's C9 chassis switched mode power supply (SMPS).*
Philips Consumer Electronics

An ac voltage is continuously applied to the power supply in all C9 chassis TV sets. Therefore, it's necessary to provide a method to degauss the picture tube. Degaussing is performed by the circuitry of triac SCR401. An ac voltage is applied across the degaussing coil, thermistor R401, and triac SCR401. When the chassis is first powered on, pulses in the secondary winding (1 and 2) are rectified by D402 and filtered with C402, This voltage turns on SCR401, allowing ac voltage to flow through the degaussing coil until thermistor R401 increases in resistance to stop the current flow, degaussing the CRT.

The differential amplifier (Q407) monitors the output voltage, compares it to a 6.2-V reference diode (Z436), and develops a dc correction voltage (Fig. 2-17). This correction voltage is inverted by the main control amplifier (Q406) and is fed back to the optoisolator (IC404). The charge to pin 1 of IC404 is proportional to the charge of the 130-V source.

If the source voltage increases, the light-emitting diode (LED) section of the optoisolator conducts more, creating more voltage at pin 4 of IC404. The increase at pin 4 adds to the charge voltage on C420, causing Q403/Q402 to turn on earlier. This action causes Q400 to turn off earlier, lowering the output voltage. The control circuitry varies the duty cycle and frequency to maintain the output at the preset level. Therefore, the output voltage is maintained by the input to the transformer and the on/off ratio of the switched mode regulator.

Startup procedure

The initial requirement for startup is to provide forward-bias voltage for Q400. When ac power is applied, base voltage is provided from the bridge-rectified 155 Vdc through the 22-Vdc zener supply, R404, Z408, and R406. The 155 Vdc is applied across the primary winding of the transformer, and collector current starts to flow when Q400 turns on. A voltage is also induced into the secondary winding (L_B) providing a positive voltage for the base drive for Q400 via pins 11, R421, D416, and C418.

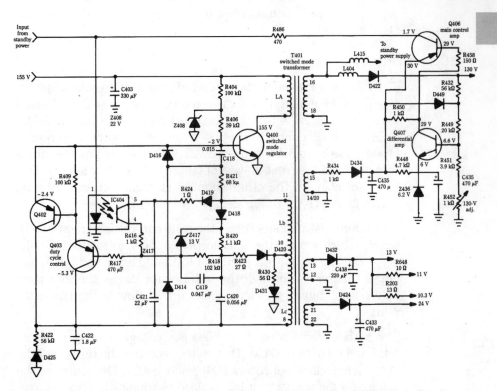

■ 2-17 *The switched-mode power supply control circuitry of Sylvania's C9 chassis.*

During this on time, a negative voltage from T401 (pin 8) is charging C422 with a negative charge. Q400 stays turned on until C420 charges positive enough to forward-bias the duty cycle control transistors (Q403/Q402). This applies negative voltage from C422 to the base of Q400, turning it off. Diode D416 blocks the startup voltage from flowing through the secondary winding (L_B). During the off period of Q400, the voltage from T401 (pin 11) becomes negative, and this negative charge on C418 is limited by the conduction of D414. By limiting the negative charge on C418, the positive charge of C418 can be accomplished more quickly. C418 controls the startup frequency during the period of time that the output voltage is building up to the operating level of 130 V.

The startup sequence repeats for several cycles until the dc voltage from the differential amplifier and main control amplifier via the optoisolator takes over control by adding or subtracting from the voltage developed by C420. During the on time of Q400, a negative voltage (–5.3 V) develops across the network of R422 and C422. This voltage provides a negative voltage for turning Q400 off when the duty cycle control transistors (Q403/Q402) turn on. When normal output voltage (130 V) has been achieved, the turn-on base voltage is then supplied by the secondary winding (L_B) via D416 and R421.

Control circuitry

The 130-V output voltage is monitored by the differential amplifier (Q407) and compared with the reference voltage (6.2 V) derived by zener diode Z436 through the divider network of R432, R449, R451, and R452 connected between the 130-V source and ground. Diode D449 is connected to the 24/27-V source and provides an alternate path of current in case of failure of the 130-V supply. This helps to maintain the other supplies at regulated levels during such a failure (refer to Fig. 2-17).

The differential amplifier drives the main control amplifier (Q406), which in turn sources the LED portion of the optoisolator (IC404), generating an output proportional to the dc error voltage. When current increases through the optoisolator diode, the increased light output causes an increase in current flow in the transistor portion.

The optoisolator feedback modifies the voltage developed across R418 and C419 from C420. That voltage controls the turn-on of the duty cycle control transistors (Q403 and Q402). The source of the voltage for the optoisolator is produced by winding L_b (pin 11), rectified by D419, and applied to pin 5 of optoisolator IC404. The output of the optoisolator is from pin 4 through R416 to the base circuit of Q403. Along with the voltage across C420, this voltage will adjust the trip point for the duty cycle control transistors proportional to the error voltage, thus advancing or delaying the turning off of Q400.

The maximum on time for Q400 is controlled by the RC time constant of R420 and C420. The induced voltage of winding L_b charges C420 via D418 and R420. The voltage developed across R418 and C419 is then added to this voltage, resulting in a shift in the turn-off point for the switched-mode transistor (Q400). When the sum of these voltages builds up enough to forward-bias Q403, the duty cycle transistors turn on and Q400 turns off.

During this off time, C420 will discharge through R423, D420, and through the lower winding (L_c). Additional control is provided by zener Z417 (13 V) and R419, which becomes a controlling factor when the voltage across L_b exceeds the zener voltage of Z417. The voltage across C420 is made up of two components, a positive portion and a negative portion. The positive portion is through R420 and D418, and the negative portion is developed by R423, D420, R430, and D421 during the off period of Q400.

At maximum load, the on time for Q400 will be at its maximum as determined by the voltages on C420. Q400 will remain on until C420 charges from its negative condition to a level sufficient to turn on Q403. Now the maximum on time occurs when the power supply is loaded at its maximum rated output. Feedback current from the optoisolator is zero, and the voltage across R418 and C419 will remain at zero as long as maximum load (short circuit) is on the output.

Normal operation

Regulation during normal operation is maintained by changing the duty cycle as well as the frequency. An output load change affects frequency, and ac line voltage changes affect the duty cycle. The frequency change occurs because the power supply is self-oscillating. The frequency will change rapidly with a change in power output, whereas a change in ac line voltage will produce a more gradual correction in output voltage.

Overload and short circuit protection

The current through the transistor portion of the optoisolator (IC404) approaches zero as the output load reaches maximum (full load). The bias voltage from pin 4 of IC404 is no longer contributing to the turn on of Q403, because the error voltage is too low to light the LED. The transistor now depends solely on the voltage across C420 for base bias. A longer time is required for Q403 to reach forward bias, resulting in an increase in conduction time for Q400.

The output voltage (130 V) will start to decrease as the load is further increased, creating an overload condition. The optoisolator feedback can no longer compensate for the output voltage change, because the light from the LED portion was already at minimum. The input to pin 1 of the optoisolator is proportional to the change in the 130-V output.

The decrease in the 130-V output causes the negative charge on C420 to become more positive during the off time of Q400, resulting in a reduced on time for Q400 proportional to the output voltage level, or inversely proportional to the output load. This process prevents the output power from exceeding the maximum design limit, and provides current protection as the output approaches a short circuit.

The current from the 155-Vdc source is also reduced when the supply is overloaded. This causes an increase in the pulse voltage at the cathode of D418 during the on time of Q400, activating the control circuit Z417 and R419. This action will increase the voltage across R418 and C419, causing Q403 to turn on earlier and reducing the on time of Q400. The turn-on time for Q400 continues to reduce as the output voltage approaches zero. Just prior to a short circuit, at a very low output voltage, the startup circuit will attempt to turn on Q400, but it will be turned off almost immediately. This will not provide any supply voltage for the 130-V output.

Under these conditions, only a small current will build up in the transformer, and normal oscillation can't be maintained due to the failure in buildup of the output voltage. Q400 is held off for the duration of time the secondary diodes are conducting until the current in the transformer reduces to zero. The cycle will repeat until the condition is corrected. With a complete short on the 130-V output and 120-Vac applied to the chassis, the on time for Q400 reduces to a series of very short spikes (approximately 10 μs every 160 μs). The on time is then controlled by the small startup current, which (together with C418) determines the cycling time. Startup cannot be obtained until the short is removed and the output voltage level is allowed to build up.

Regulator circuits

A defective regulator circuit can cause low (or no) voltage, hum bars, intermittent operation, and chassis and high-voltage shutdown. The regulator circuit can consist of transistors, SCRs, zener diodes, ICs, or a combination of these to provide accurate voltage regulation regardless of the fluctuations of the power line. Most regulator circuits have a B+ control that the technician can use to adjust the operating voltage applied to the horizontal output and low-voltage circuits.

A typical B+ regulated circuit is shown in Fig. 2-18. Here a B+ regulator and error amp transistor provides 115-V, 114-V, and 12-V regulated voltages for the TV chassis. Adjustment of R812 (115 V) places the exact operating voltage to the collector terminal of the horizontal output transistor. The F1 fuse protects the power supply from an overloaded horizontal circuit.

The line-operated low-voltage power supply can be regulated with an IC circuit (Fig. 2-19). A half-wave diode rectifier (D803) provides 135-Vdc to the regulator system, which consists of IC801 and associated components. The regulated 115-Vdc voltage provides power to the horizontal output transistor.

To quickly check the IC regulator system, measure the voltages at the input terminal (1) and at the output terminal (4) of IC801. Suspect a leaky IC801 if there is low output voltage. Excessive high voltage at pin 1 and 4 of IC801 can indicate a leaky IC regulator. Higher than normal output voltage on pin 4 can produce high-voltage shutdown. Check for a leaky D805 diode with one terminal

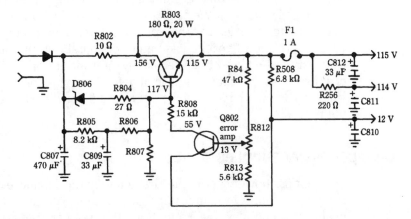

■ **2-18** *A typical low-voltage regulator circuit with a B+ regulator and error amp transistor.*

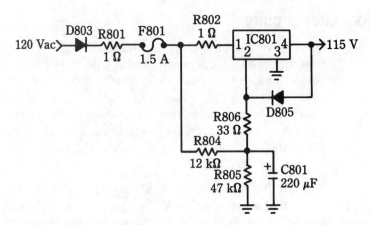

■ 2-19 *IC801 regulates the output voltage in a line-operated power supply.*

removed from the circuit board. If the chassis is hit by lightning, inspect D803 and the IC for visible component body damage.

The defective line-operated regulator can become shorted, leaky or open. A leaky or shorted IC regulator to common ground can blow the line fuse, open surge resistor and destroy silicon diodes in the power supply. Open or shorted components tied to the body of the regulator may cause the dead symptom. The open regulator can cause improper voltages to the horizontal output transistor.

The defective line voltage regulator may destroy the horizontal output transistor. These regulators can cause chassis shutdown. A shorted or leaky regulator from input to output terminals can place a high voltage on the collector terminal of the horizontal output transistor and destroy the transistor. Overloaded components or an overloaded flyback, yoke, and horizontal output transistor can destroy the line voltage regulator. Intermittent operation and chassis shutdown can result from a defective regulator. Suspect a leaky flyback with a leaky line voltage regulator and shorted output transistor.

Chopper power supplies

The chopper power supply was first found in the Japanese chassis and now is found in many American TV sets. The main components of the chopper power supply are the chopper regulator transistor, chopper transformer, regulator control transistor, x-ray latch, and overcurrent shutdown transformer (Fig. 2-20).

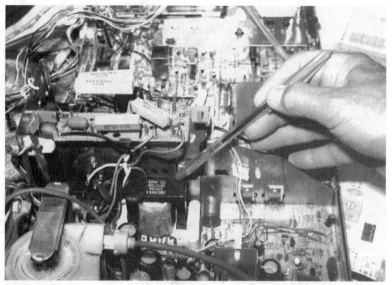

■ **2-20** *The chopper transformer located in RCA's CTC131 chassis.*

A chopper-type regulator circuit is found in RCA's CTC131 and CTC132 chassis (Fig. 2-21). Line (ac) power is supplied to a full-wave bridge rectifier circuit and a standby power transformer (T101). It is active at all times when ac power is applied.

The bridge rectifier circuit supplies an unregulated B+ of 150 Vdc to the chopper output circuit through chopper transformer T105. The standby transformer (T101) supplies power to the tuner control module and standby startup power to the chopper regulator IC (U401).

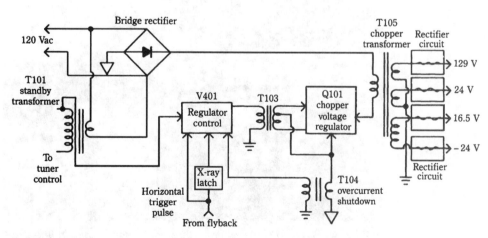

■ **2-21** *A block diagram of RCA's CTC131 chopper circuit.*

The pulse-width-modulated (PWM) chopper-regulated power supply is similar to the horizontal deflection system found in many TV chassis. The regulator free-running oscillator frequency of approximately 15 kHz is triggered by a horizontal pulse derived from the flyback, thereby locking it to the horizontal scan frequency (15734 Hz).

The output of the regulator circuit is applied to regulator drive transformer T103. Increased pulses on the secondary of T103, in turn, control the on/off state of the chopper output circuit. The chopper output circuit is powered by the unregulated "raw" B+ (150 Vdc) developed by the full-wave bridge rectifier circuit. The on/off state of the chopper output circuit causes a pulsating dc action to occur in the primary winding of the chopper transformer (T105). Increased pulses in the secondary windings of T105 are then rectified to produce a number of dc voltages used to power the TV receiver. The 129-Vdc source is fed back to the regulator control circuit where it is applied to a voltage comparator circuit. The voltage comparator output is applied to an error amp that is used to control the duty cycle of the pulse width modulator, thereby providing regulation for the B+ sources produced by the secondary windings of chopper transformer T105 (Fig. 2-22).

Suspect an open input fuse or a leaky chopper output transistor (Q101) with a no sound/no picture/no control symptom in the RCA CTC131 chassis. Sometimes a voltage isolation resistor ahead of the bridge rectifier diodes will open. Check U401 for poor voltage regulation.

■ **2-22** *A close-up view of the chopper components in the RCA CTC131 chassis.*

Flyback transformer low-voltage circuits

Today, many of a TV's low-voltage circuits are driven from separate windings on the flyback or horizontal output transformer. Several advantages are gained by tapping the flyback winding for a low-voltage source. A positive or negative voltage can be obtained by reversing the low-voltage diode polarity. Voltages can be produced without large voltage-dropping resistors in a regulator voltage network; thus, less heat is generated. Filtering is less critical because of the much higher ripple frequency. Often, small filter capacitors (less than 470 µF) are found in the flyback voltage circuits. You might find transistors and zener diode regulators in some low-voltage circuits of the flyback winding. Remember, the horizontal deflection and horizontal output circuits must function before any voltages are found at the flyback voltage power source.

A typical low-voltage flyback circuit has a half-wave rectifier diode with a capacitor filter network (Fig. 2-23). Various voltages are tapped off with small resistor-capacitor filter components. A zener diode can regulate the voltage source (12 V) that feeds the luminance and chroma circuits. In other circuits, a transistor or SCR will regulate the low-voltage source.

Another voltage source from the flyback winding can supply voltage to the screen or color output circuits (Fig. 2-24). Usually this voltage is over 200 V with a single half-wave rectifier (D701). R702

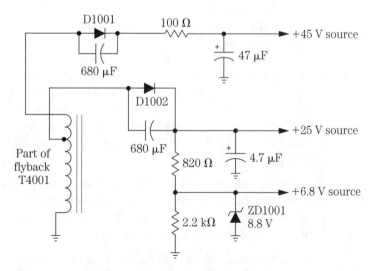

■ **2-23** *Check the secondary circuits of flyback for voltage sources feeding other TV circuits.*

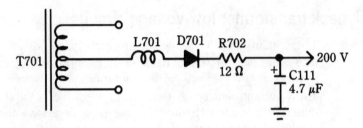

2-24 *Voltages feeding the color output transistors and screen grid circuits of the picture tube are developed in the secondary circuits.*

and C111 are an RC filter network. A negative or low brightness is a possibility when the voltage is missing at the color output stages. Sometimes only a faint color can be seen in the raster. Check both D701 and R702 for leaky or burned conditions.

Remember, when secondary flyback voltage sources feed voltage to the horizontal oscillator IC, the horizontal circuits must operate before any of these voltages are produced. When no or improperly low voltages are found in the secondary sources, the horizontal oscillator sweep IC, driver, horizontal output transistor, and flyback circuits can be defective. Any one of these circuits or a combination of components can result in a no secondary voltage source.

RCA CTC130 IHVT scan derived B+ voltages

Most of the operational B+ supplies and horizontal keying pulses are derived from the secondary windings of the integrated high-voltage transformer (IHVT) (T402). The primary winding of the IHVT is switched by the horizontal output transistor (Q402). The B+ for the primary winding is supplied by the regulated (127-V) B+ circuit. The energy developed across the primary winding of the IHVT (as a result of the switching of the horizontal output transistor) is induced into the secondary windings and is used to develop the derived operational supply voltages and horizontal keying pulses (Fig. 2-25).

The supply voltages are the high voltage, focus voltage, screen voltage, and filament voltage to the picture tube. The dc return path for the high-voltage supply is through the beam limiter circuit inside the chroma/luminance IC (U701, pin 28).

A negative-going 200-V p-p pulse at pin 7 of the IHVT is trace rectified by CR109 and filtered by C121. This provides the distributed 16-V supply used by various circuits throughout the in-

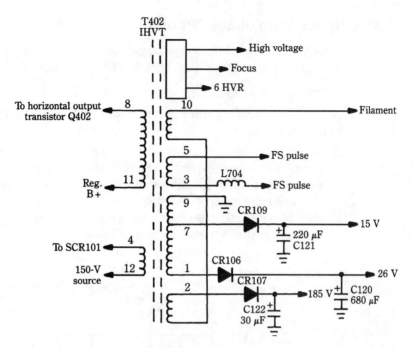

2-25 *Scan-derived voltages found in the secondary winding of the flyback transformer in an RCA CTC130 chassis.*

strument. The 16-V supply is applied to dropping resistor R325 and zener diode CR301 to develop the 12-V supply and to dropping resistor R745 to develop the 11.2-V supplies that are used extensively throughout the IF and chroma/luminance processing circuits.

A negative-going 200-V p-p pulse at pin 1 of the IHVT is trace rectified by CR106 and filtered by C120, and this provides the main distributed B+ source of 26 V. Positive-going horizontal pulses developed at pin 10 of the IHVT (used primarily for CRT filament supply) are sampled by the x-ray protection circuit. If the pulses exceed a certain level, the x-ray protection circuit is activated, defeating the horizontal oscillator circuit at U401 (pin 4), and the chassis is shut down.

A positive-going 20-V p-p horizontal pulse developed at pin 2 of the IHVT is trace rectified by CR107 and used to power the horizontal driver circuit. The output of CR107 is then applied to R114 and filtered by C105 to provide a 185-V source. In some models employing the CTC130 chassis, this voltage (185 Vdc) source also supplies B+ to the tuner.

Servicing the low-voltage circuits

Three quick tests of the low-voltage power supply can locate the defective component or dead chassis within seconds (Fig. 2-26).

☐ Take a peek at the fuse.

☐ Measure the voltage on the case of the horizontal output transistor.

☐ Measure the voltage at the output of the low-voltage power supply. The fuse can open because of an overloaded condition inside or outside of the power supply. Power line outage or lightning can also open the suspected fuse. The fuse might open with a flash or arc over a component, and when replaced, restores the set to normal operation.

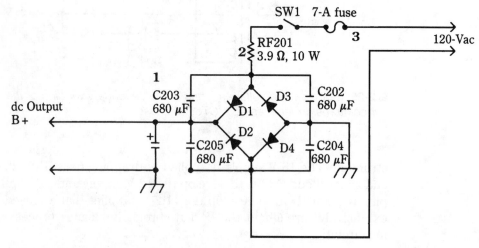

74

■ **2-26** *Check the low-voltage circuits by the numbers 1, 2, and 3 in a J.C. Penney 685-2033.*

After checking the fuse, test to see if low dc voltage is present from the low-voltage power supply. Quickly check fuseable resistor RF201 (3.9 Ω) for an open condition if there is no output voltage. With the digital multimeter (DMM) set at diode test, check all diodes in the low-voltage circuit. A leaky diode will show leakage in both directions. The conventional bridge, half-wave, or full-wave rectifier power supply can be checked within 2 minutes.

In this particular J.C. Penney model 685-2033 shown in Fig. 2-26, the 7-A fuse and RF201 were found open. Diodes D3 and D4 were shorted. The two diodes were replaced with 2.5-A silicon replacements.

Besides fuses and low-ohm resistors and diodes, look for an open switch or poor wiring connections. If the on/off switch appears erratic, replace it. Disconnect the power cord and measure for continuity with the switch in the on position. Another quick method is to plug the power cord in and measure the 120-V power line voltage across the switch terminals. Replace the defective switch if the full power line voltage is measured across the switch terminals in both on and off positions. A quick ac voltage measurement across the bridge rectifier will identify problems with an open fuse, low-ohm isolation resistor, and power cord circuits.

You can locate a separate winding of the flyback transformer in the ac circuits of the low-voltage power supply. The horizontal circuits are "kicked on" by ac applied to this winding (Fig. 2-27). When ac voltage cannot be measured across the diodes at points 1 and 2, suspect an open winding of T702 or poor board connections to the flyback winding and R705.

Blowing fuses

Suspect overloaded conditions in the low-voltage power supply or connecting circuits when the fuse blows immediately after replacement. The most common leaky components in the low-voltage circuits are the silicon rectifiers and filter capacitors. Outside of the power supply, the horizontal output transistor and damper diode cause most fuses to open.

A quick resistance measurement between the horizontal output transistor collector terminal (case) and chassis ground indicates if

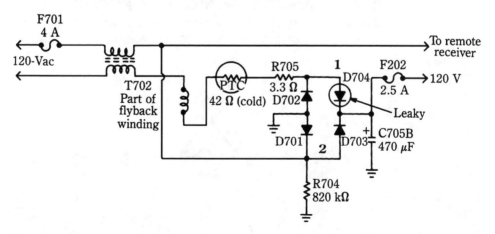

■ **2-27** *Suspect an open winding in T702, poor board connections, and an open R705 when ac voltage cannot be measured at numbers 1 and 2.*

the leaky component is outside the low-voltage power supply and in the horizontal section (Fig. 2-28). Set the DMM at the diode test and check from transistor case and chassis ground. A low resistance reading indicates a leaky output transistor. Any resistance measurement below 500 Ω indicates a leaky transistor or damper diode.

Suspect leaky diodes in the power supply if the resistance measurement is above 1 kΩ across the transistor and chassis ground. Check each diode for a low resistance in both directions. A leaky input filter capacitor can cause the fuse to open.

Another method to determine if the overloaded condition is outside the power supply is to clip a 100-W bulb across the fuse terminal (Fig. 2-29). The light bulb will be bright if there is a leaky

■ **2-28** *Take a quick diode or resistance test between the collector (body) and the chassis ground to locate a leaky output transistor.*

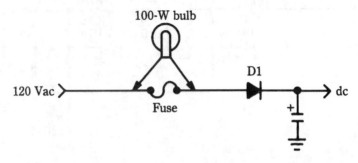

■ **2-29** *Clip a 100-watt bulb across the fuse clips to determine if overloaded condition exists, when the line fuse keeps blowing.*

component in or outside the power supply. Under normal conditions, at first the light bulb will appear bright and slowly go out due to the degaussing circuit action. Remove the horizontal output transistor and notice if the bright light goes out. This indicates an overloaded condition in the horizontal output circuit. Each circuit tied to the low-voltage power supply can be cut loose. When the light goes out, you have located the overloaded circuit outside the low-voltage power supply. Even a shorted picture tube can blow the main power fuse.

A real black area inside the glass fuse indicates a direct short ahead of the fuse. When both the line fuse and secondary fuse open, suspect a defective component in the horizontal output transistor circuits. You may find an open fuse and after replacement, the TV chassis operates without any other problems. The loose fuse in the clip-type socket can cause an intermittent or shutdown symptom.

Intermittent power supply

Most intermittent problems in the low-voltage power supply are caused by improper component connections, poor wiring connections, and defective transistors. Monitor the ac voltage applied to the rectifier at point 1 and the dc voltage output of the low-voltage power supply at point 2 (Fig. 2-30). When the voltage is missing from either or both circuits, you know in what section the defective component is located.

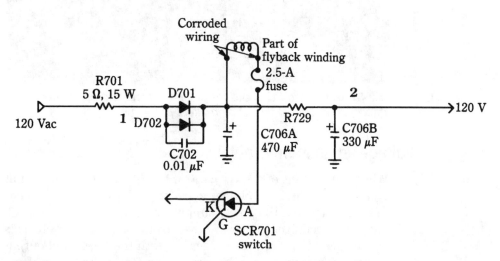

■ **2-30** *Poor or corroded terminal connections to the flyback winding caused intermittent reception in a Sharp C1950.*

Sometimes by just moving or tapping the TV chassis, the intermittent condition begins to run in and out. Do not overlook a loose fuse holder. Suspect a defective relay or relay contact in TV receivers with remote or relay-controlled ac circuits. Poor ac cord interlock contacts can also produce intermittent low-voltage conditions.

Double-sided wiring boards have a tendency to warp. This cracks the printed wiring at the feed-through eyelets in the low-voltage sources. Just touching or pushing on the board can cause the raster or color to come and go. These intermittent wiring connections can be repaired with lengths of hookup wire fed through the eyelet and connected to the nearest corresponding component connection of the same broken circuit (Fig. 2-31). Low-ohm tests between broken wiring connections can help to locate the intermittent connection.

■ **2-31** *Check the PC board for broken eyelets and component connections, which can cause intermittent problems.*

B+ voltage adjustment problems

When adjustment of the B+ control has no effect on the output voltage from the low-voltage power supply, suspect component breakdown within the power supply circuits. The B+ control will have no effect if the load is removed from the power supply. This is especially important to remember with the horizontal output transistor removed or when chassis or high-voltage shutdown occurs.

Improper adjustment of the B+ control can cause hum bars in the raster or insufficient image width. Incorrect setting of the B+ control can cause a very high voltage to be applied to the horizontal output transistor resulting in high-voltage shutdown. In some chassis, incorrect setting of the B+ control can cause the horizontal section to go out of sync (diagonal bars) instead of causing high-voltage shutdown.

Check for leaky or open regulator transistors in the regulator circuit of the low-voltage power supply with no B+ adjustment. Look for overheated zener diodes producing lower regulated voltage. Measure the resistance of each resistor in critical low-voltage regulator circuits. Always adjust the B+ control at the manufacturer's required operating voltage.

Panasonic GXLHM B+ voltage adjustment

Connect a dc voltmeter to TP91 (Fig. 2-32). Set the color pilot switch to off. Set the brightness control (R312) to minimum and the panabrite (R320) to minimum. Tune in a black-and-white signal. Cause the raster to disappear by turning the screen (R372) and subbright (R316) controls.

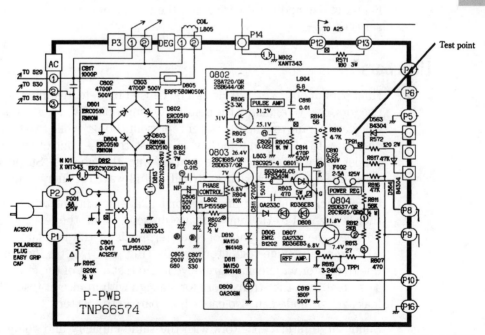

■ **2-32** *Connect a voltmeter to TP91 to set the B+ voltage in a Panasonic CTH-2560R.*
Matsushita Electric Corporation

Adjust the B+ adjustment control (R812) to 129-V I IV at TP91. To confirm the B+ voltage, check the voltage at TPD7 (12 V), TPD9 (26.8 V), terminal C7 (200 V), tuner BM (12 V), and tuner BP (5 V). Now, return the screen and subbright controls to their original positions. Tune in channel 12 and confirm the voltage at TPA8 (12 V), then change to channel 2 and confirm that the voltage is within –9 V and –21 V.

Lightning damage

Lightning or high-voltage power line surge can destroy several components in the low-voltage power supply. Usually, lightning damage occurs in the antenna lead or in the power line circuits. A heavy lightning strike can damage several sections of the TV chassis. Stripped sections of printed wiring might be found as a result of excessive lightning damage.

Check for burn or smoke marks on the chassis and around the ac wiring connected to the TV chassis. Look for blown-apart components such as transistors, ICs, and resistors. Burned silicon diodes and connecting board wiring can be signs of lightning damage. The receiver can be repaired if there is only a little damage to the power supply or antenna terminals. Total out the cost of the TV if lightning damage is excessive and the set cannot successfully be serviced.

Hum bars

Hum bars of 60 Hz or 120 Hz can be caused by defective filter capacitors, leaky regulator transistors, or improper B+ adjustment. One dark bar moving up the raster represents 60 Hz, while two hum bars is a 120-Hz symptom. Try to adjust the B+ control to the proper voltage setting before tearing into the chassis. Suspect dried-up large filter capacitors in half- or full-wave rectifier circuits.

Check each regulator transistor for leakage conditions when either 60-Hz or 120-Hz hum bars are present. Often the leakage is between the emitter and collector terminals of the regulator or APF amp in voltage regulator circuits (Fig. 2-33). After locating one transistor with leakage, remove it and test it out of the circuit. Then test the other transistors for leakage with in-circuit tests. A leaky zener diode can also cause hum bars in the raster.

Shunt large filter capacitors with the same capacity and voltage rating when hum bars are found in the raster. Always shut the

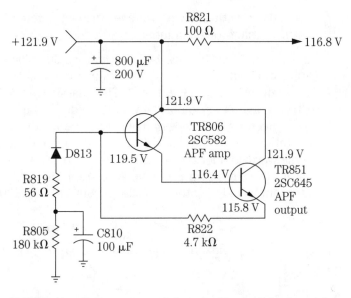

2-33 *Hum bars can result from leaky regulator transistors in the low-voltage circuits.*

chassis down and clip the filter capacitor in place to prevent damage to transistors or ICs. Clip across each filter capacitor until the defective section is located. Replace the entire filter can if more than one filter capacitor is in the same container, even if only one capacitor is defective.

Chassis and high-voltage shutdown

High-voltage shutdown can occur with too much voltage from the low-voltage power supply fed to the collector terminal of the horizontal output transistor. The chassis might shut down instantly or operate for a few minutes. Check the B+ voltage adjustment. Sometimes if the B+ adjustment is too high and the line voltage increases, the higher dc voltage can cause the chassis to shut down.

Monitor the dc voltage at the collector terminal of the horizontal output transistor. Often a 10-V increase in voltage at the collector can cause high-voltage shutdown. Lower the dc voltage with a variable ac power transformer at the power cord. Set the transformer for correct voltage at the collector terminal. If the chassis shuts down, lower the ac voltage about 10 V. Keep lowering the ac voltage as the dc voltage creeps up. If the chassis operates for 30 minutes without shutdown, notice the ac setting. If the chassis shuts down with a low ac voltage, suspect a defective horizontal output or shutdown circuit.

In an Admiral 4M10 chassis, the raster would collapse after operating for about 10 minutes. The B+ adjustment had no control over the high dc voltage (Fig. 2-34). The chassis would shut down at 131 Vdc and operate normally at 120 Vdc. Voltages were high at the collector and base terminals of the regulator transistor. After operating a few minutes, D902 (125-V zener diode) should become warm and let the voltage increase to 130 V. Replacing D901 solved the shutdown problem. After replacing D901, the B+ adjustment would vary the dc voltage from 114 V to 130 V.

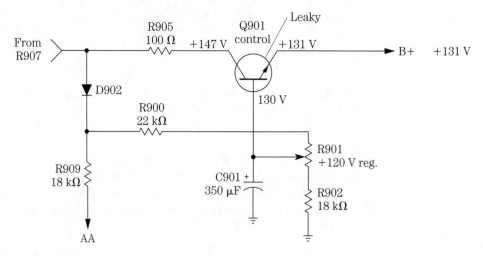

■ **2-34** *A leaky Q901 caused the raster to collapse after ten minutes of operation.*

Leaky regulator transistors can cause chassis shutdown. Make sure regular or zener diodes in the regulator circuits are not damaged. Check for a leaky regulator IC component in many of the later TV regulator circuits. While the transistor is out of the circuit, check each resistor for correct resistance.

Shutdown horizontal lines

Suspect higher than normal voltage on the collector terminal of the horizontal output transistor and low-voltage power supply when the raster displays horizontal lines. Measure the dc voltage at the low-voltage source (Fig. 2-35). In some early solid-state TV chassis, to prevent excessive high voltage at the picture tube, a shutdown circuit would make the picture go into horizontal lines

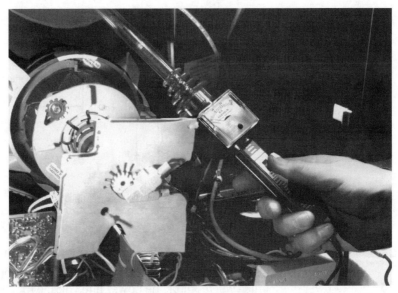

■ **2-35** *Check the high voltage at the anode of the CRT when horizontal lines are present, indicating high-voltage shutdown.*

instead of complete chassis shutdown. These horizontal lines cannot be removed with the horizontal hold control as long as the chassis is in shutdown.

To determine if the horizontal oscillator circuits are defective or if the chassis is in shutdown, check the dc voltage applied to the horizontal output transistor. If the B+ adjustment will not turn the voltage down, suspect a defective low-voltage regulator circuit. First, check all transistors and diodes for leakage. Next, measure the resistance of each critical resistor, especially those that have changed color and feel warm. Check electrolytic filter capacitors with a capacitor tester for a change in capacity. Suspect horizontal oscillator trouble if the dc voltage is normal at the horizontal output transistor.

Excessive line voltage

Excessive power line voltage can destroy components in the TV chassis and cause continued breakdowns. The ac power line voltage should never exceed 125 V (Fig. 2-36). The TV receiver was designed to work on a 117-V to 120-V ac power line. Usually, excessive power line voltage is found where the power line transformer is located nearby.

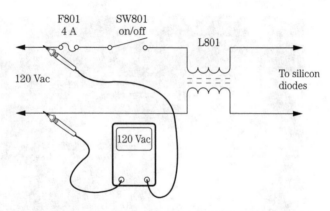

■ 2-36 *Check the ac voltage at the TV ac line input for normal 120 volts ac.*

Check the power line voltage where the TV plugs into the ac receptacle. Next, check the power line voltage at the fuse box. Most power companies will monitor the power line voltage for 3 to 5 days to determine if the voltage is high or low. Repeated service calls or chassis breakdowns can be caused by high power line voltage.

Insufficient image width

Often, insufficient width is caused by component breakdown in the horizontal output circuits. The raster can pull in from the sides if insufficient low-voltage power is applied to the horizontal output transistor. You might find the B+ control is functioning, but the dc voltage is lower than normal and cannot be raised above the operating voltage.

A leaky regulator transistor can cause an increase in the dc voltage source. Although an open transistor can lower the dc voltage, the open zener diode will let the voltage increase. Suspect an open regulator transistor or defective filter capacitor when the dc voltage cannot be returned to normal operation (Fig. 2-37). Sometimes these regulator transistors will test normal, but will open under actual working conditions.

Unusually low voltages

The 1.25-A fuse would blow in a K-Mart KMC1311G portable TV when the ac cord was plugged in. A voltage check at the horizontal output transistor indicated low voltage, then shutdown. A 100-W bulb was clipped across the fuse terminals. Low sound and no raster resulted. This indicated that the horizontal and low-voltage circuits were functioning, but there was no high voltage. A 2-A fuse was in-

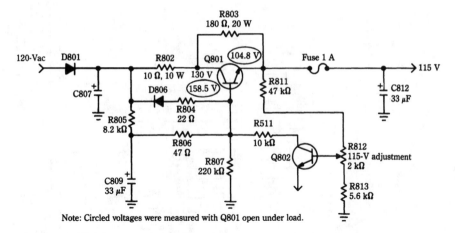

Note: Circled voltages were measured with Q801 open under load.

■ **2-37** *An open filter capacitor or regulator transistor can produce low output voltage.*

serted and it blew at once. After 2 hours, a 3-A fuse was inserted. Right away, the picture tube started arcing in the gun assembly.

Although fuses with larger amperage should never be left in a TV chassis, the TV symptoms here indicated a defective component in the high-voltage section. The larger fuse was tried as a last resort. If left operating too long, the horizontal output transistor could have been damaged. Use this method only to smoke out a defective component and only for a few seconds. Installing a new picture tube solved the repeated blown fuse problem.

Servicing the Realistic TC-1011 low-voltage power supply

To determine if the low-voltage power supply is defective, measure the dc voltage at the bridge rectifiers or filter capacitor (140 V). With low or no voltage, suspect trouble in the diodes, R602, or the line fuse (Fig. 2-38). If the dc voltage is high at the converter output transistor (normal 138.8 V), check Q601 for open conditions or no base signal. If no waveforms are found at the collector and base of Q601, check all transistors, diodes, and small resistors with the DMM.

Check diodes D605, D606, D607, and D650 with the diode test. Check resistors R606 (27 Ω), R609 (120 Ω), and R630 and R603 (47 Ω) for correct resistance. Remove one lead if the meter measurement is off. Test Q601, Q602, Q603, and Q604 in and out of the circuit. Each side of the photocoupler (IC601 and IC602) can be checked like any transistor (Fig. 2-39). The LED should show a resistance in only one direction, with the emitter and collector elements at no resistance, even with reversed test leads.

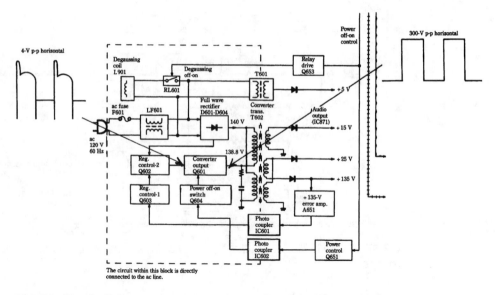

■ **2-38** *Check diodes and transistors in the Realistic TC-1011 converter-type power supply.* Radio Shack

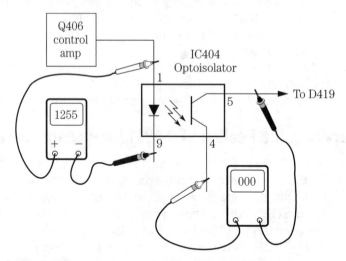

■ **2-39** *Check the optoisolator (IC404) with the diode tester of the DMM.*

Overloading of the secondary voltages (15 V, 25 V, and 135 V) with a defective component can shut down the chassis and destroy Q601. Check each voltage source for an overload. Make sure the silicon rectifier is not leaky. Take a resistance measurement from the collector of each diode to common ground. Remove one end of the diode when a power source is suspected, thus eliminating it from the circuit. Locate the defective component in that voltage source, and take another leakage test.

Servicing Sylvania's C9 switched-mode power supply

Troubleshooting power-related problems in the C9 chassis will most likely be performed within the switched-mode power supply (SMPS). Most of the voltages for the chassis are developed in the SMPS, except a 200-V source developed from the horizontal scan system. Remember, there are two different ground systems in this chassis, and the isolation transformer should be used. The chassis (signal) ground connection can be made to the IF area shield, and the hot (ac) ground can be made to TP19. Do not use any heat sinks for a common ground connection.

The most common defects in the chassis are Q400, Q402, and Q403. When Q400 becomes shorted or leaky, check D425 and D421 for burned or leaky conditions. Often, the small resistors R422 (56 Ω) and R430 (56 Ω), within each diode circuit must be replaced. First, check all transistors and diodes within the SMPS circuit (Fig. 2-40). Do not forget to check for burned bias or shunt resistors when replacing a leaky transistor or diode. Notice the hot ground is on the primary side of T401 and the common ground is on the secondary side. Check the following service notes:

☐ A blown fuse (F400) will almost always point to a defective switched-mode regulator transistor (Q400).

☐ A shorted Q400 can create a short or open circuit in the bridge diodes (D404 through D407).

☐ The switched-mode regulator transistor (Q400) will be destroyed without the function of the feedback circuit of the duty cycle control transistors (Q403 and Q402). Always check Q403 and Q402 when Q400 fails.

☐ If either of the duty cycle control transistors (Q403 or Q402) are shorted, the startup of the SMPS cannot be accomplished because the base of Q400 is always returned to a negative voltage.

☐ If the duty cycle control circuit is open, the switched-mode regulator transistor (Q400) will be destroyed as soon as the chassis is powered on.

☐ If any of the above conditions are found, verify the value of R434. R434 is the fuseable resistor for the 24–27-V source. If it is open, supply an external 24-V supply to the circuit with an external dc power supply.

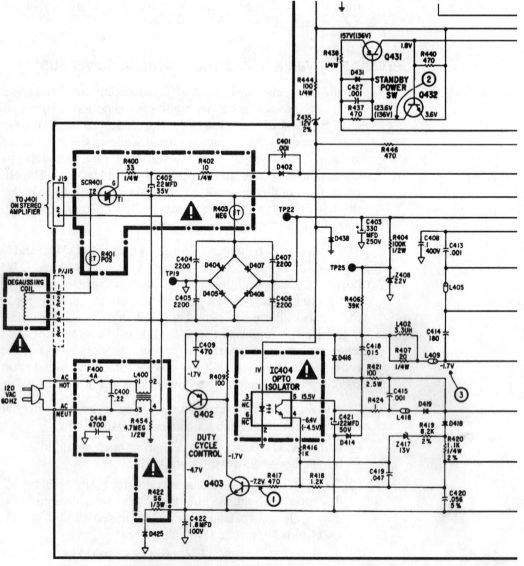

■ 2-40 *Check all transistor and diodes in the SMPS circuit of the Sylvania C9 chassis after taking critical voltage and waveform tests.*

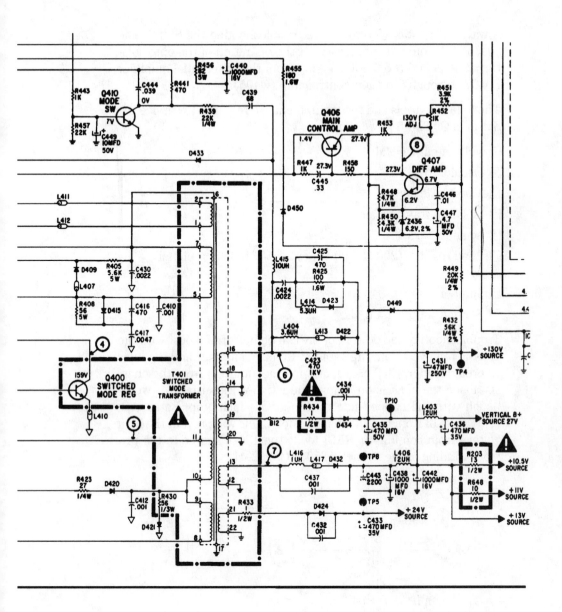

Always repair this section before troubleshooting the SMPS. The 24- to 27-V source is necessary for the operation of the optoisolator (IC404) and the feedback loop from the differential amplifier (Q407) through the main control amp (Q406).

The following tools and test equipment are required to service the Sylvania C9 SMPS power supply:

- [ ] Voltmeter or DMM
- [ ] Oscilloscope
- [ ] Variable ac supply
- [ ] Isolation transformer
- [ ] ac ammeter (5 A)
- [ ] Load resistor (200 Ω, 100 W) or a 60-W bulb and pigtail socket
- [ ] Jumper wires
- [ ] Service shop hand tools

Troubleshooting the power supply can be performed best with the chassis removed from the cabinet. Remove all plugs and modules (stereo decoder, CRT board, and microcomputer) from the chassis. Place the chassis on a nonconductive work surface or workbench with the copper side of the chassis facing up. Be sure to remove R513 from the chassis for safety. This will prevent the high voltage from coming up. R513 feeds the 130-V source to the primary winding of the flyback (T504) (Fig. 2-41).

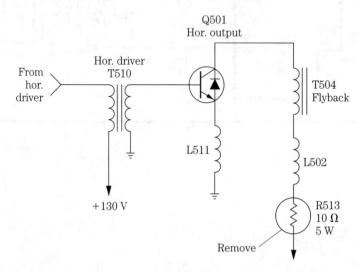

■ **2-41** *Remove one end of resistor R513 to prevent high voltage from coming up in the Sylvania C9 chassis.*

Check the following components with the ohmmeter. Be sure the chassis is removed from the ac power source. Make these checks before firing up the chassis:

☐ Check fuse F400.

☐ Check switched-mode regulator transistor Q400.

☐ Check the duty cycle control transistors (Q403 and Q402).

☐ Check the differential amp (Q407) and the optoisolator (IC404).

☐ Check the fuseable resistors in the secondary (R433 and R434).

☐ Check diodes D404–D407, D424, D432, D434, and D422.

During the following test, the power supply will be operated without a load on the 130-V supply:

1. Place a jumper wire from the collector to the emitter of Q410 (this disables the standby power supply).

2. Plug the isolation transformer into the ac outlet.

3. Plug the variable ac supply into the isolation transformer. If you have a variable isolation transformer, use it.

4. Connect a dc voltmeter to the 130-Vdc source (TP4).

5. Use the ac ground for the oscilloscope ground.

6. Connect the oscilloscope probe to the collector terminal of Q400.

7. Turn the variable ac supply to minimum.

8. Plug the ac line cord of the TV into the variable ac supply.

9. Make sure R513 is disconnected from the circuit.

10. Turn on the variable ac supply and slowly increase the ac voltage while monitoring the 130 Vdc and the ac ammeter.

There should be a rush of current until regulation is achieved, and then very little current is necessary to maintain the 130-Vdc source. If the ac ammeter does not peak and start back in the other direction, or if the 130-Vdc source continues to go beyond 130 Vdc, a defect is present in the chassis. Do not increase the ac input beyond 90 Vac during this test.

When the 130-Vdc source is higher than normal, there is an open in the feedback circuit of Q407, Q406, and IC404. With a normal working power supply and chassis, the ac ammeter will reach a peak around 375 MA and develop the 130-Vdc regulated source at approximately 40-Vac input to the chassis. Continued increase of the ac supply voltage will allow the ac current to decrease to approximately 240 MA at 60 Vac. Further increase of the ac supply voltage will begin to increase the regulated output voltage beyond the 130-V level. The ac ammeter reading will be 220 MA at 90-Vac input.

Connect a 60-W light bulb or a 200-Ω, 100-W resistor across the 130-Vdc power source for this test:

1. Connect the load resistor or light bulb between the 130-Vdc source (TP4) and the chassis ground (IF shield).
2. Place a jumper wire from the collector to the emitter of Q410 (this disables the standby power supply).
3. Plug the variable isolation transformer into the ac socket.
4. Connect a dc voltmeter to the 130-Vdc source (TP4).
5. Use the ac ground for the scope.
6. Connect the scope probe to the collector of Q400.
7. Turn the variable ac supply to minimum.
8. Plug the ac cord into the variable isolation transformer.
9. Make sure R513 is removed.
10. Slowly increase the ac voltage while monitoring the 130 Vdc and the ac ammeter.
11. If the ac ammeter does not peak and start back in the other direction, or if the 130-Vdc supply source increases beyond 130 Vdc, a defect is present in the chassis. (This does not include the horizontal output and IF area.)

An increase beyond the 130-Vdc level indicates an open in the feedback circuit of Q407, Q406, and IC404. When normal, the ammeter will reach a peak around 2.4 A and develop the 130-Vdc regulated source at approximately 90-Vac input to the chassis. Continued increase of the ac supply voltage will allow the ac current to decrease approximately 1.4 A at 120 Vac.

This next test will give you an idea of how the SMPS power supply operates with overload on the 130-Vdc line. Remember, the 130-Vdc source feeds the horizontal output transformer circuits. Any leaky component or no base drive signal on the horizontal output transistor can shut down the chassis.

1. Connect a jumper wire from the 130-Vdc source (TP4) to the chassis ground.
2. Place a jumper wire from the collector to the emitter of Q430 (this disables the standby power supply).
3. Plug the variable isolation transformer into the ac outlet.

4. Plug the ac cord into the variable isolation transformer.

5. Connect an ac voltmeter to the 130-Vdc source (TP4).

6. Connect the scope to the collector of Q400.

7. Turn the variable isolation transformer to minimum.

8. Make sure R513 is removed from the circuit.

9. Slowly raise the variable ac supply while monitoring the 130 Vdc and the ac ammeter. If the ac ammeter does not move up the scale very much in this test, the SMPS will sense this short and will not have any output voltage, indicating overcurrent shutdown.

10. The ac current does not increase a slight amount as the ac voltage approaches 120 Vac. It can read as much as 350 MA at 120 Vac, but should not exceed this value.

Realistic circuit board indications

Voltage regulators and IC pin connections in Realistic 16-410 TV chassis are indicated in Figure 2-42. The 3-pin regulator IC terminal input connection is shown with black around the input terminal, while in other ICs pin 1 and every fifth pin are indicated with black around the pin terminals. Socket pin terminal connections are indicated with black area around pin number 1. Double-check the position of terminal 1 of any regulator ICs, microprocessors, plugs, and sockets before removing them from the PC board.

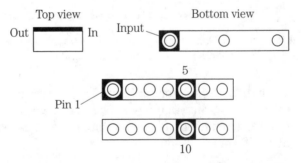

■ **2-42** *In some of the Radio Shack TV chassis, the 3-pin IC regulator and other ICs are numbered for correct identification.*

"Tough dog" power supply

Anyone who has spent hours trying to find the defective component in a TV chassis might refer to it as a "tough dog" repair problem. These problems are hated by most TV technicians, and you never forget about it by the next time the same chassis comes across the service bench. You might have one that bothers you, but it might not be a service problem for the next technician. Do not blame the manufacturer for a "tough dog" problem.

The first time a Goldstar CNR-405 chassis appeared on the service bench, it was classified as a "tough dog" repair of the power supply circuit. Not only was servicing the chassis without a schematic a great problem, several different components were found to be defective (Fig. 2-43). Now, servicing the same power supply circuit is a cinch and it takes very little service time.

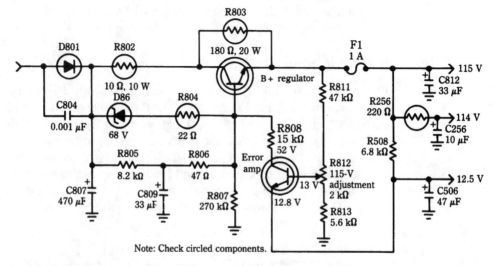

■ 2-43 *A "tough dog" GoldStar CNR-406 chassis took a lot of service time to locate all the defective components in the low-voltage circuits.*

Check the B+ regulator (2SC1629) and error amp (2SC1573P) for leakage conditions. Double-check for leakage after removing them from the circuit. The B+ regulator can be replaced with a universal ECG86 or an SK3563.

Zener diode D86 can become leaky and produce higher-than-normal output dc voltage. When both D86 and the regulator transistor become leaky, the dc source can go to 143 V, producing high-voltage shutdown. Check fuseable resistor R804. Other critical power resistors in the circuit of the low-voltage regulator transistor are R802 and R803.

Standby power supplies

The standby power supply is found in many of the new chassis to provide some means of turning the TV off and on with remote power. Most earlier TV sets used a separate power supply because standby power was needed continuously. Now standby power is only used during the time full power is not needed, making the TV set more economical to operate.

In older remote TV receivers, the standby power was turned on with relays from a separate power supply. Today, the standby power circuits are switched on with switching transistors or IC circuits. To disable the standby power circuit while servicing the TV chassis, short the collector and emitter terminals of the mode switch transistor (Fig. 2-44). Just place a clip wire across the emitter and collector terminals of the mode switch.

Sylvania C9 standby/full power circuits

Operation of the standby/full power circuit maintains low-level operation of the switched-mode power supply (SMPS). Any time the set is plugged into the ac power line, the SMPS begins running in the standby mode. This operation is accomplished by taking the square-wave pulses from T401 (pin 16) and applying them to two places (Fig. 2-45). First, the pulses are applied to diode D433 where they are rectified to provide 13 Vdc to the standby power

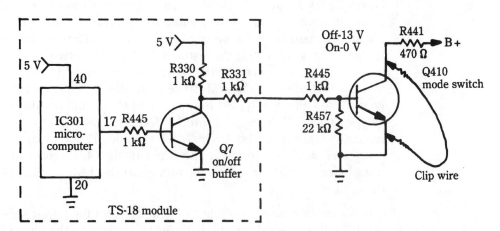

■ **2-44** *Disable the standby power supplies in a Sylvania C9 chassis with a clip wire across the collector and emitter terminals of Q410.*

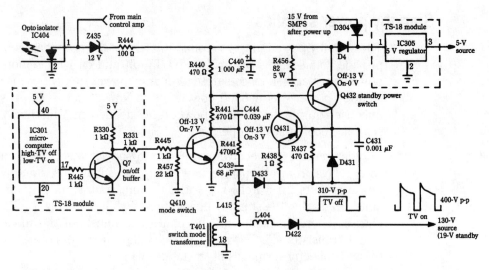

■ 2-45 *The standby power supply circuits in the Sylvania C9 chassis.*

switch (Q431/Q432). Second, the pulses are coupled across the network of C439, R439, and C444 to the base of Q432 to turn the standby power switch on.

With Q431/Q432 turned on, the rectified dc voltage (13 V) is applied via D437 to the 5-V regulator (IC305), keeping the standby circuits alive. IC305 is located on the TS-18/TS-15 module.

When the 5 V is supplied to the microcomputer, it remains active to the input commands of the remote control transmitter or customer keyboard. A high output at pin 17 of the microcomputer turns on the power on/off transistor (Q7). With Q7 turned on, the base of the mode switch (Q410) is held low, keeping it turned off and the TV turned off. With Q410 held off, the pulses from T401 (pin 16) are allowed to pass to the base of Q432, and the dc developed by D433 is coupled through Q432 via D437 to the 5-V regulator located on the TS-18/TS-15 module. The 13 V at the emitter of Q432 is also applied to zener diode Z435, which forward-biases the zener supplying voltage to the LED portion of the optoisolator.

Q410 is the central controlling factor in the standby power supply. If Q410 is turned on, the path for the pulses to the base of Q432 is grounded. Now the SMPS is allowed to come up to full power. During the full power-operation, the 13-V source from the SMPS is applied via D304 to the 5-V regulator IC as the source voltage.

Servicing standby circuits

First, take critical voltage measurements and waveform observations of each component in the standby circuitry. Place a jumper across the emitter and collector terminals of the mode switch and notice if the chassis turns on. Test each transistor and diode with in-circuit tests. If leaky or open measurements are noted, remove the collector terminal of each transistor from the circuit. Likewise, check the suspected leaky diodes in the same manner. Remember, most of the stages in the standby power circuits are either on or off.

Five new low-voltage power supply problems

Here are five problems that can occur in some common TVs.

Dead low-voltage capacitor

No B+ voltage was found upon the collector terminal of horizontal output transistor Q502 in a dead Panasonic CLT-1032R portable TV. After replacing open fuse F801 (3 amp), the fuse opened up again. Q502 tested good. The low-voltage circuits were checked with Q502 removed from the circuit. Critical voltage measurements upon the line voltage regulator (IC 801) indicated problems (Fig. 2-46). A closer peek indicated that C806 had the top blown off.

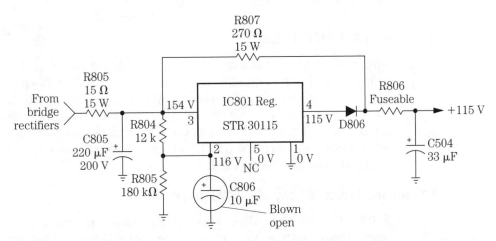

■ **2-46** *The dead-low-voltage source was caused by C806 in a Panasonic CLT-1032R portable.*

Intermittent RCA FPR2722T TV

Sometimes the chassis would operate for days before quitting, and when the chassis was touched or bumped, it came back on. The raw ac voltage was monitored at the VIPUR output transistor (Q4100) and appeared normal when the chassis ceased operation. No voltage was found at the 130-V source or horizontal output transistor. PW4700 (a diode SIP board) was removed and a crack was found in the PC wiring around L4700 and CR4701. All leads were soldered upon the SIP board, solving the intermittent and dead symptoms in several places (Fig. 2-47).

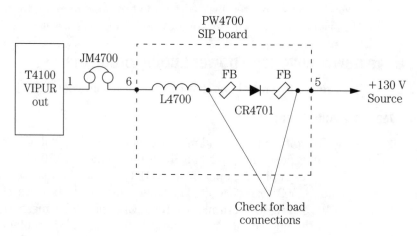

■ **2-47** *Check all the silicon diodes on SIP board in the RCA FPR2722T chassis for cracks around the terminals.*

Dead Quasar ALCD177 chassis

After replacing the main fuse, no voltage was found at the +131-V source or horizontal output transistor. R801 and F002 (a 1.5-amp secondary fuse) were also found open. After being replaced, F1002 was blown, indicating an overloaded IC regulator or horizontal output transistor. The horizontal output transistor tested normal with a leaky IC801. Replacing the STR30130 line-voltage regulator solved the dead chassis problem (Fig. 2-48).

A blown fuse in a dead Sony SCC-548D portable

F001 (a 5-amp fuse) opened again after replacement, indicating a defective low-voltage power supply or leaky horizontal output transistor. Q502 (2SD1555) was found shorted in the horizontal output circuits. The secondary fuse F602 (1.25 A) was replaced with the same results (Fig. 2-49). Low voltage was measured upon filter capacitor C602 (560 μF, 200 V), but not on fuse terminal of F602. R616 (2.2 Ω) was also found open. Replacing F601, F602, R616, and Q502 put the TV back into operation.

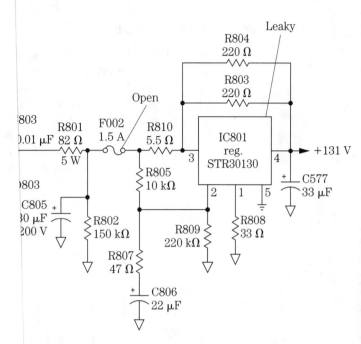

...en in a dead Quasar ALCD177 chassis. The cause was a

■ 2-49 *A shorted horizontal output transistor (Q502) in a Sony SCC548D chassis opened resistors R616, R602, and R603 in the low-voltage circuits.*

Green light on—no picture

In an RCA CTC140 chassis, the green power light would not turn off, with a no sound-no picture-no HV symptom. Voltage measurements in the VIPUR circuits were fairly normal. Replacing both CR4104 and CR4201 in the leg of the VIPUR transformer (T4100) output circuits (Fig. 2-50) solved the problem.

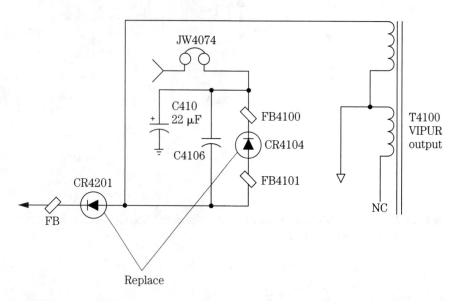

■ **2-50** *The green power light would stay on with no picture and no sound in an RCA CTC140 chassis. The problem was caused by a defective CR4201 and CR4104.*

Conclusion

Most problems within the low-voltage power supply can be located with a DMM, VTVM, and a scope. A defective low-voltage power supply can prevent one or all other sections of the TV chassis from functioning. A 100-W bulb can be used to indicate overloaded conditions in or beyond the power supply.

A poorly soldered connection or loose fuse can produce intermittent voltage from the low-voltage power supply. Hum bars can be created by a defective component in the regulator circuits, or by large filter capacitors in the dc power supply. Check all components in the low-voltage power supply when it is hit by lightning.

Chassis or high-voltage shutdown can be caused by higher-than-normal voltage from the dc power supply. Chassis shutdown can occur because of overloaded circuits connected to the rectifier low-voltage power supply of the flyback transformer winding. The horizontal and horizontal output circuits must function before secondary low-voltage is developed in the flyback low-voltage circuits.

Always remember that defective components in the power supply circuits can cause different symptoms in other connecting circuits. Do not overlook a defective filter capacitor in the power supply. These capacitors cause strange things to happen in other circuits. If in doubt, shunt or bypass each filter capacitor to help isolate power supply problems. Careful visual inspection of components and wiring can locate the defective part. Check Table 2-1 for a power supply troubleshooting chart that enables easy servicing.

■ Table 2-1 Use this power supply troubleshooting chart for easy servicing.

What to check	How to check it
Inspect line fuse.	Test in circuit with ohmmeter.
Check voltage across filter capacitor.	Test silicon diodes and isolation resistors.
Check output voltage at regulators.	Check transistor regulator and zener diodes.
Check horizontal fuse to horizontal output transistor.	Check fuse with ohmmeter if soldered in circuit.
Check voltage at horizontal output transistor.	Open isolation resistor or coil and flyback winding.
Overloaded circuits	Remove horizontal output transistor from chassis. (Voltage should be normal or a little higher with no output transistor in circuit.)

Servicing the horizontal sweep circuits

3

THE MOST TROUBLESOME SECTION OF THE SOLID-STATE TV chassis is the horizontal sweep section; it also has the most interesting circuits (Fig. 3-1). After mastering horizontal sweep circuits, most other electronic circuits are a cinch. New service techniques must be used to ensure quick and proficient servicing methods. Servicing the horizontal section can be fun—and very frustrating, at times.

Although many of the regular horizontal oscillator and output symptoms are the same, several new circuits with the same symptoms are found in today's solid-state chassis. Horizontal drifting and off-frequency symptoms can be caused by the horizontal oscillator circuits. The horizontal oscillator can be contained in one IC with both a horizontal driver and vertical circuits. You might find a horizontal and vertical frequency countdown circuit in one IC processor.

The no-raster, no-sound, no-high-voltage symptom can be caused by just about any component in the horizontal sweep or low-voltage stages. The no-raster, good-sound symptom can be caused by a defective component in the high-voltage circuits. High-voltage shutdown can be caused by the HV shutdown circuits, excessive low voltage, and defective components in the horizontal output circuits. Chassis shutdown can be caused by a breakdown in the low-voltage or horizontal sweep section.

The integrated flyback can produce arcing when self-enclosed high-voltage diodes shut down the chassis. The horizontal output transistor can be quickly destroyed by the flyback before the fuse opens. With the new integrated flyback transformer, a breakdown in one of the low-voltage circuits tied to one of the extra windings of the flyback can load down the transformer causing chassis shutdown. Today the flyback transformer provides voltage to many separate circuits and additional waveforms to other circuits.

■ **3-1** *Eighty-five percent of the problems within the TV chassis are found in the horizontal and vertical circuits.*

Horizontal oscillator problems

A dead TV chassis can be caused by an open or leaky horizontal oscillator transistor or IC. High-pitched sound can indicate the horizontal oscillator is off frequency. After several seconds, the horizontal oscillator might drift off frequency because of a defective transistor, IC, or corresponding components. If the TV comes on with horizontal lines, the horizontal hold control might need adjusting, there might be a defective horizontal oscillator circuit, or the horizontal stage has been shut down by a high-voltage shutdown circuit (Fig. 3-2). Excessive pulling at the top of the picture can be caused by improper sync or filtering of the low-voltage circuits. Check the sync or sync control circuits when the picture moves sideways.

Horizontal oscillator circuits

The horizontal oscillator stage can be checked by looking at the sawtooth waveform at the output terminal. The circuit might be one transistor in a Colpitts oscillator circuit, or an IC with oscillator and driver stage inside one component. In the latest IC circuits, a vertical-horizontal countdown IC can contain a horizontal sync separator, AFC detector, horizontal oscillator flip-flop, horizontal

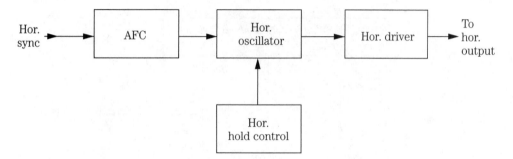

3-2 *The horizontal circuits consist of the AFC, horizontal oscillator or countdown, horizontal driver, and horizontal output stages.*

amp, and x-ray high-voltage protection, as well as vertical, luminance, and color circuits.

The horizontal frequency in a simple oscillator circuit (Fig. 3-3) is controlled by L601, C609, and C610. L601 is the horizontal hold control. If the horizontal oscillator drifts off frequency, suspect Q601, C609, C610, and L601. Spray Q601, C609, and C610 with coolant and notice if the horizontal lines flop into a picture. Replace C609, C610, and L601 with original replacement parts.

During the past 10 years, most horizontal oscillator transistors have been replaced by IC components (Fig. 3-4). Here C605 is charged and discharged repeatedly to produce the horizontal oscillator frequency. The charge time constant is controlled by R601, R603, R607, and R609. A feedback pulse from the flyback transformer passes through R605 and C605 and is turned into a sawtooth waveform feeding the AFC circuits at pin 35. The sawtooth

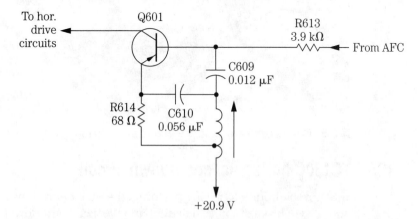

3-3 *In early low-priced TVs, a simple horizontal oscillator circuit incorporated a plastic shaft as horizontal hold control.*

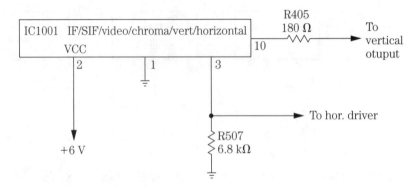

```
IC1001   IF/SIF/video/chroma/vert/horizontal        R405
                                                     180 Ω      To
           VCC                                       ─WW─        vertical
                                                  10             otuput
         2              1            3

                        ┴
                        ─
                                    R507
    +6 V                            6.8 kΩ
                                    ┴
                                    ─
```

■ 3-4 *Today's horizontal oscillator or deflection circuits may be found in one large IC, with IF/SIF/video/chroma and vertical/horizontal control circuits.*

signal is compared to the horizontal AFC circuit output and kept in phase with the horizontal pulse waveform at pin 24, feeding the horizontal drive transistor (Fig. 3-5).

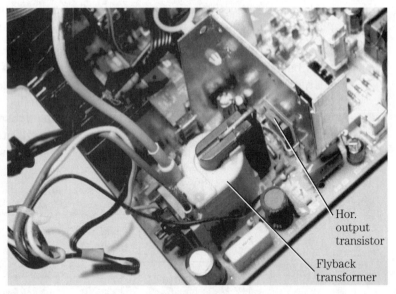

Hor. output transistor

Flyback transformer

■ 3-5 *The horizontal circuits are found close to the flyback transformer*

RCA CTC130C horizontal countdown circuit

Integrated circuit (IC) U401 performs the dual operations of horizontal synchronization (oscillator) and vertical countdown. Along with other functions, U401 provides horizontal drive for the horizontal output stage and vertical drive for the vertical output stage.

The IC (U401) is powered from the 26-V source via R116 (680 [OMEGA]) connected to pin 5 (Fig. 3-6). Internal to the IC, pin 5 is connected to a 10-V shunt regulator.

To determine if the horizontal oscillator is working, scope pin 16 of U401. If the chassis is shut down and will not start, the horizontal oscillator should be checked by injecting 26 V at pin 5 and scoping the horizontal drive at pin 16 (Fig. 3-7). Check the horizontal waveform with that shown on the schematic. If there is no output waveform, replace U401.

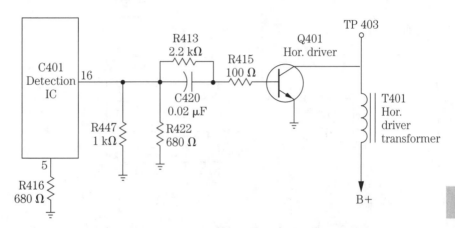

■ 3-6 *The horizontal countdown circuit in the RCA CTC130C chassis.*

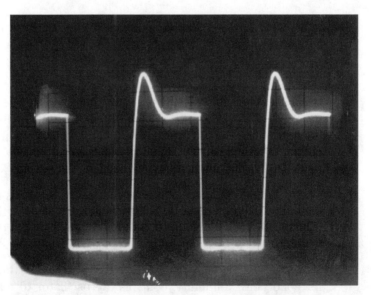

■ 3-7 *The horizontal output waveform found at pin 16 of U401 in the RCA CTC130C chassis.*

RCA CTC121 countdown IC

Integrated circuit U401 performs the functions of horizontal synchronization (horizontal oscillator) and vertical countdown. This combined circuit supplies both horizontal drive for the horizontal output stage and vertical drive for the vertical deflection amplifier (Fig. 3-8).

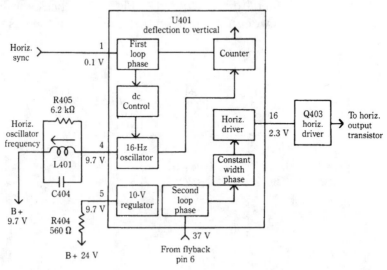

■ **3-8** *A deflection counter (U401) provides horizontal and vertical countdown to drive circuits.*

A horizontal rate output pulse is coupled from U401 (pin 16) to horizontal driver Q403. The drive stage drives the horizontal output (Q404). U401 can be checked with a pulse waveform at pin 16.

Horizontal sync is applied to U401 (pin 1). Inside the IC, horizontal sync is applied to the input of the first phase loop, and vertical sync is supplied to a vertical sync comparator for further processing. U401 is powered from an internal 10-V shunt regulator that is biased through the deflection startup circuit or 24-V supply via R404 to U401 (pin 5).

The horizontal deflection synchronization system uses two phase-lock loops and an oscillator operating at 251 kHz, exactly 16 times the horizontal rate. The frequency of the oscillator is determined by the LC network between U401 (pins 4 and 5).

The first phase-lock loop, which has a relatively long time constant for good noise immunity, controls the oscillator to maintain the control signal in correct frequency and phase with the horizontal sync signal. To compensate for load-dependent variations in the delay of the horizontal deflection stage, a second phase-lock loop is used. This loop includes a phase detector, one input of which is coupled back to the output of the first phase-lock loop via a 4-μs delay, and a second input that is coupled back to the flyback transformer via U401 (pin 6). A loop filter with a relatively fast time constant is coupled to the output of the second phase detector (pin 13) for filtering the control currents to form a control signal. The entire block diagram is shown in Fig. 3-9.

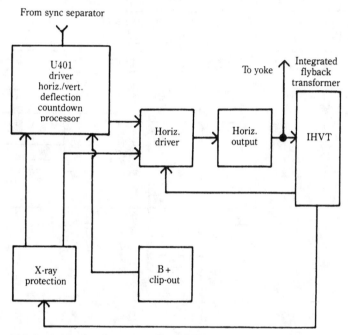

From sync separator

■ 3-9 *A block diagram of an RCA CTC121 chassis with a countdown IC.*

RCA CTC130C dual phase-control loop

The horizontal synchronization system uses a dual phase-lock loop (PLL) design with an oscillator operating at exactly 16 times the horizontal rate. The frequency of the oscillator is determined by the oscillator tank circuit (L401, C406, and R411) connected between pins 4 and 5 of U401.

The PLL circuit controls the horizontal oscillator, and maintains proper frequency and phase with the incoming horizontal sync signal (Fig. 3-10). To compensate for load-dependent variations in the delay of the horizontal deflection stage, a second PLL is used.

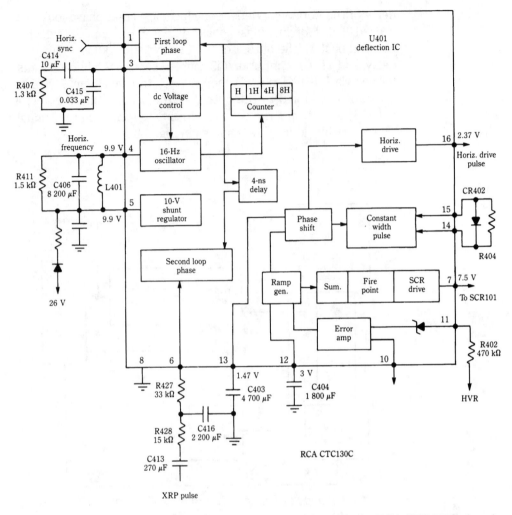

■ 3-10 *The internal section of the deflection IC (IC401) found in the RCA CTC130C chassis.*

The second PLL is coupled to the first PLL via a 4-μs delay and a second input (pin 6 of U401), which is coupled from pin 10 of the IHVT (XRP pulse). A loop filter, with a relatively fast time constant, is coupled to the output of the second phase detector (pin 13) for filtering the control current to form a control signal. This maintains the proper phase relationship between video and the horizontal yoke current. The horizontal drive pulse is found at pin 16 (Fig. 3-11).

The SCR regulator control circuit consists of an internal 6-V zener diode that senses the high-voltage resupply line (pin 11), the error amplifier controlled by the zener output, and the regulated B+

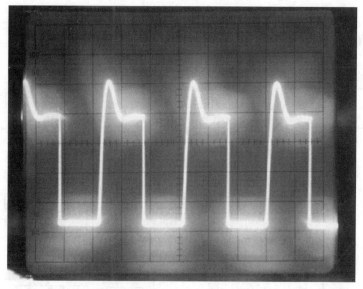

■ **3-11** *The horizontal square-wave drive pulse found at pin 16 of U401 (in an RCA CTC130C chassis).*

sensor line (pin 10). The output of the error amp controls the SCR driver stage (pin 7) gates of SCR 101 for B+ regulation control.

Horizontal oscillator IC tests

Take crucial waveform test measurements of the horizontal IC oscillator, sync, and output pulse waveforms. Check pin 10 for horizontal oscillator drive pulse to the horizontal driver transistor (Q601) (Fig. 3-12). Check for proper horizontal sync pulse from the sync separator to pin 16. Check the AFC waveform at pin 14. Improper waveforms at any of these pin numbers can indicate a defective IC.

Go a step further and take critical voltage measurements on the horizontal oscillator (pin 11) (12.5 V). Very low voltage at pin 11 can indicate a leaky IC. Check crucial voltage test points at pins 10, 11, 15, and 16. If any pin has a low voltage measurement, disconnect the pin or component tied to that pin number to see if it is loading down the suspected IC. Sometimes voltage measurements can be quite close, even if the IC is defective. Replacing the IC is the only answer.

Sylvania C9 horizontal circuit

The video IF/sync integrated circuit (IC201) generates the horizontal/vertical drive signals, a three-level sandcastle output pulse,

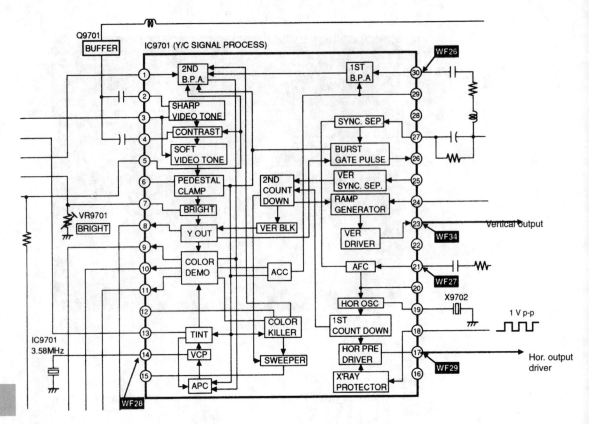

■ 3-12 *A waveform at pin 18 will indicate if the oscillator section is working.*

and coincidence voltage output and processes the video IF. Figure 3-13 shows the horizontal portion of the video IF/sync IC.

The horizontal oscillator begins running when 10.5 V is applied to pin 7 of IC201. The composite video signal, which comes from the IF processing section of IC201, exits the IC via pin 17. A 4.5-MHz filter removes any IF sound products before the composite signal is buffered by Q244 and applied to pin 25 of IC201. The horizontal oscillator is synchronized to the incoming TV signal via pin 25. When synchronization is attained, the coincidence output from IC201 (pin 22) is high (6.6 V), and when out of sync, the output at pin 22 is low (1.4 V).

The coincidence signal is applied to the tuning system to advise the tuning system that the sync is locked into the proper signal. The sync separator circuit passes the horizontal sync signals to the gating and phase 1 detector, where a dc voltage is developed and fed from pin 24 to pin 23 to control the frequency of the horizontal oscillator. Also connected to pin 23 is the horizontal frequency adjust control (R233).

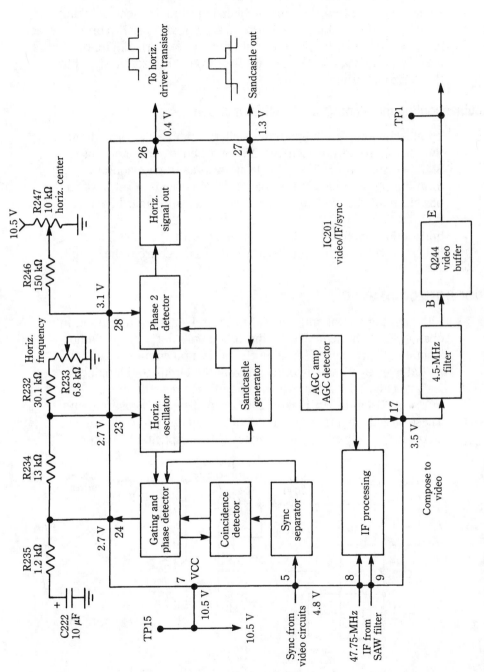

■ **3-13** A block diagram of the horizontal signal-generation system of the Sylvania C9 chassis. *Philips Consumer Electronics*

Horizontal oscillator IC tests

The output from the horizontal oscillator is fed to the sandcastle generator, which outputs on pin 27 a three-level signal that includes burst keying, horizontal blanking, and vertical blanking pulses. The horizontal pulse feedback is input via pin 27 to provide phase control for the phase 2 detector in the horizontal circuit. IC201 outputs via pin 26 the phase-controlled horizontal signal to the horizontal driver (Q500).

Troubleshooting the C9 horizontal oscillator circuit

Troubleshooting the horizontal oscillator can be performed with the chassis removed from the TV, or when troubleshooting the SMPS power supply. The video IF IC outputs the horizontal waveform during the unloaded or loaded troubleshooting setup of the power supply as explained in SMPS troubleshooting. Inject 10.5 Vdc to pin 7 of IC201 with the scope at pin 26 to determine the horizontal waveform. This test is made with the chassis power plug removed from ac and resistor R513 removed from the circuit.

Poor horizontal sync

Although horizontal and vertical sync problems are discussed in chapter 7, the horizontal sync discussed here determines if a horizontal hold problem is found in the horizontal circuits. Check the AFC diodes by removing one end from the circuit and looking for leakage (Fig. 3-14). Remove one end of each resistor above 50 kΩ in the AFC circuit to make accurate resistance measurements. These large resistors have a tendency to increase in resistance or go open. Check each small capacitor with a capacitance meter.

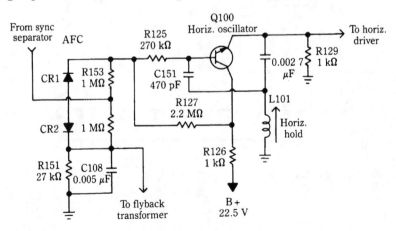

■ **3-14** *Poor horizontal sync can be caused by AFC diodes CR1 and CR2.*

Short out the AFC diode to ground to determine if the AFC circuits or the oscillator is causing the sync or off-frequency problem. Readjust the horizontal hold control to try and lock in a picture. If the picture is fairly stable, the oscillator circuits are good. Now check the AFC and sync input circuits for lack of horizontal sync. Scope the flyback waveform fed back to the horizontal oscillator in some chassis (Fig. 3-15).

In today's TV chassis, horizontal and vertical sync circuits are included inside one large IC with video, brightness, color, tint, horizontal and vertical sweep, and x-ray protector circuits. Check the input waveform of the sync separator terminal upon IC. If sync is normal here, suspect a defective IC or improper voltage source at supply pin. This signal processor IC can cause poor vertical and horizontal sync while other circuits are normal. Often, poor sync is found with a poor video picture.

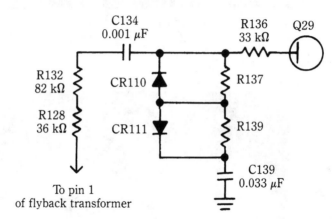

■ **3-15** *Poor horizontal sync in the RCA CTC97 chassis resulted in insufficient waveform at the base of Q29.*

Critical horizontal waveforms

The horizontal sweep waveform can be traced from horizontal oscillator or countdown IC to the collector output terminal of horizontal output transistor with the oscilloscope. Figures 3-16A to 3-16D, represent the sweep output waveform from countdown sweep IC. The drive waveform is applied to the base of driver transistor, amplified and applied to the primary winding of driver transformer. This drive signal is applied to the base terminal of the horizontal output transistor, amplified, and applied to the primary winding of the horizontal output transformer or flyback. When one of the waveforms is missing, or there is incorrect polarity and improper waveform, suspect trouble within that horizontal circuit.

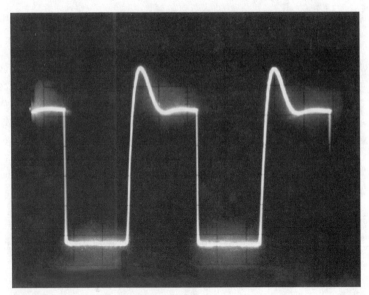

■ **3-16a** *The horizontal pulse found at the base of the horizontal driver transistor (1 V p-p to 4.5 V p-p in a 19-inch TV).*

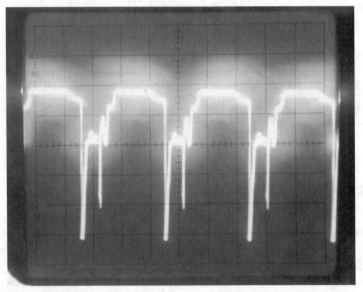

■ **3-16b** *The drive waveform found at the base terminal of horizontal output transistor. (14 V p-p to 25 V p-p in a 19-inch TV).*

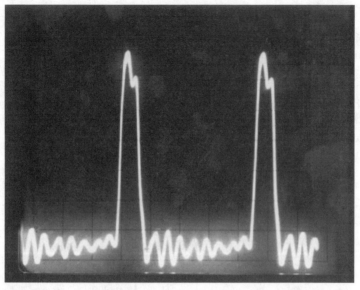

■ **3-16c** *The high voltage waveform found by placing the scope probe near the flyback transformer.*

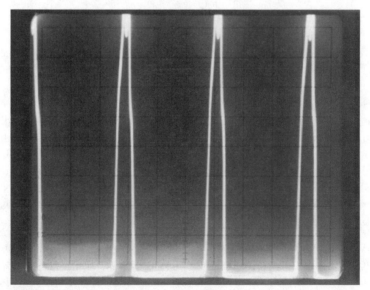

■ **3-16d** *The horizontal output waveform applied to the primary winding of the horizontal output transformer (70 V to 250 V p-p in a 19-inch TV).*

Horizontal off-frequency problems

Most TV technicians can hear the horizontal oscillator when it has drifted off frequency. Additionally, horizontal lines appear on the screen, and there is no horizontal hold control. Sometimes the oscillator will drift off after operating a few minutes. This indicates that a component is getting warm. Check the horizontal transistor or IC component for leaky conditions. Replace the resonating and coupling capacitors found in the emitter-tapped oscillator coil circuit (Fig. 3-17).

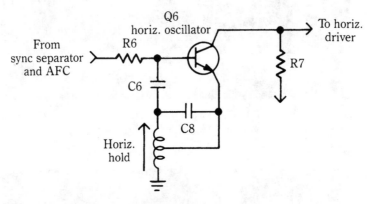

■ **3-17** *Check capacitors C6 and C8 in the horizontal coil circuit when the horizontal frequency drifts off.*

Suspect the IC, small filter capacitor, and RC time-constant circuits when the horizontal oscillator is off-frequency. First, adjust the horizontal hold control. Take a quick voltage test on the supply voltage and horizontal oscillator pins of the suspected IC. Replace the IC. Next check each small electrolytic capacitor in the circuit (Fig. 3-18). Test each bypass capacitor with the capacitance tester in the circuit. Remove one end of each resistor for correct resistance measurement.

A dried-up or open electrolytic filter capacitor in the power source feeding the horizontal circuits can cause poor horizontal lock, or an off-frequency condition. Sometimes the electrolytic capacitors tied to the horizontal oscillator circuit do not filter out the sync or hash signal, producing funny horizontal pictures. Horizontal jitters can be caused by a poor filter bypass capacitor. Suspect a defective filter capacitor when the horizontal lines cannot be straightened up after all other tests are made.

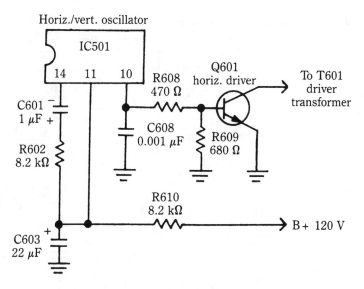

Horiz./vert. oscillator

IC501

■ **3-18** *Clip another capacitor across the suspected one and notice if the picture returns in a Sharp C1335A chassis.*

Horizontal drifting and pulling

Excessive horizontal pulling of the picture can be caused by poor filtering in the low-voltage power supply. Higher-than-normal voltage applied to the horizontal oscillator transistor IC can be caused by a defective zener diode in the power supply circuits. Horizontal flickering of the picture can result from poor voltage regulation in the holddown or regulator circuits. Pulling at the top of the picture can be caused by the malfunction of a small electrolytic bypass capacitor in the horizontal circuits.

Horizontal pulling with the frequency drifting in an Admiral M25 chassis (Fig. 3-19) was caused by a small electrolytic filter capacitor in the 19.83-V power source. When addressing this problem, scope the B+ source and shunt another capacitor across the suspected one to locate the defective capacitor. The same symptom was found in a General Electric CD chassis, except the voltage applied to the horizontal oscillator was too high (Fig. 3-20). Often, with dried-up filter capacitors the voltage source is too low. Here 34.1 V was found at the collector terminal of the horizontal oscillator transistor (Q502). Replacing the defective 22-V zener diode with a 5-W type solved the horizontal pulling symptom. Do not overlook horizontal pulling and lines caused by excessive high voltage at the picture tube.

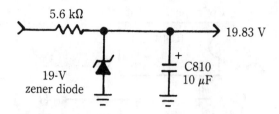

■ 3-19 *Scope the low-voltage source and shunt another capacitor across the suspected one to locate a dried-up electrolytic.*

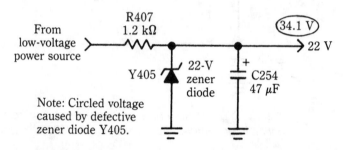

■ 3-20 *A defective Y405 caused a higher dc voltage than specs, resulting in horizontal lines in the raster.*

Cannot tune out the lines

Suspect that the horizontal oscillator is off-frequency, or that there is poor filtering in the low-voltage source when the horizontal lines cannot be adjusted out of the picture. The picture might come on with uncontrollable lines or, after several minutes, go into horizontal lines that cannot be straightened up. Determine if the horizontal line problem is in the oscillator circuits, power supply, or other sources. Monitor the voltage at the power source. If the voltage is either higher or lower than normal, check the low-voltage source feeding the horizontal oscillator circuits (Fig. 3-21). Check the low-voltage circuits for poor voltage regulation.

Check the high-voltage and shutdown circuits for excessive high voltage. Simply measure the voltage at the picture tube anode terminal. In the early high-voltage shutdown circuits, if the high voltage went too high, the horizontal circuit was shut down and all you could see were horizontal lines. Improper setting of the HV or B+ control can cause the same condition. Raise and lower the brightness control and notice if the picture goes into horizontal lines, indicating too much high voltage at the picture tube.

In the RCA CTC46 and CTC48 chassis (Fig. 3-22), high voltage at the picture tube above 30 kV can cause the shutdown circuits to

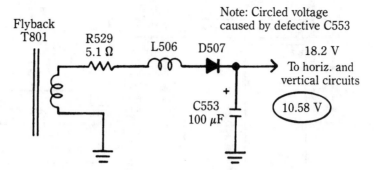

■ **3-21** *Low voltage in the scan-derived flyback source pro-duced horizontal lines in a Sony 1500.*

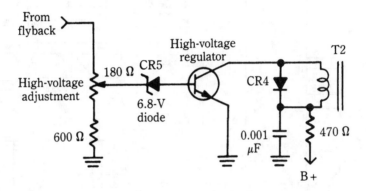

■ **3-22** *High-voltage shutdown and horizontal lines result from a defective CR5 zener diode.*

operate, causing horizontal lines in the picture. Sometimes adjusting the brightness control can raise or lower the high voltage, causing the raster to bloom with a picture and then go into horizontal lines. Often the horizontal-line symptom is caused by a leaky zener diode (CR5) or HV regulation transistor. Poor regulation transformer board connections can cause horizontal lines with excessively high voltage.

Realistic 16-261 horizontal deflection circuits

The IF/video/chroma/deflection circuits are tied to IC101. The horizontal oscillator begins to run when 7.7 V is applied to pin 20 (VCC). The horizontal oscillator countdown circuit is controlled and kept on frequency with crystal X401 (Fig. 3-23). The horizontal sync is taken from the video circuit and fed to pin 26 of the internal sync separator. The sync separator and AFC are tied to the horizontal oscillator countdown circuits.

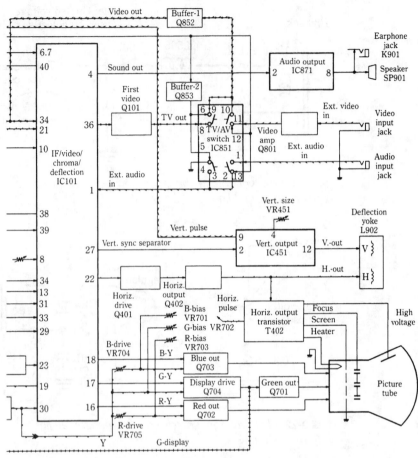

■ 3-23 *A block diagram of the Realistic 16-261 horizontal circuits.* Radio Shack

The horizontal oscillator signal is sent to the phase-shift stages and to the horizontal output terminal (pin 22). The horizontal holddown circuit ties to pin 23 of IC101. The horizontal square waveform is fed from pin 22 to the horizontal drive transistor (Q401).

To determine if the horizontal oscillator is functioning, inject 7.7 V at pin 20 and scope pin 22 for a square wave from IC101. If no waveform is found at pin 22, suspect a defective IC101. Do not plug in the TV's ac power cord for this test. You know IC101 is good when the square waveform (horizontal) is found at pin 22.

RCA CTC130C horizontal driver and output circuits

The horizontal drive pulses are found on pin 16 of countdown IC U401. The horizontal pulse is applied to the horizontal drive transistor (Q401). Remember, the master oscillator within R401 must be running to reverse the horizontal pulse at pin 16.

Collector voltage for the horizontal driver transistor is developed by the 185-V supply source in the IHVT (T402) secondary. The initial B+ required to enable the driver stage is approximately 130 V and is derived from the regulated B+ source via R402 and CR410. Remember, the horizontal output circuits must be operating before voltage (185 V) is developed for the driver transistor (Q401) (Fig. 3-24).

Horizontal output

Collector voltage for the horizontal output transistor (Q402) is drawn from the regulated B+ supply through the IHVT (R402) primary winding. Current is drawn through the primary winding and the yoke winding to ground. Q402 turns off as the beam reaches the right edge of the picture tube, starting the beginning of horizontal retrace. After Q402 turns off, current flows from the horizontal yoke and IHVT primary windings into retrace capacitor C423, where energy from the system is temporarily stored.

When the retrace capacitor (C423) is fully charged, the current flow reverses and begins flowing from C423 back into the primary winding of the IHVT and the horizontal yoke windings. Thus, the electron beam from the right edge of the picture is deflected back to the left edge to begin the next time, completing a retrace.

When C423 has been discharged, the current flow is then controlled by the damper diode (CR412), which begins drawing current from B+ to ground and storing it in the IHVT primary winding, while at the same time drawing current through the horizontal yoke winding and storing it in yoke return capacitor (C425).

The turning on of the damper diode starts the beginning of the first half of the horizontal scan period. The IHVT primary decreases toward zero when the first half of the horizontal scan continues or the current flows through the damper diode to the yoke. When the current flow reaches zero (electronic beam in the center of the picture), Q402 begins to conduct. This method reverses the current flow and begins drawing current from the IHVT primary winding and from the horizontal yoke winding to ground through R418, completing the second half of the scan.

Sylvania C9 horizontal driver and output circuits

The horizontal output square wave pulse starts at pin 26 of IC201 and is fed to the horizontal driver (Q500). The horizontal signal is amplified to drive the horizontal output stage, (Q501) (Fig. 3-25).

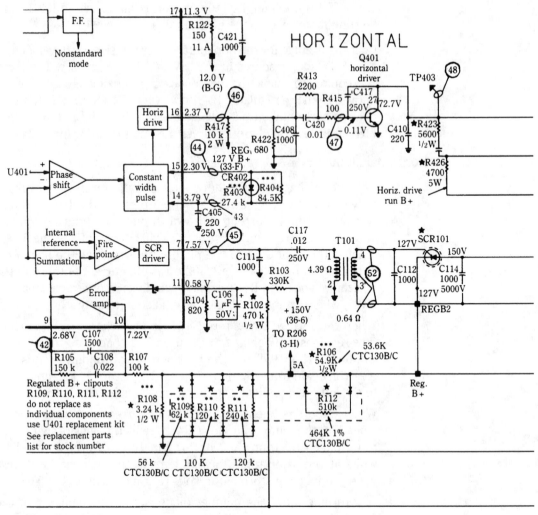

3-24 *RCA's CTC130 horizontal driver and output circuits.*

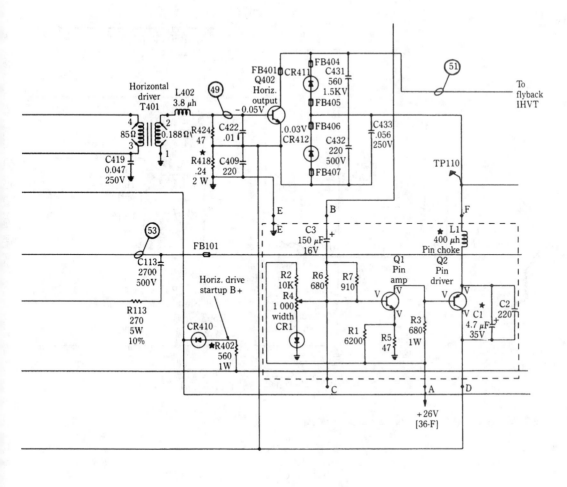

125

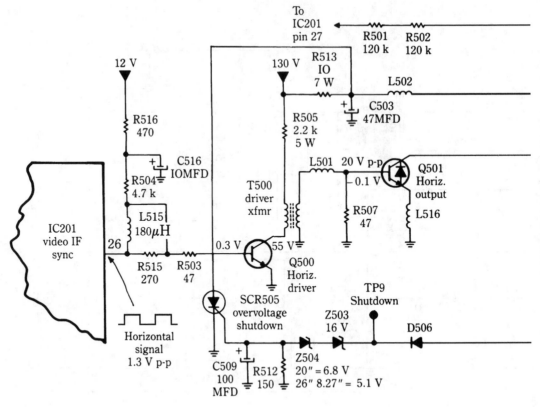

■ 3-25 *Sylvania C9 horizontal output circuits.*

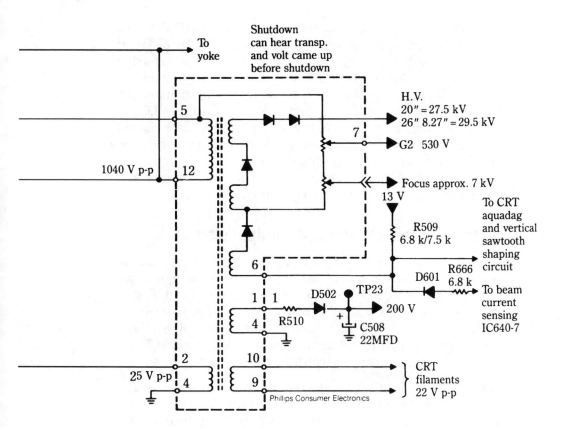

To yoke

Shutdown can hear transp. and volt came up before shutdown

5

12

1040 V p-p

6

1 1

4 R510

2

25 V p-p 4

10

9

Phillips Consumer Electronics

D502

TP23

200 V

+ C508
22MFD

H.V.
20" = 27.5 kV
26" 8.27" = 29.5 kV

7

G2 530 V

Focus approx. 7 kV

13 V

R509
6.8 k/7.5 k

D601

R666
6.8 k

To CRT
aquadag
and vertical
sawtooth
shaping
circuit

To beam
current
sensing
IC640-7

CRT
filaments
22 V p-p

This unfiltered dc voltage (12.5 V) will trigger SCR505 if it rises above the combined zener voltage (22 V) of zener diodes Z503 and Z504. Once the SCR is triggered, the 130-Vdc supply is clamped to ground through R513, which removes horizontal drive by removing the 130-Vdc source from the collectors of the horizontal driver transistor (Q500) and the horizontal output transistor (Q501). When the 130-Vdc source is clamped to ground, the switched mode power supply (SMPS) shuts down. Then ac power must be removed from the set and reapplied to reset the shut-down system. Before the horizontal driver and output circuits can be serviced, the SMPS must be restored to normal.

Servicing the Sylvania C9 horizontal output circuits

When the SMPS and horizontal oscillator are functioning, IC201 in the horizontal output circuits can be serviced. If the horizontal output transistors (Q501) become leaky, the SMPS might be shut down. Remove resistor R513 from the circuit to determine if the horizontal output circuits are shutting down the SMPS. Check both Q500 and Q501 in the circuit. Do not overlook open L516 in the emitter terminal of Q501 with a shorted transistor.

Take waveforms on the base and collector of the driver transistor (Q500). Q500 might become leaky. With no drive base signal, R505 (2.2 kΩ) will run warm. Replace a leaky Q500 if a drive pulse is found at the base and an improper waveform is found at the collector terminal. The collector voltage can be very low with leaky driver transistors or no drive pulse.

The horizontal output transistor can be checked by scoping the base and output signal of the collector (Fig. 3-26). Do not place the scope probe on the collector (metal) terminal, but lay the probe close by the horizontal output transformer. The high peak-to-peak pulse with ringing indicates that the horizontal output circuits are normal (Fig. 3-27).

The horizontal output transistor

Horizontal output transistors cause more service problems and are replaced more than any transistor in the TV chassis. The output transistor can be insulated from the metal chassis or bolted directly to a separate heat sink on the PC board (Fig. 3-28). In some chassis, the separate heat sink and chassis can be hot or above chassis ground (Fig. 3-29). All voltage measurements made around the horizontal circuits must be made from the hot chassis—not chassis ground.

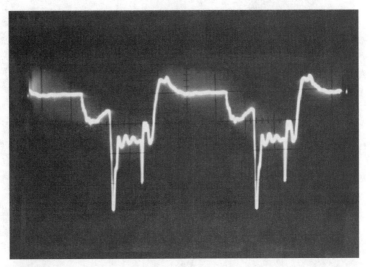

■ **3-26** *The scope waveform found at the base terminal of the horizontal output transistor.*

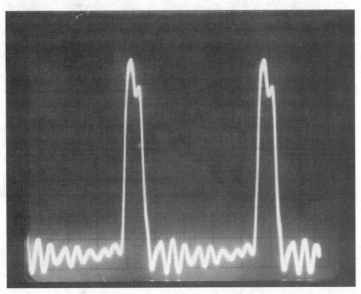

■ **3-27** *The horizontal output waveform with the probe held near the flyback transformer.*

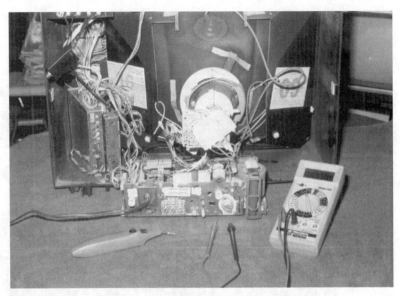

■ **3-28** *The horizontal output transistor is bolted to the metal chassis or heat sink in the RCA CTC108 chassis.*

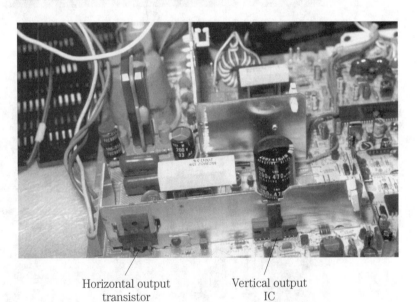

Horizontal output
transistor

Vertical output
IC

■ **3-29** *The horizontal output transistor is insulated above the heat sink and may have a hot ground.*

The defective horizontal output transistor can appear leaky, shorted, or open. A leaky output transistor can blow the fuse or trip the circuit breaker. Take a low resistance measurement between the collector terminal (body) and chassis ground to check for a leaky transistor. Only a few ohms should register on the DMM. Most output transistors leak or short between the collector and emitter terminals. The output transistor should be removed from the circuit for accurate leakage tests. Because the driver transformer secondary terminals are across the base and emitter terminals, an accurate leakage test between base and emitter cannot be made (Fig. 3-30).

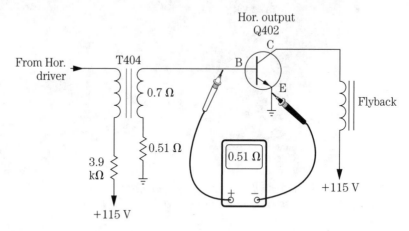

■ **3-30** *A resistance or leakage test between base and common ground may not be accurate with a transformer winding in the circuit.*

Open output transistor tests between the base and collector terminals can be accurate, but not between the base and emitter terminals while the transistor is in the circuit. Remove the horizontal output transistor from its socket for leak or open tests. In addition to the secondary driver winding in the base circuits, the damper diode in the collector circuit can produce inaccurate measurements.

The overheated horizontal output transistor can be caused by a shorted flyback transformer, open damper diode, or insufficient drive voltage. Suspect a leaky flyback when the transistor heats up at once and shuts down. The leaky flyback can destroy the output transistor each time the transistor is replaced. Because the damper diode is shunted from the collector terminal to ground, a quick resistance check with the DMM from the collector terminal (body) to chassis ground can indicate an open by low ohmmeter measurement (Fig. 3-31).

You might find the damper diode inside the same case as the horizontal output transistor (Fig. 3-32). This type of transistor is found in the latest TV chassis. Check the schematic diagram when you find the output transistor has some leakage. If the transistor is checked in the circuit, a low resistance can exist between the emitter and base

■ **3-31** *Check the horizontal output transistor from the collector (leaky) to common ground with the diode tester of the DMM.*

terminals. A low resistance can be measured between the emitter and collector terminals, indicating a leaky transistor under normal tests. In this case, however, the resistance measurement is of the internal damper diode in one direction. Remove the transistor from the circuit and make another leakage test.

A quick voltage measurement at the collector terminal (body) of the horizontal output transistor can determine if the transistor is defective. Low voltage can indicate a leaky output transistor, or no drive voltage at the base terminal. Higher-than-normal voltage at the collector terminal can indicate that the transistor is open. No voltage measurement at the collector terminal can indicate an open horizontal protection fuse, isolation resistor, or flyback primary winding.

Always place silicone grease on both sides of the insulator and transistor when installing a new output transistor. Wipe off all old grease and dust from the metal heat sink. Snug up the mounting screws, but not too tight, because sometimes the corner of the insulator can break through, causing voltage leakage between the transistor and heat sink.

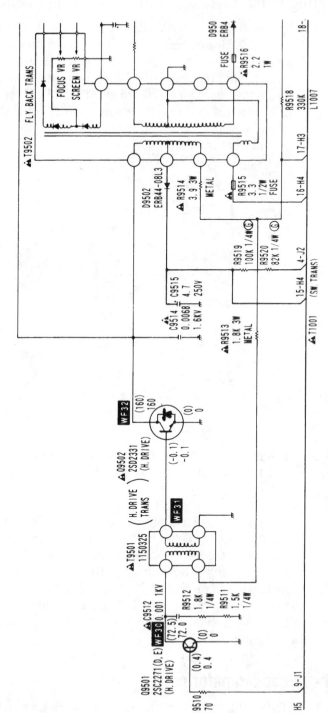

3-32 Notice that the damper diode is located inside the horizontal output transistor (Q9502) 25D2331.

The horizontal output transistor

Testing horizontal output transistors with damper diodes

In many of the new TV chassis, the damper diode is found inside the output transistor. Do not be alarmed if you have a low ohm measurement between the emitter and collector terminals. Usually when a low resistance is found between the emitter and collector, the transistor is discarded (Fig. 3-33). However, the SK9119/ECG89 horizontal output transistor has a 480-Ω resistance between the emitter and collector terminals, with the positive lead at the emitter terminal.

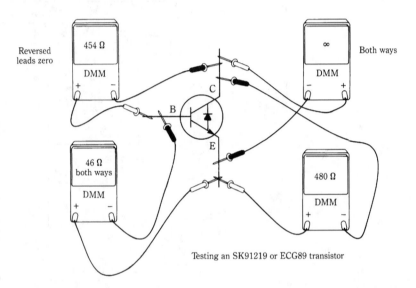

Testing an SK91219 or ECG89 transistor

■ **3-33** *Typical ohmmeter measurements across the three elements when the damper diode is located inside the output transistor (SK9119 or ECG89).*

Actually, you are measuring the resistance of a silicon diode inside the transistor. When a low resistance measurement is found with reverse test leads, replace the output transistor. Notice the 50-Ω resistance from the emitter to the base terminal. This measurement will be found in both directions with the DMM diode test. Do not throw away the transistor. If the reading in both directions is below 11 Ω, the transistor is shorted between the base and emitter terminals.

The flyback transformer

Today, the flyback does a lot more work than the old tube transformer used to do. The integrated high-voltage transformer (IHVT) can have several different voltage windings for low-voltage sources and high-voltage diodes molded inside the same plastic housing (Fig. 3-34). Instead of a separate tripler unit with the low-voltage flyback, the integrated transformer has diodes built inside the windings (Fig. 3-35).

The integrated flyback can run warm, pop and crack, and cause chassis shutdown. Like the diodes and capacitors in the tripler unit, the high-voltage diodes break down inside the transformer. Excessive arcover can cause overheating or arcing of the plastic material. Sometimes the flyback arcs over to the metal core area.

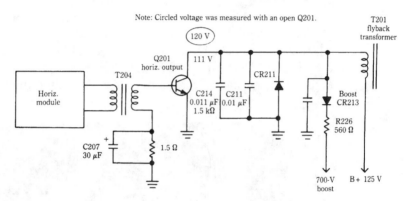

3-34 *Collector voltage that is very low can indicate a leaky output transistor. A very high voltage can indicate an open output transistor.*

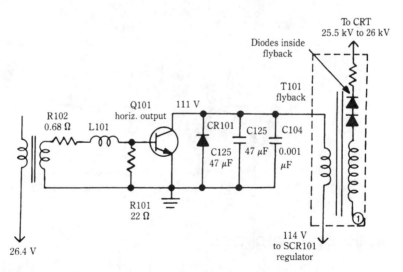

3-35 *No resistance can be measured in the secondary winding of the IHVT flyback with high-voltage diodes in the circuit.*

The integrated transformer should operate fairly cool, unless overloaded by a leaky horizontal output transistor or internal high-voltage diodes. An overload in one of the secondary voltages can cause the transformer to shut down the chassis.

Most flyback transformers should be replaced with the original part number. Although the American-made transformer can be replaced with universal types, imported transformers must be replaced with the original (Fig. 3-36). The Japanese component might include the screen and focus controls as one unit.

Often the horizontal output transistor is damaged by a defective flyback transformer. Use the variac transformer tests to locate a defective flyback and prevent damaging another output transistor. Before replacing the flyback, check the output transistor and the damper diode. Make sure leaky secondary components are not overloading the flyback. Sufficient drive and regulated low voltage should be found at the base and collector terminals of the output transistor.

■ 3-36 *The high-voltage diodes or secondary winding can arc over to the metal core of the flyback, making a loud popping and cracking noise.*

Basic horizontal circuits

The horizontal circuits are made up of the oscillator, driver, output, and flyback transformer. All stages consisted of transistors in the early circuit, or one IC and two transistors in the latest circuits (Fig. 3-37). This type of horizontal circuit can easily be serviced by taking waveforms at the countdown oscillator (TP406) (pin 16), collector terminal of driver transistor Q403 (TP408), base terminal of Q404 (TP404), and collector Q404 (TP407) (Fig. 3-38).

■ **3-37** *The screen and focus controls may be part of the output transformer replacement.*

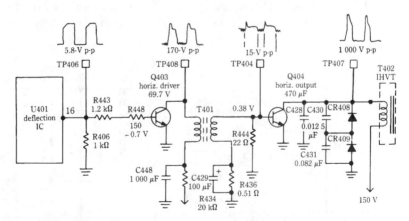

■ **3-38** *Critical waveforms and voltage measurements at the various test points identify most horizontal circuit problems in the RCA CTC121 chassis.*

The horizontal driver transistor (Q403) provides high gain for the weak pulse from U401. The driver transistor can be tested in the circuit for leakage or open conditions. Voltage measurements at the collector terminal can indicate if the transistor is leaky or open. Excessively low voltage at the collector terminal can indicate a leaky transistor. High collector voltage can indicate the transistor is open, or that there is improper drive voltage at the base terminal. Often R434 will become quite warm with either condition. Shorted turns in the primary side can be caused by an overheated horizontal driver transistor, indicated by a weak output waveform. Most defective driver transistors can be located with critical voltage and waveform tests. The universal transistor can be used effectively in the driver circuits.

The peak-to-peak waveform tests at the base and collector terminals of the horizontal output transistor can indicate if the stage is weak or dead. Remember, when taking a base waveform with the transistor out of the socket, the waveform will look different from that of a normally mounted transistor. Often the leaky horizontal output transistor or flyback can load down the base waveform with the transistor in the circuit. At least with the transistor out, the horizontal driver output and base waveforms should be close to normal if the preceding circuits are working.

Holddown capacitors

In the early Zenith, Admiral, and Magnavox chassis, a four-legged collector-shunt capacitor could open, producing excessive high voltage or no voltage applied to the horizontal output transistor (Fig. 3-39). These tuning capacitors had a tendency to open at one end. Often, the horizontal output transistor was destroyed in the

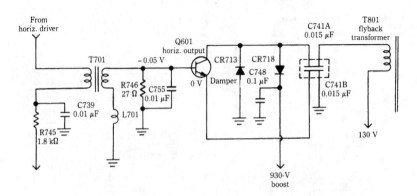

■ **3-39** A typical output safety circuit using the four-legged collector-shunt capacitor.

process. Always replace this holddown capacitor when a leaky output transistor is found.

If the capacitor does open the high voltage will increase, and it might shut down the chassis. Besides the higher-than-normal B+ voltage applied to the horizontal output transistor, check the hold-down capacitor for excessive high voltage. High-voltage arc-over can occur at the picture tube or in the flyback circuits when the shutdown high-voltage circuits do not function due to a defective capacitor. This capacitor should be replaced with one of the exact capacitance and operating voltage. The voltage can be from 1.2 kV to 2 kV (Fig. 3-40).

In some chassis, the tuning capacitor can go open and destroy the flyback and the output transistor. It's best, when replacing a fly-back transformer (in the older models), to replace the holddown capacitor and output transistor. Often the fuse or isolation resistor goes open because of a leaky transformer and output transistor (Fig. 3-41). Check the damper diode and boost diodes when you find a leaky output transistor.

The damper diode

A leaky damper diode can blow the fuse or trip the circuit breaker. The open damper diode can cause the horizontal output transistor to run warm and be destroyed. The diode can be checked with the diode test of the DMM or VOM (Fig. 3-42). Low resistance mea-

Focus control

Screen control

Flyback

Holddown or safety capacitors

■ **3-40** *The voltage of the holddown or safety capacitor can range from 1.2 kV to 2 kV with a 5% tolerance.*

■ **3-41** *Check the safety capacitors when the output transistors keep shorting out.*

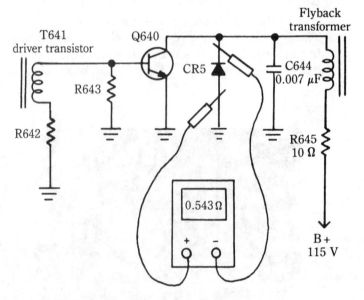

■ **3-42** *Replace the defective damper diode with one of 1.8 kV or 2 kV rating. A normal damper diode measurement.*

surements should be found in one direction, while infinite resistance is found with reversed test leads. A leaky damper diode will have a low resistance in both directions. Replace the damper diode with a 1.8-kV to 2-kV rated unit. Do not replace it with an ordinary power diode.

Critical safety components

Replace (with original replacement parts) defective safety components in the horizontal circuits. Check the manufacturer's schematic for a star, diamond, or letters alongside each safety component. Critical components such as the horizontal driver transformer, flyback, safety capacitors, bypass capacitors, metal oxide resistors, the yoke, and the horizontal output transistor should be replaced with originals or exact replacements. Always replace holddown capacitors and capacitors in the flyback and yoke circuits with exact replacements. These critical capacitors have a higher voltage measurement than ordinary bypass capacitors.

The dead chassis

Most dead chassis symptoms occur because of trouble in the horizontal output and low-voltage power supply circuits. Check for low voltage at the collector terminal of the output transistor and the low-voltage supply. Measure the low voltage at the fuse terminal protecting the horizontal circuits. No voltage found at the fuse or collector terminal indicates power supply problems. Remove the horizontal output transistor if the fuse keeps blowing to determine if horizontal circuits are defective.

Measure the high voltage at the anode connection of the CRT if normal voltage is found at the output transistor. Suspect a defective yoke if there is low high-voltage current at the picture tube. Check for an open yoke return capacitor when there is no high voltage. An open flyback transformer or capacitor in the flyback circuits can cause a dead chassis. Usually, a leaky flyback or capacitor will lower the output transistor collector voltage and can open the fuse.

Suspect an open horizontal output transistor with higher-than-normal voltage at the collector terminal. The transistor should be removed and tested out of the circuit. Low voltage at the collector terminal can indicate a leaky output transistor or insufficient drive voltage. Scope the base terminal of the output transistor. If you find insufficient waveform at the base terminal, check the horizontal drive and oscillator circuits for correct waveforms.

RCA CTC121 chassis

The most common components that break down in the CTC121 chassis are shown in Fig. 3-43. Check Q404 and CR408 for leaky conditions. Low voltage at Q404 or chassis shutdown can be caused with a leaky T402. Improper drive voltage at the base terminal can be caused by a leaky Q403, open T401, or burned R434. After replacing a leaky Q404, make sure the base resistor (R436) is not open. Sometimes Q404 can have a dead short, taking out R436, which is in series with T401. Check for open horizontal winding of the deflection yoke or capacitor C447 with a dead chassis.

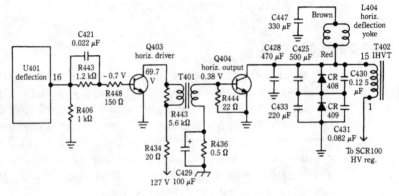

■ **3-43** *The horizontal output circuit of the RCA CTC121 chassis.*

Montgomery Ward GGY16215A

The dead chassis can be caused by high-voltage shutdown. Sometimes the chassis shuts down at once and the technician does not hear the high voltage come up. In a Montgomery Ward GGY16215A, the chassis was dead with high-voltage shutdown. Nothing, not even a click, could be heard from the flyback or speaker. One terminal of SCR430 was removed from the shutdown circuit (Fig. 3-44). The high voltage and sound came on. SC435, SC433, and SC434 were all found to be leaky and were replaced. Be sure to check the feedback pulse from the flyback at L434 to the shutdown circuits.

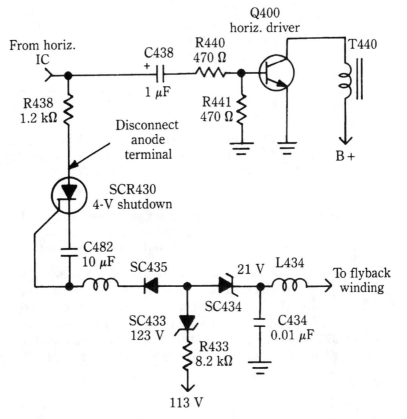

■ **3-44** *SC435, SC434, and SC433 were suspected in a Wards GGY16215A.*

Goldstar CR407

Voltage measurements (Fig. 3-45) at the collector terminal of the horizontal output and driver transistor were too high (160.5 V). No drive waveform was found at the base of the output or driver transistor. The horizontal drive signal was traced back to the deflection IC (IC401). The voltage at pin 11 was only 1.5 V. High voltage was found on one side of R508 (160.5 V) with only 1.5 V on the other. The resistance measurement between pin 11 and ground was 54 Ω. Replacing a leaky HA11235 deflection IC solved the problem.

Dead Sylvania 26C9 switched-mode power supply

The fuse was blown in a Sylvania C9 chassis with a leaky Q400 switched-mode regulator transistor. Q403 and Q402 were also

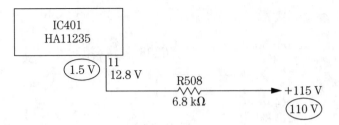

3-45 *A leaky deflection IC (IC401) in a Goldstar CR407 caused the dead chassis. A 54-ohm resistance between pin 11 and chassis ground indicates a leaky IC.*

found to be leaky. After checking the diodes and small resistors in the primary circuit, D425 and D421 were found to be shorted. R430 and R432 were burned, and so were replaced (Fig. 3-46).

The chassis was fired up, but had no output voltage (130 V). R531 was removed and a 75-W bulb was used as the load on the power supply. All components in the secondary were checked for open or leaky diodes and resistors. R434 (1 Ω) was open and D434 was leaky. The voltages on Q406 and Q407 were found to be normal when 24 V were injected at the collector of D434. All other components were good.

144

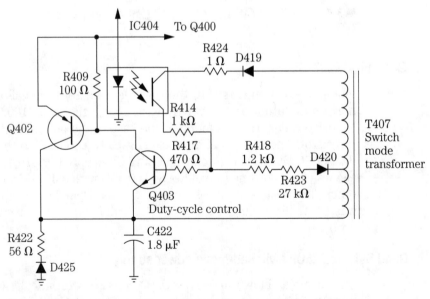

3-46 *Q403 and Q402 were found leaky in the switched-mode power supply (SMPS) of the Sylvania 26C9 chassis.*

Q400 was replaced again with a new fuse. The chassis was normal. The switched-mode transistor was fine after replacing defective components R434 and D434 in the 24-V power source.

Miscellaneous dead checks

Poor pincushion and regulator transformer terminals can cause a dead chassis. If the connections are intermittent, the raster will narrow down, with low high voltage at the picture tube. Improper connections between the collector bolt and metal body of the output transistor can cause a dead raster with no voltage at the collector terminal. Sometimes if the mounting bolts are too tight, the transistor can short through the insulation and ground out the collector voltage. Be careful when taking voltage measurements at the horizontal output circuits in some chassis where the output circuits are above ground. Always measure the collector voltage at the isolated ground terminal or chassis. Otherwise, low-voltage measurements will not be accurate.

RCA CTC121 horizontal circuits

The horizontal deflection synchronizing IC (U401) contains the horizontal oscillator and vertical countdown circuit (refer to Fig. 3-43). The countdown circuit provides a signal to the horizontal driver and output stages. A horizontal rate output pulse is coupled from U401 (pin 16) to horizontal driver Q403. To obtain horizontal drive pulses from U401, the master oscillator (counter) must be working.

Horizontal driver

Horizontal drive pulses from U401 (pin 16) are applied to horizontal driver Q403. B+ for Q403 is supplied from horizontal driver supply rectifiers CR109 and CR-110 (not shown), which rectify an IHVT-derived horizontal retrace pulse. (During initial receiver turn-on, before the IHVT-derived supply voltages are developed, Q403 is supplied initial B+ from the regulated 127-V line through R434.)

Horizontal output

Horizontal drive pulses from Q403 are coupled through T401 to the base of horizontal output Q404. Q404 is biased on when the beam is at about midscreen. Current flows through the horizontal yoke winding and Q404 to ground. When the beam reaches the right side of the screen, Q404 is turned off and the current in the yoke is directed into C430. At the same time, current flows into

C430 from the regulated B+ via the IHVT primary winding. Due to resonance, the current then reverses and flows back through the horizontal yoke winding. This action deflects the electronic beam back to the left side of the screen, completing the retrace. Current also flows through the IHVT primary into the 127-V regulated B+.

When the voltage on C430 reaches zero, damper CR408 turns on and starts the first half of the scan. Current also flows from ground via CR409 and CR408 into the IHVT primary and regulated B+. As these currents decrease to zero, Q404 takes over conduction and starts the second half of the scan.

No raster, no sound, no high voltage

With a no-raster, no-sound symptom, check the high voltage at the picture tube with a high-voltage probe. If the high voltage is present without a raster, troubleshoot the video and picture tube circuits. Go directly to the horizontal output stages if there is no high voltage. Quickly check each horizontal stage with the scope. Start at the base of the output transistor and go backwards towards the horizontal IC to locate the lost signal (Fig. 3-47). When the signal is located, troubleshoot the preceding stage with voltage and resistance measurements.

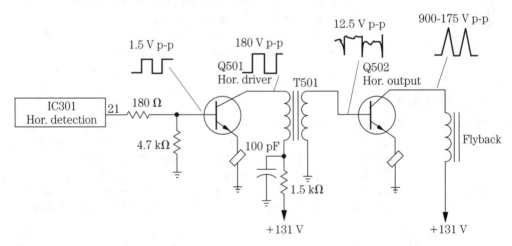

■ **3-47** *Typical waveforms found in the horizontal circuit.*

Sharp C1335A with shorted output

Most horizontal problems are found in the horizontal output circuits. Check the voltage at the flyback transformer and the collector terminal of the horizontal output transistor. Scope the input waveform at the base terminal. Remove the output transistor and test it out of the circuit for open or leakage conditions.

In Fig. 3-48, Q602 was leaky on all terminals and was replaced with a 2SD869, 2SD870, or a universal ECG89 replacement transistor. R613 resistor was also found to be open. The chassis will "motorboat" with R613 open. Also, check the secondary driver transistor for an open winding when there is a leaky output transistor.

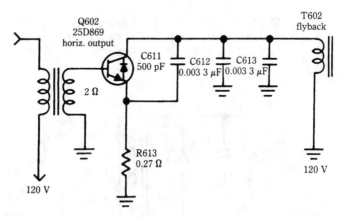

■ **3-48** *Q602 was found leaky in a Sharp C1335A chassis caused by a shorted internal damper diode.*

Open output transistor in Magnavox T995-02

Often, higher-than-normal low voltage can be measured at the collector terminal of the output transistor with an open transistor. Lower voltage at the collector terminal can indicate a leaky output transistor, improper drive pulse, or leaky output transformer. Here (Fig. 3-49), the low voltage had increased to 135 V, indicating a leaky output Q1. Q1 was replaced with a GE-259 transistor. Always replace C4 when there is a defective output transistor. Check the condition of the damper diode D2 before replacing Q1.

Leaky tripler in Zenith 17FC45

With a leaky flyback or tripler unit, the voltage at the collector terminal of the output transistor will be lower than normal (Fig. 3-50). First, check the output transistor for leaky conditions. Make sure a drive waveform is found on the base terminal with the scope. Remove the input lead to the tripler unit. If the collector voltage goes back to normal (144 V), suspect a leaky tripler unit. Check the flyback transformer when the voltage remains the same after removing the tripler input lead.

No raster, no sound, no high voltage

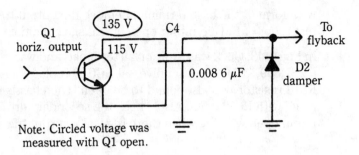

Note: Circled voltage was measured with Q1 open.

■ **3-49** *Replace C4, a four-legged safety capacitor, and check the damper diode D2 if you find a shorted output transistor Q1 in a Magnavox T995-02.*

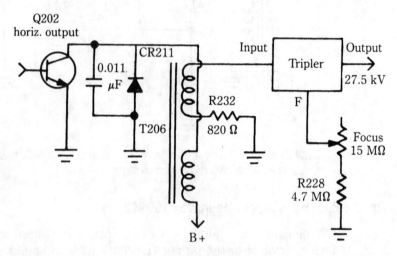

■ **3-50** *A leaky tripler unit in a Zenith 17FC45 chassis loaded down the output transistor.*

No drive pulse in J.C. Penney 685-2012

Go directly to the collector terminal of the driver transistor if there is no drive pulse at the base terminal of the output transistor. Check the waveform on the base and collector terminals. If a waveform is located at the base, suspect a leaky transistor or improper voltage at the collector terminal (Fig. 3-51). Measure the collector voltage. Zero voltage at the collector terminal can indicate R518 and the primary winding of T502 are open, or that there is no supply voltage. Low voltage at the collector can indicate a leaky driver transistor or improper drive waveform.

Check TR503 for leaky conditions with an in-circuit transistor test. Remove the transistor and make another test out of the circuit.

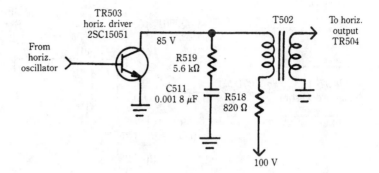

3-51 *TR503 was found leaky in a J.C. Penney 685-2012 with R518 (820 ohms) burned open.*

R518 might run quite warm with a leaky transistor. Measure R518 for a change in resistance. Shorted turns in the primary winding of T502 can provide insufficient drive voltage at the base terminal of the output transistor. Most driver transistors can be replaced with universal replacements.

Realistic 16-410 improper drive pulse

Check the horizontal circuits by taking quick waveforms of each stage. Start at the horizontal oscillator, sweep, or countdown IC terminal that contains the drive signal fed to the horizontal driver transistor. Sometimes a large IC may contain a countdown horizontal signal with a drive amplifier inside the large IC.

In the Memorex 16-410 color TV/VCR portable, IC9701 contains the horizontal amp pre-driver, which feeds a drive signal to Q1001. The drive signal is taken from pin 17 of IC9701 to driver transistor Q9501 and driver transformer (T9501) to the base of the horizontal output transistor (Fig. 3-52). No waveform was found at pin 17, indicating no horizontal drive voltage. Low supply voltage (VCC) indicated a leaky IC9701.

No oscillator waveform in RCA CTC111A chassis

Check the output of the oscillator transistor or IC with the scope. No waveform or an improper waveform can indicate a defective transistor, IC, or surrounding component. Measure the collector voltage of the transistor and the supply voltage terminal of the IC. Low voltage can indicate a leaky transistor or IC.

Check all voltages tied to the IC deflection circuit (U401). Notice if any voltages are very far off from normal (Fig. 3-53). In this RCA chassis, zero voltage was found at terminal 5 of U401. Terminal 5

No raster, no sound, no high voltage

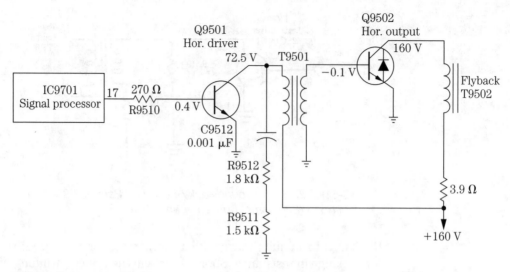

■ **3-52** *Schematic diagram of the horizontal output circuits within the Realistic 16-410 portable TV/VCR.*

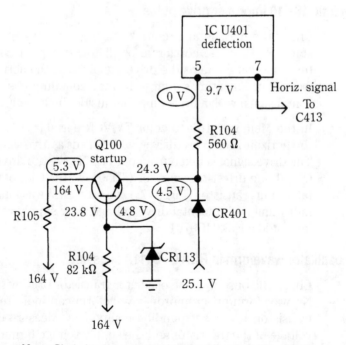

Note: Circled voltages were found with a defective CR401.

■ **3-53** *No horizontal drive signal at pin 7 was found in an RCA CTC111 chassis caused by a leaky diode CR401.*

was unsoldered from the PC wiring. A low resistance measurement was not found from the terminal to chassis ground. A resistance measurement across CR401 indicated a leaky diode. The leaky CR401 prevented Q100 from starting and prevented IC401 from applying a sawtooth voltage to the driver circuits.

Bad connection in a GE AB-B chassis

Besides locating defective components in the horizontal section, do not overlook possible bad board connections. A poorly soldered connection where the component ties into the PC wiring or a broken wiring connection can cause the loss of horizontal sweep. Critical voltage measurements with scope waveforms can locate the defective circuit. A break in the low-voltage power supply (35.1 V) of the General Electric chassis (Fig. 3-54) was located when there was no voltage at the collector terminal of the horizontal driver transistor.

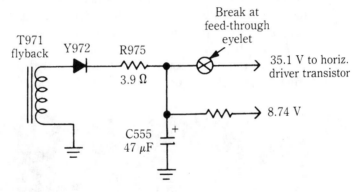

■ **3-54** *A break in a soldered eyelet of the PC board resulted in a loss of voltage at the driver transistor in a GE AB chassis.*

No raster, normal sound

In the older TV chassis, you could have normal sound with a defective horizontal section. Today, with the horizontal circuits and the sound powered from a low-voltage source connected to the secondary windings of the flyback transformer, the horizontal circuits must function to receive good sound. Check the high-voltage circuits for possible trouble when there is normal sound and no raster. A defective picture tube or its circuit can prevent a normal raster. If the high voltage and picture voltages are normal, suspect the video circuits.

Intermittent raster

The intermittent raster can originate in the horizontal or high-voltage circuits. A defective gun assembly of the CRT or picture tube circuit can produce an intermittent picture. The intermittent raster can be caused by a poor regulated low-voltage power supply. Although high voltage might be present, intermittent video components can cause an intermittent raster. Poor board or wiring connections cause many intermittent problems.

The intermittent raster that changes constantly is fairly easy to troubleshoot to locate the defective component. A TV chassis that might take hours to quit working or only "dies" once a week is very difficult to diagnose. Besides voltage and waveform monitoring, the intermittent component might be found with coolant and heat applications. Moving the board or prodding components on the board might turn up a poor board connection.

Transistors and IC components produce most of the intermittent problems in the TV chassis. The horizontal output transistor and the high-voltage regulator are likely suspects (Fig. 3-55). Of course, any transistor in the horizontal circuits can break down. The suspected transistor can be monitored with the scope and with voltage measurements. Monitor the low-voltage supply at the

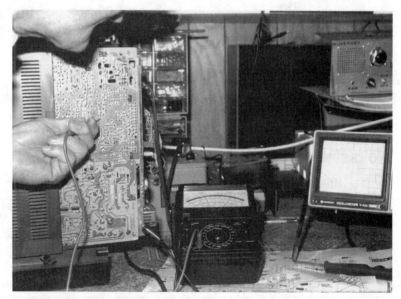

■ **3-55** *Check the waveform at the base and collector terminals of the horizontal output transistor with a scope to determine if the output transistor is leaky or open.*

horizontal fuse for intermittent power problems. Poor transistor socket connections can cause intermittent problems. Bad component-to-board connections or bad double-board feed-through eyelets can be located by pressing up and down on the chassis.

With difficult intermittents, the whole back side of the TV should be covered with a blanket to add more heat, so the intermittent will act up. Raising the power line voltage slightly can help. Monitor the low-voltage supply at the fuse or where dc voltage enters the flyback transformer. Connect the scope lead to the base terminal of the horizontal output transistor (Fig. 3-56). Insert the high-voltage probe to monitor the high voltage. If the raster disappears, note which monitor indicates a poor measurement. No voltage at the low-voltage meter indicates trouble in the low-voltage power supply circuits. If the waveform is not found at the base terminal of the horizontal output transistor, troubleshoot the horizontal stages. The high voltage will disappear with horizontal problems. If the base waveform is normal and the high voltage is low or absent, suspect a defective picture tube or the video and high-voltage circuits. In case the high voltage increases rapidly, suspect high-voltage shutdown. Concentrate on locating the defective component after isolating the intermittent to a given section of the chassis.

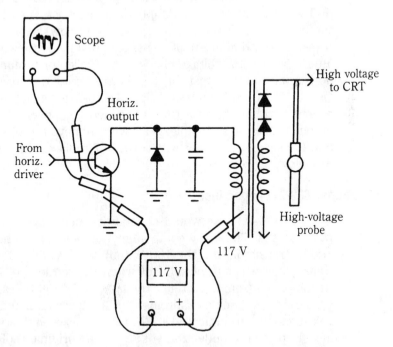

■ **3-56** *Monitor the intermittent chassis with the scope and critical voltage measurements.*

Keeps blowing the fuse

Suspect a defective component in the horizontal or low-voltage circuits when the fuse blows at once. If the fuse takes a few seconds to open, suspect a tripler unit or a component in the high-voltage circuits. A leaky power supply diode or filter capacitor can cause the fuse to repeatedly open. Remove the horizontal circuit fuse or output transistor to determine if the component is in the low-voltage or horizontal circuits.

Check the damper diode and horizontal output transistor when the horizontal fuse will not hold. Measure the resistance between the collector (body) terminal of the output transistor to chassis. Remove the transistor with a low measurement and test it out of the circuit. A leaky flyback transformer can cause the case of the horizontal output transistor to run warm before the fuse is blown. The leaky flyback test should be made with the universal line power transformer to prevent damaging a new output transistor. Isolate a possible shorted yoke by removing the plug or the red wire connected to the yoke plug. It's best to remove the red lead if the yoke plug has a shorting wire for the B+ circuits.

Suspect a leaky tripler unit if the fuse does not blow until after 4 or 5 seconds have passed. Simply disconnect the input lead from the tripler unit. If the fuse does not open now, suspect a leaky tripler. In an Admiral 2M10C chassis, F102 (1-A) fuse would open after 4 seconds. A voltage check at the collector terminal of the horizontal output transistor was normal until the fuse opened (Fig. 3-57). A 100-W bulb was clipped across the open fuse and appeared fairly bright with the overloaded component. When the output transistor and tripler lead were removed, the light went out. Replacing the tripler with a GE-537 replacement solved the fuse blowing problem.

J.C. Penney 682-2114 open fuse

The F910 (1-A) fuse would constantly open when a new fuse was inserted. The leaky component was beyond the horizontal output transistor with the 100-W bulb test. At first the flyback transformer was suspected. But when the red lead of the yoke was removed from the circuit, the light bulb brightness dimmed (Fig. 3-58). A quick measurement of the horizontal and vertical yoke connections indicated a 2.7-Ω short between the two windings. Replace the deflection yoke with the original replacement part.

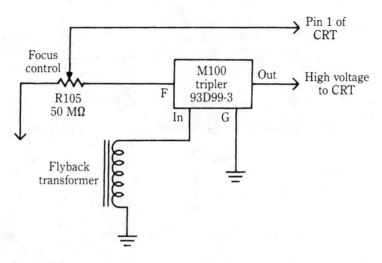

■ 3-57 *After a while the fuse will blow if there is a leaky tripler unit in the horizontal output circuits.*

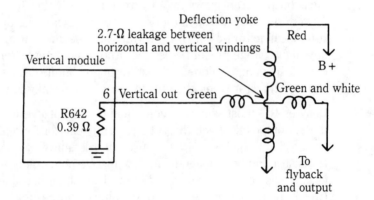

■ 3-58 *The deflection yoke leakage in a J.C. Penney 682-2114 chassis, caused the line fuse to blow.*

RCA CTC59 15-minute trip

The circuit breaker would trip after 15 to 20 minutes of operation, and sometimes when pushed in would kick right out. A 100-W bulb was clipped across the open circuit breaker. The power supply and horizontal oscillator modules were replaced without any luck (Fig. 3-59). When removing the mounting screws of the trace switch (SCR101), the light went out. Replacing the intermittent SCR solved the intermittent raster.

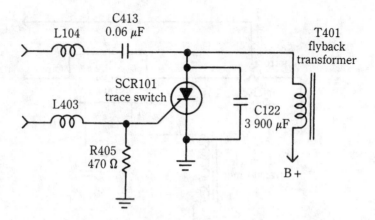

■ 3-59 *A leaky trace switch (SCR101) caused the chassis to shut down in an RCA CTC59.*

Keeps destroying the output transistors

Most output transistors are destroyed by a leaky flyback transformer, an open damper diode, or open or leaky holddown and tuning capacitors. If the output transistor becomes warm without any high voltage present, suspect either component. Remove one end of the damper diode and test for open conditions. Replace the tuning capacitor when the output transistor is leaky or shorted.

Besides the most common components, do not overlook a leaky deflection yoke. Check the yoke return capacitor for leaky conditions. Check the low-voltage source. If it is quite high, the output transistor can be damaged with a defective high-voltage regulator circuit. Disconnect the boost diode and check for possible leakage. Check the diodes in the secondary voltage sources for leakage before inserting a new output transistor.

The new output transistor might be defective. Before installation, check each transistor or diode. The output transistor can be damaged because of lower breakdown voltage. When the original transistor is not available, install one with a 1200-V or 1500-V rating. Many service technicians use the higher-voltage-rating output transistors so that they will last longer. If the case is warm when the transistor is operating, either insufficient drive or too high an operating voltage is applied to the output transistor. Checking the leaky output transformer takes a little longer.

To prevent damaging another output transistor, use a variable line transformer at the power cord (Fig. 3-60). Bring the line voltage up from 60 V to 80 V and check the voltage at the output transis-

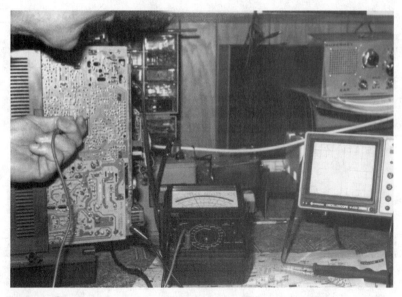

■ **3-60** *Connect the TV chassis to variable line transformer and vary to about 60 to 80 Vac to check horizontal circuits before shutdown.*

tor. Because only two-thirds of the line voltage is used, the dc voltage should be down one-third from normal. Suspect a leaky flyback when there is low dc voltage and under 400 peak-to-peak waveforms at the collector terminal of the output transistor. Remove possible leaky secondary components on the transformer winding with an integrated flyback. Simply remove each lead of the diodes furnishing power to other circuits on the secondary windings and notice if the dc voltage returns.

Red hot output transistors

When the horizontal output transistor runs red hot and sometimes ends up with a dead short, suspect an improper drive waveform, a faulty driver transformer, and a bad small filter capacitor. Determine if the waveform drive voltage is the same as shown upon the schematic. Solder all terminals upon the driver transistor. Check the primary resistance to see if it is the same as in the schematic. The Sams Photofacts lists the primary and secondary resistance, while some manufacturers do not.

If the output transistor is still hot, replace the small electrolytic capacitor (usually a 1-μF, 4.7-μF, and 10-μF at 250 volts) located on the primary of the horizontal driver transformer (Fig. 3-61). Also, replace the filter capacitor connected to the primary of the flyback in

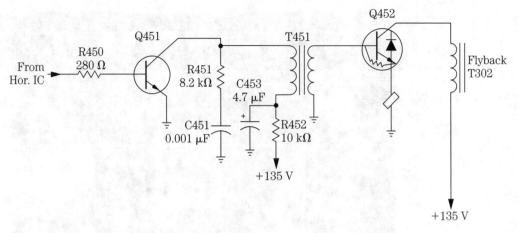

3-61 *Check small electrolytic capacitors (1 µF to 10 µF) in the horizontal driver circuits when the horizontal output transistor runs red hot.*

the B+ source. Replace the driver transformer after all other parts have been replaced. Do not overlook a leaky flyback transformer.

Checking the flyback in an RCA CTC111L chassis

To determine if the flyback transformer is defective in almost any RCA chassis with an integrated transformer, the following method can be used:

1. Short capacitor C113 with test leads (Fig. 3-62).
2. Connect a jumper across the gate and cathode terminals of SCR100.
3. Adjust the variable line transformer to 60 Vdc at the cathode terminal of SCR100.
4. The remote chassis will not function unless a jumper lead is attached across the relay terminals (large red and white tracer).
5. Check the waveform at the base of the horizontal output transistor (15-V p-p) (Fig. 3-63).
6. Check the horizontal output transistor waveform (body) (450-V p-p). The waveform should be clean and without ringing. If ringing is found, suspect a shorted high-voltage transistor. Defective high-voltage diodes (inside transformer) can cause noise spikes. Distortion or retrace spikes denote mistuning or a cracked high-voltage transistor core.
7. If the waveform is clean, raise the ac voltage to a normal B+ 130 V.

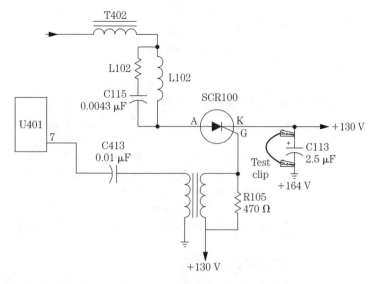

■ 3-62 *Testing the flyback in an RCA CTC111L chassis.*

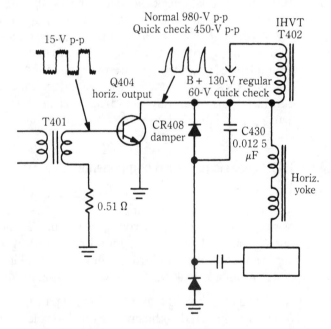

■ 3-63 *Check the waveforms at the base and collector terminals with the scope of the horizontal output transistor.*

8. If heavy current is pulled, suspect a leaky horizontal output transformer.

9. Check the flyback secondary voltages and diodes CR106, CR107, CR109, CR110, and CR112 for possible leakage.

10. Do not overlook the deflection yoke, pincushion, and centering circuits.

11. Replace a leaky flyback transformer with the correct part number.

A GE 19QB chassis transistor blows

Always replace the tuning capacitor in the chassis when the horizontal output transistor is found shorted. If a new transistor is installed without replacing the capacitor, it will blow again. Replace C234 with a 0.0047-μF unit at 1.2 kV or 1.8 kV. These capacitors can appear normal, even when they open up inside the ceramic case. Make sure the tuning capacitor replacement has the correct operating voltage. This also applies to the damper diode when it is replaced.

Additional damaged parts

With a leaky horizontal output transformer or with arc-over found in the TV chassis, check for several parts that might be damaged. Often the chassis comes in with a no-sound, no-raster symptom. You might find a leaky horizontal output transistor, or damper diode, or an open tuning capacitor and flyback transformer in the same chassis. High-voltage arc-over can cause a chain of events that damages small diodes in the secondary winding as well as the output transistor and flyback. Look a little closer than usual, because you might find several components damaged under these conditions.

RCA CTC111H chassis damaged components

This particular chassis (Fig. 3-64) came in with a no-sound, no-high-voltage, no-raster symptom. Q404 was found shorted and was replaced. No collector voltage was found at the horizontal output transistor. SCR100 was found open and was replaced. The chassis was fired up and Q404 went out. Checking the output circuit, CR408 was leaky and CR409 had a 2-Ω leakage.

After replacing both diodes with original parts, a line voltage transformer was inserted in the line. As the line voltage was raised, Q401 was pulling heavy current. Right away the flyback transformer was suspected (T402). The yoke assembly was disconnected without any results. Next, the pincushion circuits were checked and transistor Q405 was found to be leaky. The dc voltage came up at the output transistor, indicating the overload circuit was lifted. The blown 5-A fuse, Q404, SCR100, Q405, CR408, and CR409 were replaced to put the chassis in operation.

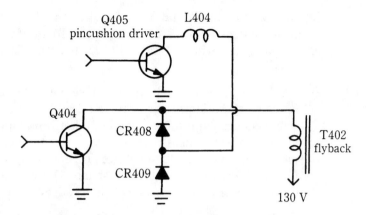

■ 3-64 *Take critical voltage and resistance measurements to locate the leaky component in the horizontal output circuits.*

Variac transformer tests

The variac, or step-up (or down) power line transformer, is the ideal test instrument for locating intermittents with defective flyback transformers and in high-voltage shutdown cases. With intermittents, the voltage can be raised above normal power-line voltage to apply more voltage or heat to the intermittent component. To prevent damage to another horizontal output transistor, the dc voltage can be lowered until a leaky flyback or excessive high low-voltage source is located. The high-voltage shutdown circuits or other components causing the shutdown can be found by applying lower voltage to the horizontal circuits.

Sometimes when raising the variac line voltage to around 90 volts, the chassis may operate with narrow width, indicating a defective line voltage regulator. Monitor the voltage at the output of line voltage regulator as the line voltage is raised. Monitor the base terminal of the output transistor with the oscilloscope. Suspect the line voltage regulator when voltage is raised higher and the chassis shuts down. Also, suspect the regulator when the chassis is dead and will only operate with about 87-90 volts applied from the variable line transformer.

High voltage too high

Excessive high voltage in the TV chassis can cause the chassis to shut down, turn the raster into horizontal lines, arcover, and destroy components in the horizontal and high-voltage circuits. In the early HEW holddown circuits, the excessive high voltage would affect the horizontal oscillator circuit and cause the raster to go into horizontal lines or off frequency. Today, if the high voltage is too high, the high-voltage shutdown circuits shut the chassis down.

Higher-than-normal high voltage can cause the raster to expand or go out when the brightness control is raised in some TV chassis. The chassis might intermittently go into horizontal lines with adjustment of the brightness control. First check the high voltage at the picture tube with the high-voltage probe. Adjust the B+ source or high-voltage control for lower voltage. The chassis might be shutting down with too high a low-voltage adjustment.

Some TV chassis will shut down quicker than others. Connect the chassis to a universal line transformer. Adjust the line voltage until the chassis shuts down. Under normal operation, you might find a few chassis will shut down at 122 or 125 Vac, while others can operate at up to 140 Vac. Monitor the high voltage with the high-voltage probe. Suspect a defective shutdown circuit if the high voltage shuts the chassis down too early. Check the power line voltage at the customer's house for higher-than-normal line voltage which might shut the chassis down at certain times of the day.

162

High-voltage shutdown

The chassis might shut down with a high power line voltage, with a defective component in the chassis, or by action of the high-voltage shutdown circuits. Use a universal power line transformer to help solve the shutdown problem. Slowly raise the ac voltage and notice at what voltage the chassis shuts down. Start all over and keep the line voltage lower than shutdown to determine what section is causing the chassis to shut down. A 100-W bulb in series with the open fuse can help in locating the shutdown component.

To determine if the chassis is shut down by another component or if it is actually high-voltage shutdown, use the universal power line transformer or remove a diode in the shutdown circuit. Often, one end of a diode or transistor can be easily removed from the shutdown circuit (Fig. 3-65). Monitor the high voltage with a probe and monitor the dc voltage source to the horizontal output transistor with a voltmeter. Be very careful at this point. Hold on to the ac plug as the chassis is plugged in. Do not leave the chassis on too long, so as to prevent damage to the picture tube, high-voltage, and horizontal circuits. Notice the high-voltage and low-voltage source measurements. If high-voltage shutdown is causing the

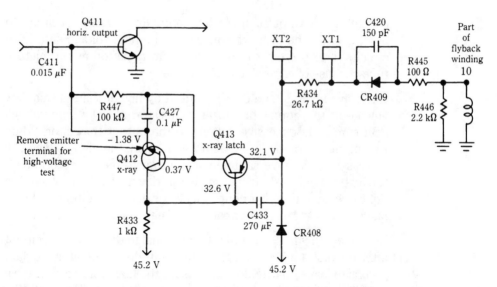

■ 3-65 *Remove one end of the diode or transistor from the circuit to determine if the high-voltage shutdown circuits are shutting down the chassis.*

problem, the chassis will remain on during this test. After making repairs, do not forget to replace the diode or transistor lead to the shutdown circuit.

Suspect a defective component in the low-voltage source with higher-than-normal dc voltage. If the low-voltage source is normal and the high voltage is too high, suspect a high-voltage regulator SCR, or the tuning or holddown capacitor in the horizontal circuits. Look for a defect in the high-voltage shutdown circuits when the high voltage is normal and the chassis shuts down.

High-voltage shutdown circuits

Most high-voltage shutdown circuits operate the same way, although some can be quite complicated and use several transistors and diodes. When the high voltage becomes too high, the shutdown circuit turns the chassis off. If the high-voltage shutdown circuits have a defective component, causing overloading, the chassis might shut down by itself. Disconnect the shutdown circuits momentarily to see if the high voltage at the picture tube is normal.

Some high-voltage shutdown circuits use a silicon controlled rectifier (SCR) to kill the horizontal oscillator stage. Other circuits can kill the drive pulse at the horizontal drive transistor or at the defective IC. The primary concern of the high-voltage shutdown circuits is to protect the operator from radiation at the picture tube because of excessive high voltage.

Excessive high voltage can be caused by higher-than-normal low voltage at the horizontal output transistor. A leaky high-voltage SCR can apply excessive voltage to the output transistor. When the tuning or holddown capacitor goes open, the high voltage will increase. Use a universal line voltage transformer to determine if the high voltage is shutting the chassis down or if the problem is a defective chassis component. A leaky or defective integrated horizontal output transformer can cause high-voltage shutdown.

Reconnect the diode or transistor terminal to the shutdown circuit after repairs. A defeated high-voltage shutdown circuit is too dangerous to leave unattended. Make sure the high-voltage shutdown circuit is working by shorting out a component. In Fig. 3-66, the shutdown circuits can be tested by shorting test points XT2 and XT1 together. Most chassis have an easy method to check the high-voltage shutdown circuits. The chassis should not operate with these two test points shorted. If the chassis does not shut down, suspect a faulty high-voltage shutdown circuit.

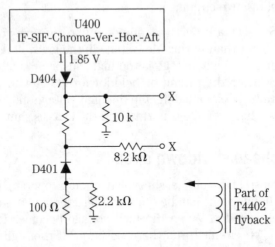

■ 3-66 *Short the two X connections to see if the shutdown circuits will shut down the chassis.*

RCA CTC101A high-voltage shutdown

With the variac transformer in the power line, this chassis would not shut down until 30 kV was measured at the CRT. One end of the holddown diode (CR419) was removed from the circuit to verify high-voltage shutdown. The collector voltage at Q100 was 139 V and should be about 121.5 V. In this model, a new regulator resistor trim component kit was installed to lower the dc voltage. CR115, R113, R114, R115, R116, and R117 were removed from the circuit. New components from the regulator kit, CR115, R113, and R117 were installed (kit number 146399).

Sanyo 91C64 high-voltage shutdown

The chassis and sound came on for only a few seconds and shut down. To determine if the chassis was shut down by high voltage, disconnect the collector terminal of shutdown transistor Q403. Use solder wick to remove the terminal from the wiring. Measure the high voltage if the chassis stays on (25 kV to 27 kV). If the high voltage is above 28 kV, suspect high-voltage shutdown.

Check the voltage at the horizontal output transistor (normally 114 V). Try to lower the dc voltage with low-voltage control. If the dc voltage is above 120 V to 150 V, the chassis will shut down. Suspect a poor low-voltage regulator circuit when the low voltage cannot be adjusted or is too high. Notice if the high voltage varies with the low-voltage regulator control.

If the low voltage and high voltage can be adjusted and are normal with a 120-Vac line, suspect trouble in the high-voltage shutdown circuit (Fig. 3-67). Check for zero voltage at pin 8 of LA1461. Any voltage measurement at pin 8 indicates a defective high-voltage shutdown circuit.

Test Q403 in the circuit and replace it if it is leaky. Check all diodes, especially D404 (33-V zener), D405, D415, and D406. Remove one end of each diode for accurate leakage tests. Make sure 12 V is found on the anode side of D405. Check all 1% resistors in the shutdown circuits. Double-check the emitter voltage on Q403. It should be lower than the base terminal.

RCA CTC97 blast on shutdown

Suspect a leaky flyback transformer when the sound blasts on and then shutdown occurs. Often the horizontal output transistor is shorted with a leaky SCR101 high-voltage regulator. Remove the collector lead of Q38 to see if the chassis has high-voltage shutdown. If the high voltage does not come up and the chassis shuts down, insert a universal transformer in the power line. Slowly increase the ac voltage to 60-V or 80-Vac. Feel the case of the hori-

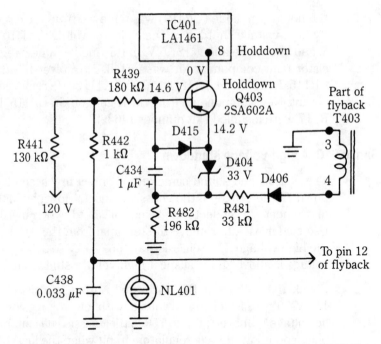

■ 3-67 *Check components C434 and leaky IC401, D415, D404, and D406 when chassis shutdown occurs because of a defective shutdown circuit in a Sanyo 91C64 chassis.*

zontal output transistor. Shut the chassis down if it gets warm. Replace the leaky flyback with a 145722.

Tic-tic noise

If, with one ear close to the flyback transformer, a tic-tic noise is heard, the horizontal circuit might not have sufficient drive voltage to the output transistor, or it could be a defective flyback transformer, leaky tripler unit, or loading down of the flyback in secondary circuits. In the modular chassis, remove each module except the horizontal oscillator to locate the defective module loading down the low-voltage circuits in the secondary voltage source of the flyback. Remove the transformer input lead to the tripler unit with a tic-tic noise to determine if the tripler is loading down the output transformer. Check the waveform at the base terminal of the output transistor to determine if horizontal circuits are normal (Fig. 3-68).

Take an in-circuit diode test with the DMM of each low-voltage diode tied to the secondary low-voltage source before suspecting the flyback transformer. A leaky diode or component tied to one of

■ **3-68** *Check the drive waveforms in the horizontal circuits with the scope, all the way up to the flyback.*

the low-voltage sources can load down the secondary circuits. Often, with a leaky secondary component, the high voltage will come up with sound and die down in a few seconds.

Defective deflection yoke

A leaky horizontal yoke winding to the vertical winding can cause the fuse to blow in older solid-state chassis. Shorted turns in the horizontal deflection yoke can cause shutdown in the latest TV chassis. Often, the chassis might operate for a few seconds before the defective yoke shuts down the chassis. The raster might come up, or there might be only a quick flash in the picture tube before shutdown.

Usually, the yoke loads down the flyback and horizontal output transistor with some of the high voltage with the variac transformer in the power line. The output transistor might run warm with an overloaded, leaky yoke assembly. Remove the red lead from the yoke assembly. The voltage will increase right away if the the yoke is defective. Of course, the high voltage will be quite low with the yoke out of the circuit. Remove the yoke assembly and check for leakage between the horizontal and vertical windings. It's very difficult to locate a shorted turn or two with the ohmmeter because the horizontal windings have a very low resistance. If the yoke loads down the voltage at the collector terminal of the horizontal output transistor, replace it. Do not overlook a leaky pincushion circuit tied to the same spot as the yoke assembly.

Chassis shutdown

Practically any component in the horizontal circuit can produce chassis shutdown. The horizontal output transistor, high-voltage regulator, flyback, and IC oscillator are the most likely components. The chassis might shut down with poor component terminal connections and loose transistor sockets. Chassis shutdown can occur at once, or several hours later. Check for arcing sounds in the flyback with chassis shutdown. The defective component might be intermittent.

Approach the shutdown problem as an intermittent or high-voltage problem. Use a variable line voltage transformer in IHVT circuits. Take scope waveforms with critical voltage measurements. Isolate the high-voltage shutdown circuits temporarily to a defective chassis component by removing the lead of the transistor or diode from the high-voltage shutdown circuits.

RCA CTC93 chassis shutdown

After an hour or so, this RCA chassis would quit operating (Fig. 3-69). At the collector terminal of the horizontal output transistor, no voltage was measured when the chassis shut down. The open high-voltage regulator (SCR600) was located with voltage tests at the collector and anode terminals when the chassis shut down.

Poor socket connections in an RCA CTC87A chassis would intermittently shut it down. Either replace the socket or solder the SCR terminal pins to the socket temporarily.

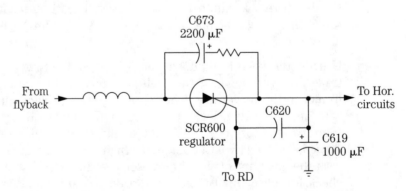

■ **3-69** *Regulator SCR600 caused the chassis to be intermittent in an RCA CTC93 chassis.*

Poor focus

Poor focus can result from a low high voltage at the CRT. Check the anode and focus voltage at the picture tube with the high-voltage probe. Intermittent focus problems can be caused by poor picture tube pin connections. A defective focus control might produce erratic focus problems (Fig. 3-70). Isolate the focus problem to the focus or horizontal circuit with low high voltage. A defective IHVT can cause poor-focus problems.

■ **3-70** *Check the focus control for erratic focus and replace it with a new one if necessary.*

Poor width

Check the high-voltage regulator circuits for poor picture width in the latest TV chassis. The high-voltage regulator transistor, SCRs, and zener diodes in the regulator circuits can produce insufficient width. Poor solder connections at pincushion, regulator, and driver transformers can result in poor width. Low voltage applied to the horizontal output circuits can cause poor width. Open bypass or coupling capacitors in the horizontal and high-voltage circuits also can cause poor width.

Excessive width

Although most width problems result in a narrow picture, sometimes a chassis comes in with excessive width. Most of the new solid-state chassis do not have a width control. Check the B+ or high-voltage adjustment for correct high voltage at the CRT. Excessive low voltage can increase the width of the raster. This can be caused by a leaky low-voltage regulator transistor, or an open zener diode. Excessive width can be caused by a high B+ boost voltage. Check the small filter capacitor in the boost circuit.

Poor high-voltage regulator

Improper voltage applied to the horizontal output transistor can result from a poor high-voltage regulation circuit. Leaky or intermittent high-voltage regulator SCRs can produce poor regulation (Fig. 3-71). High regulated voltage applied to the horizontal output transistor can cause chassis shutdown in the IHVT chassis. Check the high-voltage regulation circuits for poor voltage fed to the horizontal output stage. Replace the high-voltage regulator module for poor regulation in a modular chassis (Fig. 3-72). A regulator transistor or zener diode in the regulator circuit can cause poor high-voltage regulation.

Regulator

SCR

■ **3-71** *Suspect a high voltage regulator when voltage is missing at the horizontal output transistor.*

HV regulator module

■ **3-72** *Replace the high-voltage regulator module when there is no voltage at the horizontal output transistor.*

RCA CTC68 cannot adjust high voltage

The horizontal frequency was off and the raster would go into horizontal lines intermittently in this chassis. The high-voltage control could change the high-voltage setting, but the high voltage would constantly shift high and low. Open Q401 was replaced and it immediately burned out (Fig. 3-73). Replacing open CR404 and Q401 solved the poor high-voltage regulation problem.

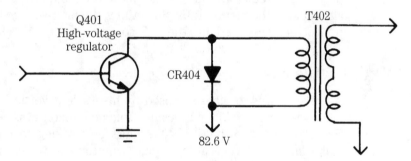

■ **3-73** *The high-voltage adjustment will not vary the voltage with a leaky or shorted CR404.*

Improper B+ or high-voltage adjustments

You might find both a B+ and high-voltage adjustment control in the early TV chassis. Some of the latest TV chassis do not have either control. Adjust for the correct voltage at the horizontal fuse or collector terminal of the horizontal output transistor with the B+ control. Set the high-voltage control with correct voltage at the anode socket of the picture tube. These voltages are listed on the schematic.

Lower the B+ voltage if the line voltage is high at the customer's home. This often happens with rural power lines. Now readjust the high-voltage control. Suspect a defective high-voltage regulator circuit when the B+ voltage has to be lowered with higher-than-normal voltage at the collector terminal of the horizontal output transistor. The wrong zener diode replaced in a high-voltage regulator circuit can produce improper high voltage. Double-check the voltage of a replaced zener diode. Lower the B+ voltage as far as possible (while retaining adequate raster width) in extreme high-voltage areas.

Flashing horizontal pictures

Check the flyback transformer for poor soldered connections or loose wires when there is flashing in the picture. Often the whole raster will flash off and on. The raster might flash across the screen and go dead. Poor PC wiring or board connections can cause a flashing picture. Inspect the yoke socket for poorly soldered or loose wire connections. A poorly soldered input terminal connection from the flyback to the tripler can cause flashing in the horizontal circuits. Discharge the CRT and refasten the high-voltage cable to the picture tube. Sometimes the high-voltage cable is not correctly inserted into the CRT socket.

Vertical white line

A horizontal white line is caused by insufficient vertical sweep, while the vertical white line results from no horizontal sweep. The flyback and high-voltage circuits must be working to receive a white line. Most vertical white-line symptoms are caused by an open yoke, yoke terminals, or yoke return capacitor. First check the horizontal winding and connections of the yoke plugs into a socket. Do not overlook an open pincushion transformer in series with the horizontal yoke return winding.

RCA CTC76A white line

Only the up and down white line was found with no horizontal sweep in this RCA chassis. After the yoke winding and terminals were tested, the yoke return wire was traced to the pin-phase transformer, R5, or L401 (Fig. 3-74). A poor soldered connection of L401 to PC wiring resulted in the vertical white-line symptom.

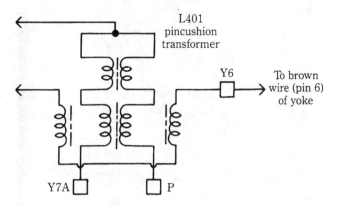

■ 3-74 *Check the pincushion terminal connections for bowing or a white line in the picture.*

Horizontal foldover problems

Horizontal foldover can be caused by a defective damper diode, safety or holddown capacitors, capacitors in pincushion circuits, and the pincushion transformer. A leaky horizontal output transistor, trace or retrace SCR, and flyback can cause foldover. Open bypass capacitors in the high-voltage regulator or SCR horizontal circuits can cause some type of horizontal linearity or foldover. Open or poorly soldered terminals of reactors and regulation transformer can cause horizontal foldover (Fig. 3-75).

Funny horizontal noises

There are many different horizontal noises you can hear that are caused by a component in the horizontal circuits. A high-pitched, high-voltage squeal or a ringing noise can be caused by loose particles or by vibration in the flyback transformer. Sometimes the flyback bolts need to be tightened to eliminate the noise. Pinning the flyback winding core with toothpicks and glue might solve the vibration or singing problem. Replace an IHVT flyback that suffers from loose particles (Fig. 3-76).

The tic-tic noise in the flyback indicates the chassis is in shutdown, or that there is no horizontal sweep. Small spitting noises can be caused with high-voltage arcover inside the flyback or the lead to the focus control (Fig. 3-77). A high-pitched, intermittent whistle noise can be caused by a vibrating ferrite bead (FB) component over the leads of a diode or transistor element in the horizontal circuits. The firing noise with a high-pitched squeal can be caused with arcover in a focus control.

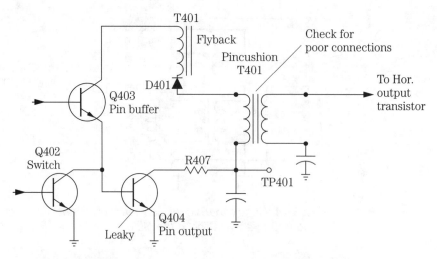

■ 3-75 *Solder all pincushion transformer terminals and check the pin output transistor for leakage in late-model pincushion circuits.*

■ 3-76 *A defective IHVT flyback can cause squealing and ringing noises.*

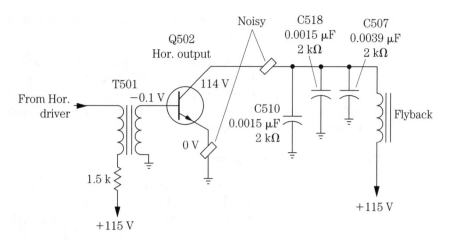

■ 3-77 *Ferrite beads within the horizontal output circuits can come loose and cause noise in the chassis.*

Noise lines in the raster

A firing flyback transformer can cause noise lines in the picture even if you cannot hear the internal arcing noise. Poorly soldered connections on the flyback terminals can cause line noise in the raster. Make sure the picture tube shield is grounded to the metal TV chassis. Check the high-voltage cable to the CRT for breaks or for a loose screw in the anode clip. Notice if the high-voltage lead is clipped properly into the picture tube.

Excessive arcing inside the focus control can cause firing lines in the raster. Simply rotate the focus control and notice if lines disappear. Replace the defective focus assembly. Horizontal lines that look like auto ignition noise can be caused by open filter capacitors in the B+ voltage regulation circuits. Remove the antenna lead to determine if the noise is picked up or occurs within the TV chassis.

Arcover and firing

High-voltage arcover can occur at the CRT, flyback, or tripler unit. Often, you can see the arcover happen. Sometimes the high-voltage arcover at the picture tube anode will show lines of firing down to the banded area of the CRT. Replace arcing flyback and tripler units. Arcover can occur at spark-gap assemblies, indicating excessive voltage or a defective component nearby.

Usually, arcover inside the picture tube gun assembly indicates an open tube filament or a broken assembly. After replacing the defective component causing the arcover, check the B+ and high voltage. Service the high-voltage regulator or low-voltage circuits producing excessive voltage at the picture tube.

RCA CTC92L dark lines in picture

Two dark lines appeared at the top and bottom of the screen and rolled upward. It looked like 120-Hz hum. All filter capacitors were shunted and the results were the same. CR305 was found to be leaky in the startup diode circuit at T201 (Fig. 3-78). Any one of the startup diodes can cause the dark bars in the raster. Replacing the start diode solved the noisy line problem.

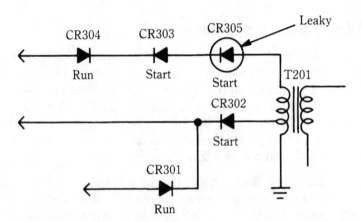

■ **3-78** *Check for a leaky startup diode, which will cause two dark lines to roll up the screen.*

Jail bars

In the tube chassis, vertical bars to the left or right of the raster indicate Barkhausen oscillations. A defective flyback or horizontal output transistor can cause vertical firing lines in the raster. Check the horizontal output and high-voltage components for vertical lines in the raster.

Sony KV-2643R jail bars

Several dark vertical bars to the left of the raster were found in this chassis (Fig. 3-79). Sometimes if the contrast and brightness

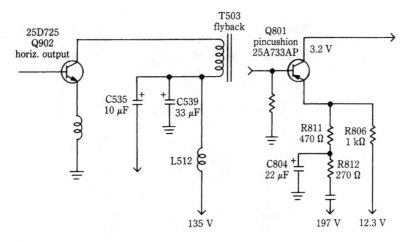

■ 3-79 *Open electrolytic capacitors cause jail bars to the left side of picture.*

were adjusted, the bars could barely be seen. Resistor R812 was found burned in the pincushion circuits. Replacing both electrolytic capacitors (C534 and C539) in the horizontal output circuit solved the problem. Both of these capacitors are mounted close to the flyback transformer area.

Pie-crust lines

The pie-crust lines often appear to the left side of the picture. These lines can be intermittent and can vary with the brightness control. The pie-crust effect can take place with open resistors across clamp diodes, or poor filtering in the horizontal stages.

J.C. Penney 685-2041 horizontal pie-crust lines

These unusual pie-crust horizontal lines were caused by a defective electrolytic capacitor in the horizontal oscillator circuits (Fig. 3-80). The 3.3-μF capacitor was replaced by one with a 250-V rating.

Horizontal motorboating

A putt-putt noise in the horizontal output circuit can be caused by a defective output transistor or corresponding component. Check for an open emitter or base resistor in the horizontal output transistor circuit. Replace the output transistor when no emitter resistors are located in the circuit. These transistors might test normal and still oscillate under load.

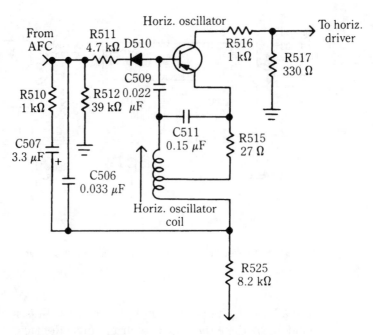

■ 3-80 *Replace C507 in a J.C. Penney 685-2041 with horizontal pie-crust lines.*

Montgomery Ward C1935 motorboating

The neon dial lights would blink, demonstrating unstable B+ voltage. At first the trouble suspected was poor low-voltage regulation. When Q602 was removed from the circuit, the motorboating sound quit. Replacing the horizontal output transistor (2SD870) did not solve the problem (Fig. 3-81). Installing a new emitter resistor (R615) solved the motorboating condition. Evidently Q602 had shorted, taking out the emitter resistor.

Poor board connections

Poor board or component terminal connections can produce a dead chassis, intermittent horizontal section, and high-voltage troubles. A poor solder connection around a regulator or pincushion transformer can cause the raster to narrow. Poorly soldered terminals of any horizontal transistor can cause an intermittent picture. Check all eyelet feed-through terminals that can cause improper voltage to a horizontal circuit.

PC wiring might break at a flyback transformer lug connection, killing the high voltage. Sometimes the wiring will break at feed-

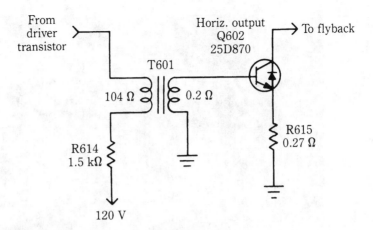

From driver transistor

Horiz. output
Q602
25D870

To flyback

T601

104 Ω   0.2 Ω

R615
0.27 Ω

R614
1.5 kΩ

120 V

■ **3-81** *Check R615 for motorboating in the output circuit with the damper diode inside the output transistor.*

through eyelets. The PC board might break where the flyback transformer is mounted if the chassis is dropped or knocked off the TV stand. Often, heavy components such as transformers will cause the board to crack when the TV is dropped. Repair the PC wiring with short lengths of hookup wire across each broken area. Check for broken PC wiring where the wiring joins a large terminal post. Check for PC wiring breaks at the various controls when the set is dropped on the front side.

Soldering IC terminals

Look for cracked board areas around large IC processors. Repair the cracked wiring with hookup wire. Sometimes the intermittent symptom might be caused by poor pin connections of the IC. When the chassis is carried or moved around, the contacts might break.

Solder each pin with a small soldering iron. The battery-operated iron is ideal for small delicate connections (Fig. 3-82). Use very small solder (0.015) for IC work. Do not hold the iron on too long, just long enough to make a good flowing joint.

SCR horizontal circuits

In the early modular chassis, two SCR components were used in the sweep circuit in place of a regular horizontal output transistor. The same SCR components were found in the RCA chassis starting with the CTC40, CTC46, CTC48, CTC58, and CTC68 chassis (Fig. 3-83). You will find the same sweep circuits in Philco, Coronado, J.C. Penney, and others during this same period. Although the

■ **3-82** *Solder the terminal pins around IC terminals with a low-wattage soldering iron.*

■ **3-83** *Two SCRs developed the drive to the flyback transformer in early TV chassis.*

SCR sweep circuits were fairly reliable, most defective components were common to all chassis.

Retrace and trace switch SCRs provide a horizontal sweep pulse to the flyback transformer. Often SCR101 is found open and SCR102 becomes leaky. CR402 and CR401 provide damping action for SCR102 and SCR101 respectively (Fig. 3-84). If CR402 goes open, SCR102 will become shorted. Replace the small insulator with silicone grease when installing a new SCR component.

Take a voltage check at the case of each SCR with a dead symptom. Normal voltage at SCR102 and no or low voltage at SCR101 can indicate a normal retrace switch with a defective trace component. Low voltage at SCR101 can be caused by a leaky tripler unit. Remove the tripler input lead for further tests. A quick scope test at SCR101 will indicate if the normal drive pulse is applied to the flyback. Remove both SCRs and test them out of the circuit with no drive pulse. Disconnect one end of CR402 and CR401 and test for leakage.

High-voltage problems

Excessive high voltage will cause these chassis (Fig. 3-85) to go into horizontal lines. The horizontal might sync in when the brightness is raised and lowered. Try to adjust the high voltage with R412. If not, troubleshoot the high-voltage regulator circuits. Check Q401, CR402, and CR404 for leakage. Remove one lead of each diode for accurate leakage tests. Check the primary of T402 for a correct resistance of 37 Ω.

Poor horizontal and vertical sync can occur with 32- to 35-kV high voltage at the picture tube. Excessive high voltage can be caused by poorly soldered terminals on T402. Poor terminal connections on T402 and T401 can produce intermittent and high-voltage problems. Replace CR402 with a 13-V, 1-W zener, diode. If CR404 opens, Q401 is destroyed. Q401 can be replaced with a universal SK3024 transistor when the original cannot be found.

Servicing the RCA CTC110A chassis

Many of the latest RCA chassis can be serviced in the same manner as the CTC110A chassis. For that matter, several different brands of TVs can be serviced with the following service procedures. After blown fuses have been replaced in sets with a no-high-voltage, no-raster, no-sound symptom, determine if the chassis has high-voltage shutdown or a defective component. With this particular chassis, the flyback circuit must perform or no voltage is applied to the horizontal oscillator and other circuits.

Use a universal power-line transformer to set the voltage at about 60 to 80 Vac. Notice if the chassis has high-voltage shutdown. An-

182

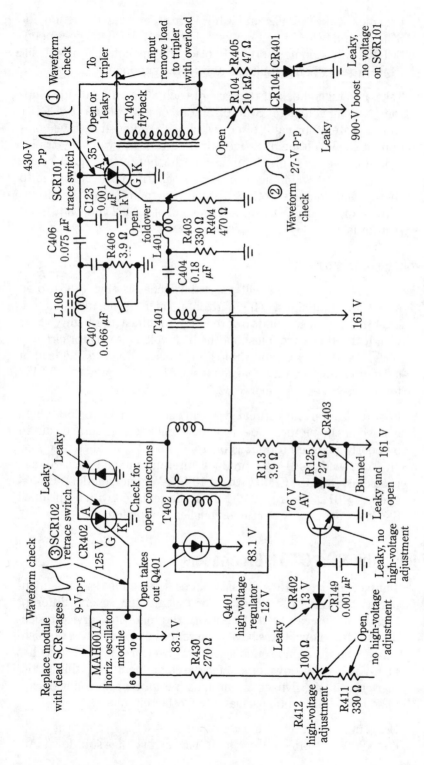

■ 3-84 The various problems caused by the defective components are found marked on the schematic.

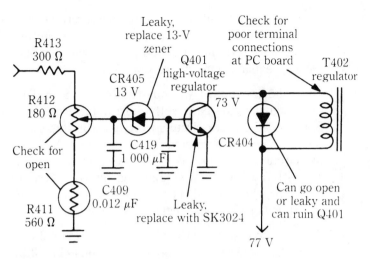

■ 3-85 *Check the various components in the high-voltage regulator circuits.*

other method is to remove the emitter terminal of the x-ray latch transistor (Q414) to prevent high-voltage shutdown. If the high voltage does not come up, suspect a defective horizontal circuit.

Take a waveform check at the output transistor. Without a drive pulse, measure the voltage at the collector terminal of the output transistor (Q412). A normal voltage should be around 121 V (Fig. 3-86). Lower voltage can indicate a leaky output transistor or damper diode. Remove Q412 from the circuit and test both components for leakage. If the voltage is high, suspect an improper

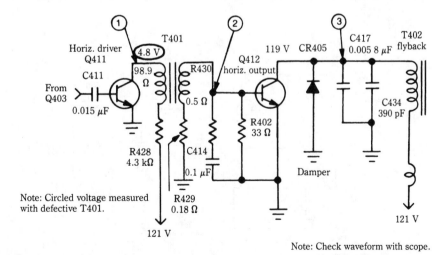

■ 3-86 *Check the voltage source and waveform at collector terminal of Q412.*

drive signal or the high-voltage regulator circuit. In this particular case, 159.7 V was found at the regulator with output of 132.5 V at the collector terminal of the output transistor, indicating the high-voltage regulator circuits were working.

Low collector voltage was found at Q411 (4.8 V). This indicated no drive signal at the base of the horizontal driver or a leaky Q411. Because the flyback circuits were not working, the 19.9 V feeding the horizontal was missing. To determine if the horizontal or output circuits were defective, the external voltage supply must be connected to the 19.9 V supply. The low-voltage source must function to provide drive signal to the horizontal output circuits. A universal power source is ideal in this situation.

Remove the power cord and connect the external voltage source to the 19.9-V source in the horizontal circuits (Fig. 3-87). Adjust the external voltage to 19 V or 20 V. The horizontal output transistor (Q412) should be removed for these tests. Check the waveform at the base of the horizontal driver transistor (Q411) with the scope. Now plug the chassis into the power line. Signal-trace the waveform from the horizontal switch through to the base of the output transistor. Suspect problems in the output circuits if a normal waveform is found at the base of Q412.

In this particular case, the drive signal was good at the base of Q411 and no signal was found at the collector terminal. Q411 was checked for leakage out of the circuit. Q411 was replaced and still no waveform was found. Both R428 and the primary winding of T401 were checked for correct resistance. When the chassis was

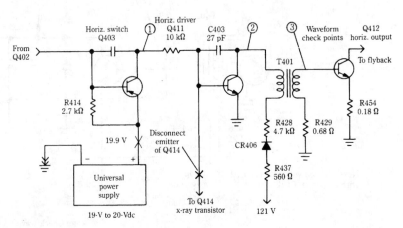

■ 3-87 *Connect an external power supply to determine if horizontal circuits are working.*

reconnected, the waveforms were good at the base of the output transistor and collapsed in 30 minutes. The primary winding of T401 had an internal short, yielding a resistance of 0.05 Ω instead of 98.9 Ω. Replacing the shorted driver transformer solved the no-raster, no-high-voltage symptom.

The TV chassis with a flyback transformer (IHVT) must be serviced in a slightly different manner than the standard output circuit. First, determine if the chassis is shut down with high voltage or a bad component in the chassis. Use a universal power line transformer or remove a leg of a transistor or diode in high-voltage shutdown circuit to determine if high-voltage shutdown is the problem. Next, determine if the horizontal circuits are functioning with an external power source and monitor the circuits with the scope.

Fire damage

Critical components that cause fires in the TV chassis are the deflection yoke, flyback transformer, degaussing coils, overheated resistors, large power resistors, and transformers. Replace these defective components with parts having the exact part number as the originals. A stuck relay or relay transistor circuit can cause the relay to stay on and apply power to the TV set. A defective relay with weak release spring or broken leaf assembly may let the relay on and not shut off. A shorted yoke or flyback transformer can cause the windings to burn other components nearby.

Red-hot resistors close to the PC board or liquid spilled down inside TV can cause the PC board to undergo a slow burning process. Shorted diodes in the low-voltage power supply can overload power transformers and produce fires. Always replace line and secondary fuses with replacements of the exact amperage. Do not place larger amperage fuses that may not open with a leaky component in the voltage sources.

Six actual horizontal sweep problems

Here are six typical problems with horizontal sweep circuits.

Dead-shorted output transistors

After replacing the fuse, the chassis groaned and out went the fuse again. The horizontal output transistor was checked to common ground and showed a 0.13-Ω short (Fig. 3-88). Q401 was replaced and the variable isolation transformer was slowly raised to 65

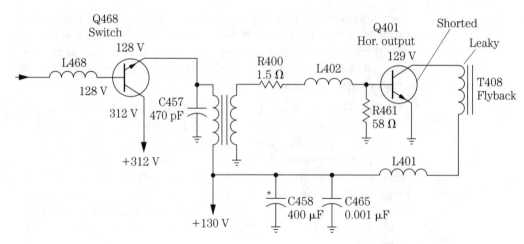

■ 3-88 *A shorted Q401 was caused by a defective flyback (T408) in a Sylvania 25C504 chassis.*

volts. The horizontal output transistor became warm, with very little drive waveform. After several attempts, it was found that the flyback transformer (T408) was knocking out Q401. Replacing the fuse, Q401, and T408 solved the Sylvania 25C504 problem.

Destroys output transistor in a few hours

After the horizontal output transistor was replaced in an Emerson MS1980 TV, the chassis shut down in about 3 hours. Again Q401 was checked, and it was found to be shorted. Replacing the transistor once again and slowly raising the line voltage with the variable isolation transformer did not cause the chassis to shut down.

Critical voltage and waveform tests were made on Q401 and the horizontal driver transistor. The collector voltage upon the driver transistor was too high, according to the schematic. C435, (1 μF, 160 V) an electrolytic capacitor, was replaced, which eliminated the damaging of the high-output transistor. Replace small-capacity capacitors in the collector circuit when it keeps destroying the output transistor.

Blown fuse—shorted output transistor

The main line fuse in a Samsung K-25 chassis would open after replacement, indicating an overload in the power supply or horizontal output circuits. Q402 (25D870) horizontal output transistor was found leaky. After replacing Q402, the chassis still blew the main fuse. A resistance measurement to collector of Q412 and common ground indicated 0.27 ohms (Fig. 3-89).

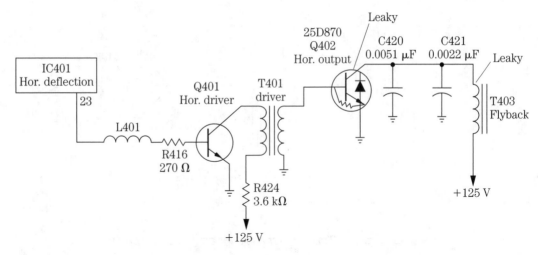

■ 3-89 *Replacing Q402, C420, and T403 solved the horizontal problems in a Samsung K-25 chassis.*

The output transistor was removed and tested normal. Again a resistance test was made from the flyback and collector terminal with the same resistance. C420 and C421 checked normal. T403 was found to be shorted. Before the repair was finished, the main fuse, line voltage regulator (IC801), Q402, and T403 were replaced.

Dead, no start—RCA CTC109A

The TV chassis (Fig. 3-90) seemed to want to start, but only a buzz could be heard in the flyback transformer. High voltage was found on the collector terminal of Q412 (137 V). The high-voltage regulator was checked, with 178 V at the anode terminal of SCR101 and 137 V at the gate and cathode terminals. Although the voltage was high, the high-voltage regulator SCR was functioning with a voltage drop of 41 V.

Often, higher voltage found at the output collector terminal is caused by a defective high-voltage regulator circuit or open output transistors. Because the regulator circuit was working, Q412 was checked for open and leakage tests. Remove Q412 from the chassis, because T401's secondary winding will show a short between the base and emitter terminals. Replacing the open output transistor (Q412) between the emitter and base terminals cured the dead chassis.

No regulator voltage—RCA CTC115A

No voltage was measured at the output transistor in a 9-inch portable color TV. Voltage measurements of 117.6 V were found at the gate

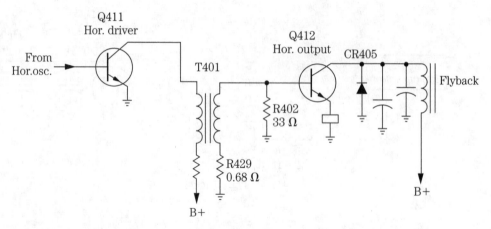

3-90 *Remove Q412 from the chassis for an accurate open test.*

and cathode terminals of the high-voltage regulator (SCR101), and no voltage was found at the anode terminal. Because U401 triggers SCR101, voltage measurements were found low at pins 5 and 17 (0.5 V). After pins 5 and 17 were removed from the PC wiring, a leakage of 54 Ω was found from pin 5 to chassis ground. Replacing R401 with a 149253 IC kit solved the no-picture, no-sound regulator problem (Fig. 3-91).

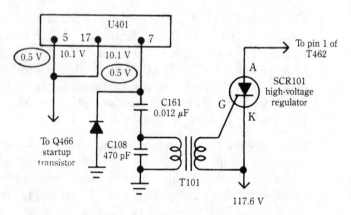

Note: Circled voltages were measured with a leaky U401.

3-91 *A leaky ICU401 caused chassis shutdown in the RCA CTC115A chassis.*

Output transistor keeps blowing

After replacing the horizontal output transistor (Q402) in the RCA CTC149 chassis, the fuse would blow and the horizontal output transistor was shorted. After applying an external voltage source to the deflection IC, no vertical or horizontal output pulses were noted. After several hours of frustration, the countdown crystal (Y3301) was found to be defective.

The new horizontal output transistor

The new horizontal output transistor found in the latest TV circuits resemble the same type on the schematic, except it has a TO-218 mounting. This output transistor looks like the TO vertical transistors. Q4400 (179743) found in the RCA CTC140 chassis is an npn, silicon, 8-amp, 1500-V replacement (Fig. 3-92). Although the universal replacements may dissipate 125 watts, they seem to run warmer than the older output transistors. See Table 3-1 for the horizontal troubleshooting chart.

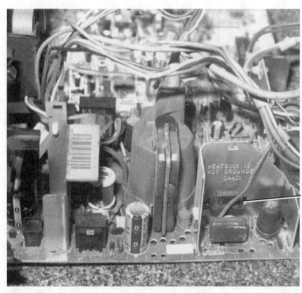

New horizontal output transistor

■ **3-92** *The new horizontal output transistors may look the same as the originals on the schematic, but they have a TO-218 body.*

■ Table 3-1 Check the horizontal sweep circuits troubleshooting chart.

What to check	How to check it
No horizontal sweep.	Check voltage to flyback and horizontal output transistor.
Voltage normal, no sweep.	Check waveform on base of output transistor.
No waveform on horizontal driver transistor.	Check output waveform on sweep IC.
Check voltage on horizontal sweep IC.	Check low voltage source (VCC) to sweep IC.
Is sweep IC defective?	Inject dc voltage to IC sweep and monitor output with a scope.
Check circuit diagram for correct voltage applied to sweep IC.	Unplug ac cord for this test.

190

High-voltage circuit problems

4

ALTHOUGH THE HIGH-VOLTAGE CIRCUITS DO NOT PRODUCE quite as many problems as in the old tube chassis, some new and old service problems still arise. Because the tripler unit replaces the high-voltage rectifier, less flyback transformer replacement is found in the new TV chassis. The tripler unit, which had many service problems at first, was replaced with the integrated flyback transformer with high-voltage diodes molded inside the flyback transformer (Fig. 4-1). These transformers have produced various service problems.

Besides the high-voltage rectifier molded inside the flyback transformer, additional windings are found on the same core, providing low voltage to many dc circuits (Fig. 4-2). You might find a separate filament winding for the CRT on the flyback transformer. These additional windings produce chassis shutdown when the circuits are overloaded by a leaky component in the connecting circuits. Do not overlook a high-voltage shutdown symptom that might occur in the flyback secondary circuits.

A defective tripler unit and integrated horizontal output transformer can destroy horizontal output transistors. The tripler or flyback can arc over, producing a loud arcing noise. Sometimes the defective integrated flyback only has a tic-tic noise with chassis shutdown. Poor focusing of the CRT can be caused by a defective flyback or tripler unit. Servicing the new high-voltage circuits might require different troubleshooting techniques, but they are rather easy to service after a few have crossed your service bench.

RCA CTC130C high-voltage circuits

In the RCA CTC130C chassis, the primary winding of the horizontal output transformer (T402) provides voltage to the collector terminal of the horizontal output transistor (Q402) and a B+ of

■ **4-1** *The modern integrated horizontal output transformer (IHVT) contains high-voltage diodes molded inside the flyback.*

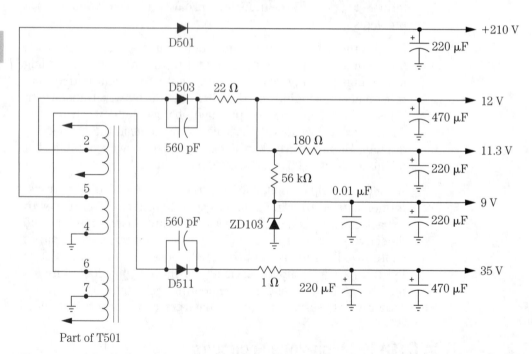

■ **4-2** *The IHVT can supply voltages to the screen grid and focus grid, plus providing several different low-voltage sources from separate windings on the secondary of the flyback.*

150-V regulated (Fig. 4-3). The horizontal deflection yoke connects to the primary winding of T402 through C425 and C434. Very seldom does the primary winding of the IHVT go open. The continuity can be checked with the ohmmeter.

The secondary high-voltage winding consists of the high voltage applied to the picture tube (25.8 kV), the focus voltage (5 kV to 7 kV), and HVR. The screen voltage is taken from one side of the focus divider circuit. Most problems develop in the high-voltage winding when the high-voltage capacitor and diodes arc over inside the molded winding. High-voltage arc-over of the CRT anode lead can occur with a crack in the anode lead or at the anode button. Sometimes the anode lead clip is not firmly clipped into the anode button and falls off during transport. Firmly hook the anode end into the

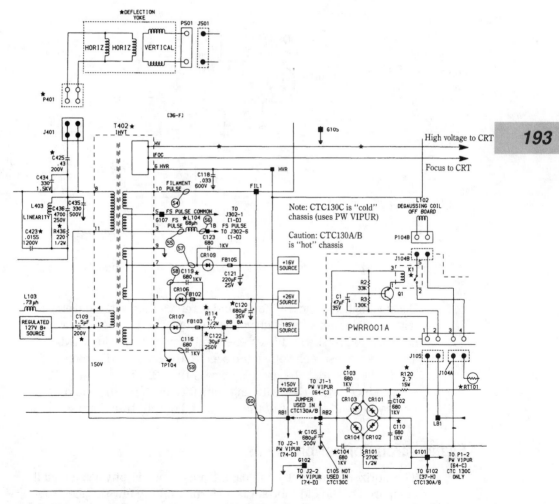

■ **4-3** *The high-voltage circuits in the RCA CTC130C chassis.* Thomson Consumer Electronics

CRT, then flip down the rubber-covered ring on the picture tube. Improper grounding can cause arcing lines in the picture.

Radio Shack 16-410 13" TV/VCR high-voltage circuits

The low-voltage source (+112 V) is applied to the primary winding of flyback (T9502) and on to the collector terminal of horizontal output transistor (Q9502). R9514 (3.93Ω) isolates the B+ circuit from the flyback primary winding (Fig. 4-4). When adequate drive voltage is applied to the base terminal of Q9502, high voltage is developed in the secondary circuits of the transformer.

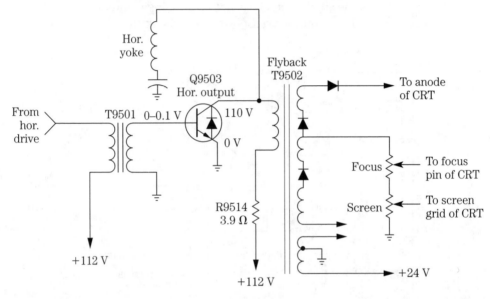

■ **4-4** *The high-voltage circuits in the Radio Shack's 16-410 13-inch TV and VCR chassis.* Radio Shack

Besides high voltage, the focus and screen voltages are developed with a divider network. High voltage is developed in the same winding with high-voltage diodes enclosed inside the molded flyback. The focus and screen voltages are fed to the focus and screen grids inside the picture tube. In this chassis the 12, 15, 32, and +112 V are produced with the switching transformer T001, instead of secondary derived voltages of the horizontal output transformer.

Flyback transformer problems

The horizontal output or flyback transformer is physically small compared to the old tube set transformers. Very few service problems are found with the flyback transformer tied into a tripler unit

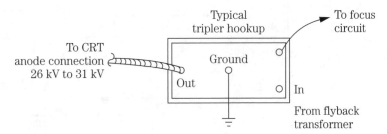

To CRT
anode connection
26 kV to 31 kV

Typical
tripler hookup

To focus
circuit

Ground

Out

In

From flyback
transformer

■ **4-5** *The high-voltage tripler contains capacitors and diodes to provide a higher dc voltage for the anode terminal of the CRT.*

(Fig. 4-5). These transformers operate at a very low voltage compared to the integrated flyback. This reduces breakdown problems. The RF output voltage is usually under 10 kV, while the integrated transformer can be over 30 kV. The tripler unit produces more breakdowns than the transformer.

The integrated transformer with several low-voltage windings and molded high-voltage diodes can cause a lot of service problems. Suspect a shorted flyback transformer when the horizontal output transformer is arcing or begins to run warm. The insulation between windings can break down, producing high-voltage arc-over in the new transformers. A breakdown of the high-voltage diodes molded inside can cause arcing or a warm flyback.

With the new high-voltage shutdown circuits, the chassis can come on with a blast in the sound and then shut down. In some transformers, all you can hear is a tic-tic noise. An intermittent squealing noise can indicate a defective flyback or a whining noise can be caused by loose particles inside the transformer or by loose mounting bolts. Replace the noisy transformer, because most ceramic cores are held together with metal tabs.

The tripler flyback transformer secondary winding can be checked with the low-ohm scale of the VOM or DMM (Fig. 4-6). The resistance of the high-voltage winding can vary from 55 Ω to 350 Ω. This depends on the output voltage and manufacturer. Although the resistance measurement does not indicate a shorted turn, the measurement does tell us if the winding is open. In the integrated transformer, winding continuity cannot be measured because of high-voltage diodes in the circuit.

The output lead of the tripler transformer can be checked by creating a 0.25-inch arc between a screwdriver blade and the input lead removed from the tripler unit. A high-voltage measurement at the anode CRT lead can indicate high voltage with the integrated transformer. If the integrated transformer is in shutdown, the transformer's condition must be checked with a universal line volt-

195

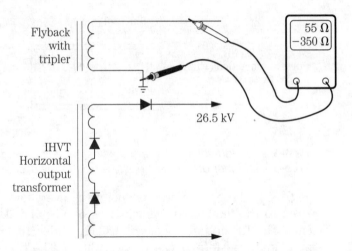

■ **4-6** *Disconnect the input lead from the tripler and measure the secondary HV winding to see if it's open.*

age or variac transformer. All circuits in the secondary winding of the horizontal output transformer are discussed in this chapter.

A defective flyback can destroy the horizontal output transistor or produce a high-voltage shutdown. The shorted flyback can cause an excessively bright raster with heavy retrace lines due to a leaky path to the core of the transformer. Because there is no high voltage, the defective integrated flyback can cause pulsating blooming of the raster.

A defective flyback in an RCA CTC97A chassis produced a pulsating and blooming raster that was out of focus. Only 12 kV were measured with the high-voltage probe. Normal voltage was measured on the horizontal output transistor, indicating improper high voltage at the picture tube.

The high-voltage probe

A high-voltage measurement at the anode connection of the CRT indicates that the high-voltage and horizontal section is working. The ground lead of the high-voltage probe must be grounded to the chassis or you will receive a jolt. High-voltage shock can be injurious to your health. Never use a low-voltage VOM or DMM in high-voltage tests. Be extra careful when working around high-voltage circuits (Fig. 4-7).

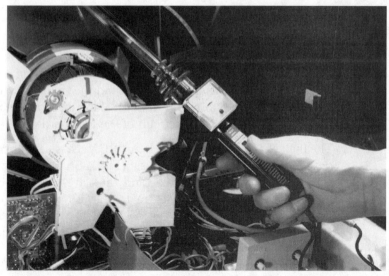

■ **4-7** *Monitor the high voltage with the high-voltage probe at the anode terminal of the CRT. Make sure a ground wire is clipped to the tube ground wire.*

The VTVM can be used as a high-voltage meter when a special high-voltage probe is attached instead of the regular voltage probe. Again, clip the ground wire to the metal TV chassis. Always choose a high-voltage probe that will measure up to 35 kV, because many of the new TV chassis generate about 31 kV. Discharge the CRT at the anode terminal before working in the high-voltage circuits. Remove the anode lead. Isolate it on heavy insulation or a cardboard box to prevent arcing when the high-voltage measurement is low. A leaky picture tube or circuits might be pulling down the high-voltage measurement (Table 4-1).

■ **Table 4-1 The high voltage level, picture tube, and model numbers of several different TV chassis.**

Make	Model or chassis	High voltage
Bohsei	13B081R	19.9 kV
J.C. Penney	685-2048	22.5 kV
Panasonic	GXLHM	29.5 kV
RCA	CTC130C	25.8 kV
RCA	CTC117	24.2 kV
Zenith	L1740W	26.5 kV

Removing and replacing the flyback

Make sure the flyback transformer is defective before installing a new one, because replacements take time to remove and install. Mark down all color-coded leads to each terminal on a separate piece of paper. List each component and the tie points that solder directly to the transformer terminals. Many integrated transformers solder directly into the circuit of the PC board.

The flyback transformer in the latest RCA chassis mounts into soldering eyelets on the fiber chassis (Fig. 4-8). Long lugs from the transformer fit into the eyelet and then are soldered. The best method is to use a solderwick material with a 150-W (or higher wattage) soldering gun. If the solderwick mesh does not have enough soldering paste, apply extra soldering paste to help suck up the melted solder. A desoldering tool can be used to pick up the melted solder.

Start about 0.5-inches down the braid of the solderwick and apply heat to the first eyelet. Slowly go around the terminal with the iron tip pressed tightly against the wick material. Make a complete turn around the eyelet picking up excess solder. If the wick material is full, start down another inch and pick up the remaining solder from the eyelet. Keep heat applied as the solderwick goes around the terminal. If excess solder is left in the hole, the terminal is hard to remove and it might crack the board or wiring when it is removed. Disconnect each terminal with the same method.

■ **4-8** *Remove excess solder from the output transformer eyelet or pin connection with large iron and solder wick.*

Be careful not to misplace or break components on the top side of the board when removing the transformer. A screwdriver blade slipped under the metal or plastic body of the transformer can help to loosen up each terminal. Most transformers will pop right up if all excess solder is removed. Suck out or remove all excess solder from each eyelet before installing the new transformer. Cut off the focus and screen cables, as well as any other wires found on top of the chassis connected to the transformer. Leave leads long enough to identify the color code of each wire.

Route the filament wires down at the bottom side of the transformer and through the back metal brace area. Improper voltage might be placed on the filament wire if it is left high around the plastic belt of the flyback. Sometimes this can burn out the picture tube. Solder the filament wires before soldering the bottom transformer lugs.

Keep the transformer tight against the PC board while soldering the eyelet terminals. Make sure each eyelet is full of solder all around the terminal. Now check each terminal to the PC wiring for a broken connection. Sometimes the wiring will break right at the eyelet. Take a continuity check with the low-ohm scale of a VOM or DMM from the transformer terminal to the nearest component tied to each terminal of the transformer (Fig. 4-9). This check can save a few hours of troubleshooting if the transformer does not work after installation.

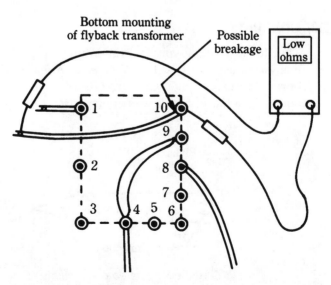

■ 4-9 *After installing a new flyback, check each terminal with the corresponding foil wire to make sure no foil is broken at the transformer connection.*

Picture tube filament circuit

In the tube chassis, the picture tube filaments are taken from a power transformer or are in series with the power line. Today, the filament voltage is taken from a separate winding on the flyback transformer. No light at the end of the picture tube can indicate an open tube filament or transformer winding.

The raster was out in a K-Mart SK1310A with no light at the end of the CRT. The picture tube filaments were checked across pins 6 and 7 with the socket removed (Fig. 4-10). The tube was normal except transformer and lead continuity was found open at pin 10. Cleaning up and soldering the connection solved the problem.

Only a very dim light could be seen in the end of the picture tube in a Montgomery Ward GAl-12994C with normal high voltage (Fig. 4-11). Because RF filament voltage cannot be measured accurately on the picture tube socket, the socket was removed and continuity checks were made. Practically a dead short existed across pins 9 and 10.

A normal filament winding measurement should be around 2.8 Ω with voltage-dropping resistor R115 in the circuit. Diodes D102 and D103 were replaced, because they contained white burn marks on the body and measured a dead short across each diode.

Tripler problems

The defective high-voltage tripler unit (Fig. 4-12) can cause many problems within the TV chassis. A leaky tripler can keep blowing fuses, or might cause the fuse to open after several seconds of operation and keep tripping circuit breakers. Very little

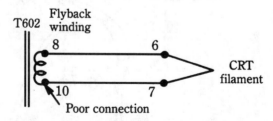

■ **4-10** *Poor soldered connections at the flyback transformer or poor pin contacts in the CRT socket can produce intermittent or no filament voltage in the CRT.*

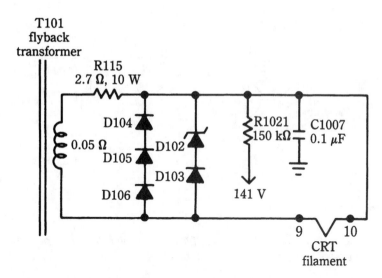

4-11 *Overheated diodes in a Wards GAI-I2994C caused a very low filament light.*

4-12 *Several different types of tripler units are found in the various high-voltage circuits.*

or low high voltage can result from a defective tripler unit. A defective tripler can load down the horizontal circuits, destroying the horizontal output transistor or causing low voltage at the collector terminal.

Feel the outside case of the suspected tripler. After pulling the power plug, replace it if it is running quite warm. All tripler units should run cool. Intermittent or constant arc-over can occur inside the tripler. Suspect a defective tripler when the flyback has a tic-tic noise with no high-voltage output. Firing lines can appear on the raster with an internally arcing tripler unit.

A defective tripler can be isolated by disconnecting the input terminal from the flyback (Fig. 4-13). If the tripler unit was causing

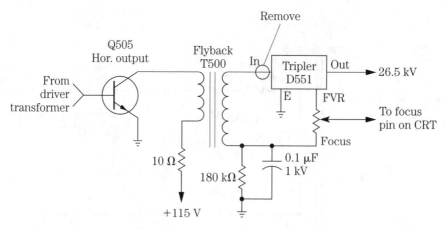

■ 4-13 *Remove the "in" lead from the tripler to determine if the unit is loading down the horizontal output circuits.*

the fuse or circuit breaker to open, the voltage will be restored at the collector terminal of the horizontal oscillator transistor with no shutdown. A normal 0.25-inch arc can be drawn from the input transformer lead with the blade of a screwdriver. This indicates horizontal circuits are normal. Do not touch the metal blade to the chassis.

Keeps blowing fuses

The horizontal output transistor and fuse was found blown in an RCA CTC92. After replacing both components, the fuse blew again. To determine if the horizontal circuits are causing the fuse to open, remove the horizontal output transistor (Fig. 4-14). Clip a 100-W bulb in place of the fuse. The light will stay bright with an overloaded component. Check the output transistor. If the light comes on bright, check for a leaky tripler. Remove the input terminal of the flyback transformer to the tripler. Install a new tripler unit if the light goes out.

Loading down—Magnavox T995-02

Only 87 V were measured on the collector terminal of the horizontal output transistor before the circuit breaker tripped in a Magnavox TV chassis. The leaky horizontal output transistor was replaced with the same results. After the flyback input lead was removed from the tripler, the voltage was normal at 114 V without shutdown. Replace the tripler component with the original part number or the correct universal replacement.

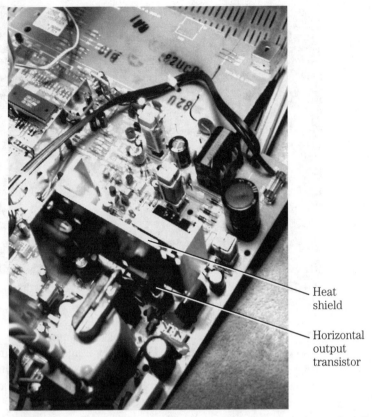

Heat
shield

Horizontal
output
transistor

■ **4-14** *Remove the horizontal output transistor to determine if horizontal circuits are blowing the fuse.*

Poor focus

A defective tripler or components in the high-voltage circuit can produce poor focus problems. The leaky flyback transformer with enclosed high-voltage capacitors can have low high voltage resulting in poor focus voltage. A defective focus control or bad high-ohm dropping resistors can cause poor focusing. Check the focus pin at the tube socket for poor connections.

First measure the high voltage at the anode terminal of the CRT. Poor focus will result if the high voltage is low. Normal high voltage with poor focus can be caused by a poor focus assembly or picture tube. Check the focus voltage at the focus pin of the CRT socket. Place the end of a paper clip inside the focus socket and measure the focus voltage from it. Low focus voltage can be caused by a poor socket or spark gap assembly inside CRT socket (Fig. 4-15).

■ **4-15** *Check all the pins on the CRT socket for corrosion, especially on the focus and filament terminals.*

The average focus voltage should vary between 3.5 kV and 6.5 kV. With the latest picture tubes, the focus voltage can vary from 6 kV to 9 kV (Fig. 4-16). Check the focus voltage with the high-voltage probe. The normal focus control should vary the voltage at least 1 kV at the CRT focus terminal.

When the focus voltage compares to that found in the schematic, suspect a defective picture tube. Check the picture tube in the tube checker. Measure critical voltages upon the picture tube terminals. A weak or gassy picture tube can cause poor focus. Suspect a focus spark gap or focus control when the focus voltage at the CRT socket will not change the focus. Blow out the socket and the focus spark gap assembly. Do not overlook poor heater terminals.

High-voltage arcing

Excessive high-voltage arcing can cause high-voltage shutdown with excessive high voltage at the picture tube. Use the variable line transformer to help locate the arcing shutdown problem. The leaky tripler can arc between plastic case and chassis. Insulate the tripler with a pasteboard box or tube carton to see if the chassis is operating normally. Do not try to apply putty or high-voltage insulation. Simply replace the leaky tripler. It might break down later under hot and humid weather conditions if it is repaired with putty or insulation spray.

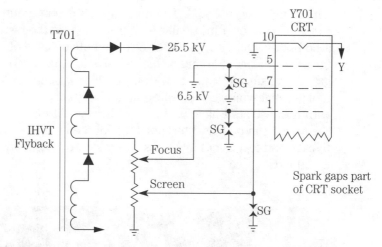

■ 4-16 *The focus voltage should vary from 3.5 kV to 6.5 kV, and in newer TVs up to 9 kV.*

Poor focus with high-voltage arcover can occur in a defective focus control or spark gap. Remove one terminal of the spark gap to determine if it is defective. In a Philco C-2518KW, excessive arcing could be heard across the spark gap mounted on top of the tripler unit (Fig. 4-17). The spark gap was removed and cleaned up. The

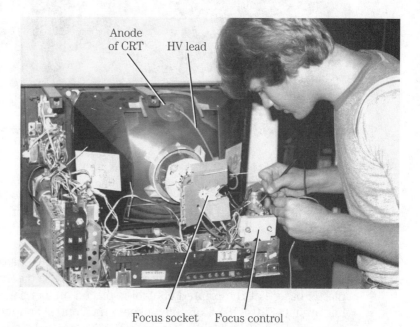

■ 4-17 *Check for arcing within the flyback, tripler, high-voltage lead, and picture-tube anode terminal.*

tripler unit was replaced. The integrated horizontal output transformer can arcover internally, with firing lines in the raster or with a no-high-voltage symptom. Check the outside case of the transformer for signs of arcing or melted plastic (Fig. 4-18). The integrated transformer can run warm, with high-voltage arcover of the molded high-voltage diodes. Sometimes tapping the transformer with the insulated handle of a screwdriver will make it act up. If the arcing can be heard but not seen, darken the area around the flyback and focus-screen units so the defective component can be identified.

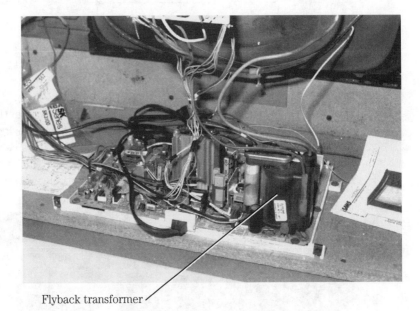

Flyback transformer

■ **4-18** *Check for signs of arcover on the flyback transformer mold case area.*

Arcing in the focus control

Thin white lines or dark zigging lines going across the screen horizontally can be caused by high-voltage arcing or poor connections to the focus controls. Some of the RCA focus and screen controls are located in one component that mounts on the TV chassis. Improper soldering of the focus control leads or a cracked ceramic or plastic base that the printed control is placed on can cause fine-line arcing (Fig. 4-19). Carefully inspect the control and feel it for warm sections. The inside elements of both controls can be ordered and replaced.

Focus-screen assembly

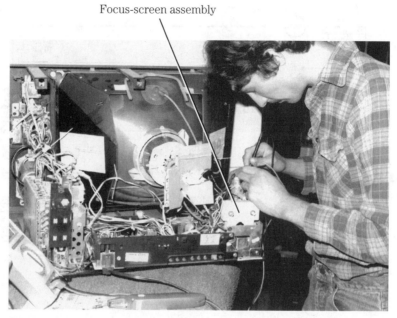

■ **4-19** *Check the focus and screen control assembly for cracks and corroded terminals that may arc over.*

Secondary flyback circuits

In the latest horizontal circuits, startup and secondary power supplies taken from the flyback transformer are found in practically every manufacturer's TV chassis. This means the horizontal circuits must function before voltage is applied to the various circuits (Fig. 4-20). A breakdown of components in any power source can cause the chassis to shut down. In most cases, secondary voltage source shutdown occurs a little later than high-voltage shutdown. Usually there are a few minutes of lag, with the sound blasting on and high voltage coming up before shutdown.

Try to isolate the defective power source before shutdown, if possible. When a normal picture and raster is noted for a few seconds and there is no sound, suspect the sound circuits are loading down the flyback transformer. A blast on of sound with normal high voltage and no picture before shutdown can be caused by the luminance or vertical circuits. Check for high-voltage shutdown if the chassis immediately goes into shutdown.

Isolate the shutdown problem with a variable line or variac transformer. Slowly bring up the line voltage and notice what circuits are working. Check the horizontal and low-voltage circuits for no

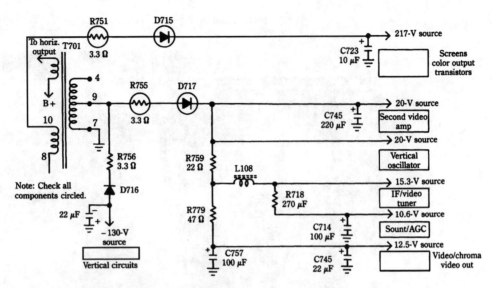

■ 4-20 *The horizontal output transistor may be overloaded if there is a shorted or leaky component in the secondary voltage sources.*

high voltage. All horizontal and high-voltage circuits must be functioning before any other voltage source is developed. Often, the high voltage might be lower, with an improper width symptom and an overloaded secondary voltage source.

Go directly to the secondary power source when circuits are not operating or appear dead. Be careful not to cut off any dc voltage connected to the horizontal circuits. Take a measurement of each voltage source. A low measurement can turn up the defective or overloaded circuit. Shut the chassis down and remove this circuit from the flyback circuit. Simply remove one end of the diode rectifier to disconnect it from the horizontal output transformer. Remember, these voltages are lower than normal, as the universal line transformer is adjusted lower before chassis shutdown.

You have located the overloaded circuit if voltage on the circuits drastically increases. Check for an increase in voltage at the collector terminal of the horizontal output transistor. Now raise the line voltage to 120 Vac. Locate the defective component in the overloaded circuit before connecting into the circuit.

Only a tic-tic noise could be heard at the flyback transformer in an RCA CTC85 chassis. The chassis would come on with sound and then shut down. Only a horizontal white line could be seen before shutdown. Because a vertical sweep collapse was noticed, the vertical module was replaced. With the module out of the circuit, the horizontal and high voltage were normal. Replacing the leaky vertical module fixed the problem.

Quick high-voltage circuit checks

With a no-raster, no-sound symptom (Fig. 4-21), check the high voltage at the picture tube (1). If high voltage is normal, notice if the picture tube filaments are lit. Now go to the video circuits if the CRT and high voltage are normal. If no high voltage is measured, take a voltage measurement at the collector terminal of the horizontal output transistor (2).

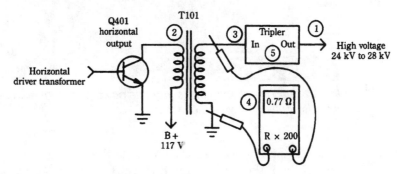

■ **4-21** *Check the high-voltage circuits by the numbers; use high-voltage and resistance measurements.*

Check the tripler circuits if the horizontal voltage at the output transistor is a little low or normal. Remove the tripler input wire from the flyback when it is suspected of loading down the secondary circuits (3). Fire up the chassis and draw a small 0.25-inch arc from the flyback input cable. Take a resistance continuity check of the flyback winding if there is no voltage (4). Replace the defective tripler if the flyback seems normal (5).

When the line fuse or circuit breaker keeps tripping, remove the horizontal circuit fuse or horizontal output transistor. Now if the fuse or circuit breaker holds, suspect a defective horizontal or high-voltage circuit. Replace the horizontal output transistor and remove the input terminal to the tripler unit. If the fuse or circuit breaker is normal, replace the leaky tripler unit. Always use a variable power transformer as B+ for these tests (Fig. 4-22).

Check the high-voltage circuits with the integrated horizontal output transformer in the very same manner (Fig. 4-23) (1). If the transformer has a tic-tic noise and is in chassis shutdown, connect a variable line transformer to the chassis (2). Slowly bring up the

■ **4-22** *Plug the chassis into a variable power transformer to determine what is shutting down the chassis.*

ac voltage and notice at what voltage the chassis, shutdown occurs. Check the chassis for high-voltage shutdown with higher than normal low or high voltage. Check the voltage at the collector terminal of the horizontal output transistor (3). Determine if the secondary voltage circuits are overloading the flyback in chassis shutdown. Disconnect each voltage source until the overloaded circuit responds with normal voltage at the horizontal output transistor (4). Do not overlook a defective horizontal output transformer.

Voltages supplied to CRT

High voltage is supplied to the anode terminal of the CRT from the horizontal output transformer. Focus and screen voltages are fed to the focus and screen grids of picture tube from an HV divider network consisting of separate focus and screen controls. In some TV chassis, high boost voltage is supplied to the CRT from a lower winding of the flyback. One or two turns of large heater wire is wound around the horizontal output transformer core to develop heater or filament voltage, which is then fed to the heater pins of the CRT. Make sure all the above voltages are found on the picture tube before replacement.

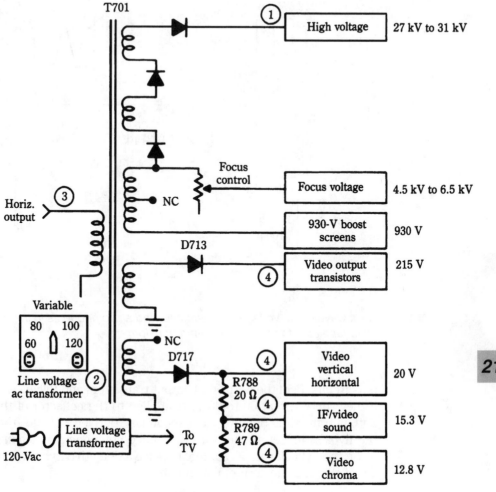

■ 4-23 *Check the horizontal output transformer circuits by the number to locate a defective component.*

Check components before replacing flyback

Before replacing the horizontal output transformer, make sure that proper driver voltage is applied to the base of horizontal output transistor. Test and, if necessary, replace the output transistor. Check for proper collector voltage at the output transistor. Remove the red lead of the yoke assembly to determine if the yoke is defective. Check continuity of the primary winding or flyback (Fig. 4-24). Determine with the variable power line transformer if the output transistor and flyback are loading down the horizontal out-

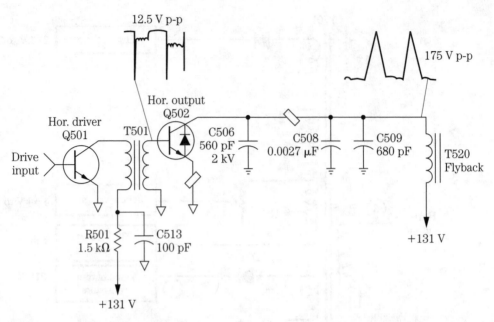

12.5 V p-p

175 V p-p

Hor. output
Q502

Hor. driver
Q501

T501

Drive
input

C506
560 pF
2 kV

C508
0.0027 μF

C509
680 pF

T520
Flyback

R501
1.5 kΩ

C513
100 pF

+131 V

+131 V

■ **4-24** *Check waveforms, voltages, resistances, and all components in the horizontal circuits before replacing the horizontal output transformer.*

put circuits. Feel the horizontal output transistors and flyback see if they run warm at 60 to 90 volts applied to the collector of the output transistor. Remove each secondary derived voltage from the flyback to determine if a low-voltage component is loading down the flyback voltage source. Replace a defective flyback with a part that has the correct part number.

Panasonic GXLHM chassis high-voltage adjustment

The following procedure is used to make high-voltage adjustments in a Panasonic GXLHM chassis.

1. Connect a high-voltage meter (50-kV rating, electrostatic type) to the second anode of the picture tube.

2. Set the color pilot switch off.

3. Set the brightness control (R312) and parabrite control (R320) at minimum.

4. Tune in a black-and-white signal and cause the raster to disappear by tuning the screen control (R372) and subbright (R316) controls.

5. Adjust the high voltage to 29.5 kV ±1.0 kV.

6. If the high voltage is more than 30.5 kV, connect a ceramic 820 PK, 2-kV capacitor (C560) between pin 9 of FBT and the chassis ground. Make sure the high voltage comes within tolerance.

7. Return the screen and subbright controls to their original positions. In many present-day TVs, the high voltage is not adjustable.

Realistic 16-261 x-ray and high-voltage warnings

Be sure all service personnel are aware of the procedures and instructions covering x-ray radiation. The only potential source of x-rays in the solid-state TV chassis is the picture tube. The CRT does not emit measurable x-ray radiation if the high voltage is as specified in the high-voltage check instructions.

It is only when the high voltage is excessive that radiation is capable of penetrating the shell of the picture tube, including the leaded glass material. The important precaution is to keep the high voltage below the maximum level specified.

Have an accurate high-voltage meter handy at all times. The calibration of this meter should be checked periodically. High voltage should always be kept at the rated value, but no higher. Operation at higher voltages can cause a failure of the picture tube or high-voltage circuitry. Also, under certain conditions, it might produce radiation in excess of safe levels.

When the high-voltage regulator is operating properly, there is no possibility of a radiation problem. Every time a color chassis is serviced, the brightness should be tested while monitoring the high voltage with a meter to be certain that the high voltage does not exceed the specified value. When the brightness is raised, usually the high voltage will go down, and when brightness is reduced the high voltage will go up.

When troubleshooting and taking test measurements on a TV with excessive high voltage, avoid getting too close to the chassis. Do not operate it longer than is necessary to locate the cause of the excessive high voltage.

RCA CTC130C x-ray protection and overcurrent shutdown

To prevent catastrophic failures, the chassis uses a shutdown circuit that disables the set if the high voltage or power demand of the horizontal output circuit increases above predetermined limits. There are two different input circuits in the shutdown circuit.

The first is the sample pulse (XRP) from the IHVT, and the second is overcurrent sensing from the horizontal output transistor emitter resistor (R418). The XRP sample pulse is detected by diode CR409, and under normal conditions develops about 24 Vdc. This voltage is divided down to about 18 V by R430 and R416, and applied to a 20-V zener diode (CR406). If the voltage remains below 20 V, the zener diode (CR406) will not conduct (Fig. 4-25).

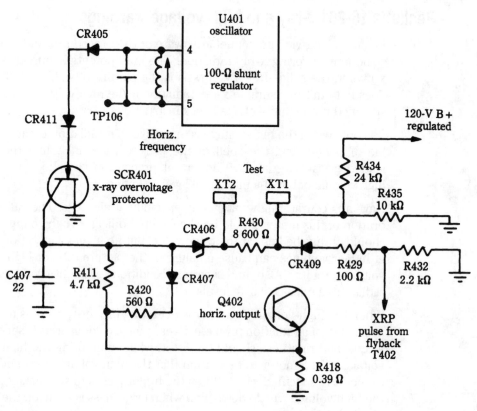

■ **4-25** *The typical x-ray circuit found in many RCA circuits.*

During a malfunction where the voltage exceeds 20 V, the zener diode (CR406) will conduct and turn on the shutdown SCR (SCR 401). The SCR connects the diode of CR409 to ground. CR408 becomes forward-biased and, in turn, forward-biases CR405. This action loads down the horizontal oscillator to the point where the oscillator is no longer functional, shutting down the chassis. The SCR continues to conduct because of raw B+ flowing through R115, R116, L401, CR405, and CR411 into SCR401.

The SCR will turn off when the ac switch is opened. The charge on C105 (raw B+ filter) bleeds off in sets with the remote feature turn off by shunting the anode current of CR405 to ground through R117. The second input to the shutdown circuit monitors the current through the horizontal output transistor. Any trouble that causes the current to increase above normal limits activates the shutdown circuit.

RCA CTC130C x-ray protection test

When service has been performed on the horizontal oscillator, horizontal deflection, high voltage, flyback transformer replacement, or B+ regulator system, the x-ray circuit must be tested for proper operation.

1. Apply 120 Vac with a variable transformer for accurate input voltage.
2. Allow the chassis to warm up for a few minutes and adjust all controls for normal viewing.
3. Locate stakes XT1 and XT2 on the PW rear control (RC) circuit board and on the schematic.
4. Momentarily short stake XT1 to stake XT2 with a short test lead. The chassis should shut down at once. If the chassis does not shut down, check and repair the x-ray shutdown circuits.

Sylvania C9 overvoltage protection

An overvoltage protection device (SCR505) monitors the voltage from pin 2 of the IFT, which is rectified by D506. This unfiltered dc voltage (12.5 V) will trigger SCR505 if it rises above the combined zener voltage (22 V) of zener diodes Z503 and Z504 (Fig. 4-26). Once the SCR is triggered, the 130-Vdc supply is clamped to ground through R513 which removes horizontal drive by removing the 130-Vdc source from the collectors of the horizontal driver transistor (Q500) and horizontal output transistor (Q501). When the 130-Vdc source is clamped to ground, the SMPS shuts down. All ac power must be removed from the chassis and then reapplied to reset the shutdown system. In some TV chassis the horizontal pulse is taken from a separate winding on the flyback transformer, and in others it is taken from the filament winding of the picture tube.

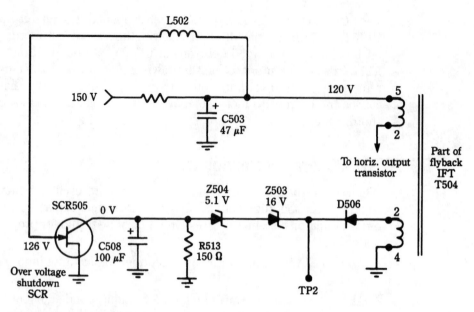

4-26 *Sylvania's 26C9 high-voltage shutdown circuits.*

Panasonic GXLHM chassis
horizontal oscillator disable circuit

The horizontal oscillator disable circuit protects against excessive high voltage and CRT beam current. If for any reason high voltage or CRT beam current exceeds a predetermined level, the circuit operates to shift the horizontal oscillator frequency and limit high voltage. The active components are contained in IC531 (Fig. 4-27).

Negative ABL voltage, which is a direct result of CRT current, is applied to pin 5 of IC531 and can be measured at TPD2. Horizontal retrace pulses from pin 3 of the flyback transformer are rectified by diode D531. The resultant dc (+) voltage at the cathode of D531 (directly proportional to high-voltage variations) is applied to the disable amp stage in IC531 via pin 4. The pin 2 output of IC531's dc amp stage is applied to the horizontal oscillator stage in IC401. During normal operation IC531 is in a static condition and has no effect on horizontal oscillator operation.

The horizontal oscillator stage in IC401 is dc operated. Adjustment of the horizontal hold control sets the dc voltage at IC401 (pin 3) and thus sets the horizontal frequency.

If the horizontal retrace pulse amplitude (representative of high voltage) increases, the positive dc voltage at the cathode of D531

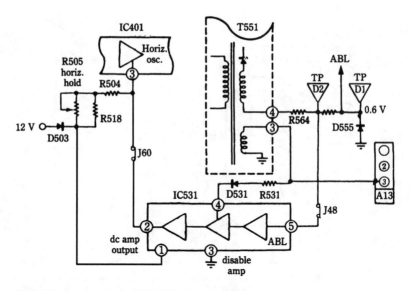

also increases. Thus, more positive voltage is applied to the disable amp stage in IC531. When this voltage reaches a predetermined level, a dc output from pin 2 of IC531 increases the dc voltage at pin 3 of IC401. This increases the horizontal oscillator frequency and limits high voltage.

Increased CRT beam current produces more negative ABL voltage at pin 5 of IC531, which brings the disable amp stage closer to its condition point. Thus, excessive levels of high voltage or CRT beam current produces an output for disable IC531 to assume operation within safe limits.

Panasonic GXLHM chassis holddown circuit

The positive dc voltage, which is supplied from the cathode of D531 for monitoring the high voltage, is applied to pin 4 of IC531 and to the base of Q903 through R907 (Fig. 4-28). The voltage at the emitter of Q903 is regulated by zener diode D901. Under normal conditions, the voltage applied across the base and emitter of Q903 is not sufficient to cause emitter current to flow, and it holds the transistor off.

If the high voltage increases over the specified voltage, the positive dc voltage that is supplied from the cathode of D531 also increases. The voltage through D531 is parted to R907 and R906,

217

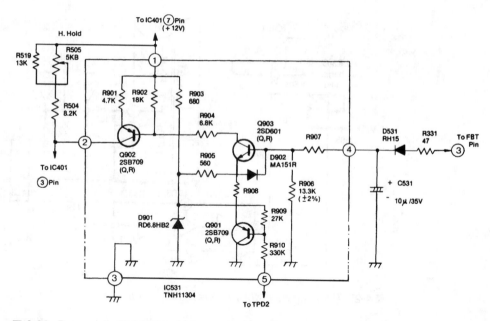

■ 4-28 *Panasonic GXLHM holddown circuits.* Matsushita Electric Corporation

and is then applied to the base of Q903. If the voltage between the base and emitter is more than 0.7 V, this transistor turns on.

Now the collector voltage of Q903, which is connected to pin 3 of IC401, turns down, causing an increase in the horizontal oscillator frequency, loss of horizontal sync, and a lowering of the high voltage. Simultaneously, pin 5 of IC531 is connected to TPD2. Voltage of TPD2 decreases when beam current increases. If the beam current increases, voltage of the base and emitter of Q901 decreases. The emitter of Q901 is connected to the emitter of Q903 through R908. When beam current increases, Q903 turns on earlier than at small beam current, and loss of horizontal sync comes early.

If the horizontal oscillator is off-frequency, the cause must be determined in either the horizontal oscillator circuit or the disable circuit. Check the horizontal oscillator disable circuit flowchart (Table 4-2). Scope the horizontal output transformer winding for a waveform when the shutdown circuits do not perform, and check all diodes and small resistors (Fig. 4-29).

■ Table 4-2 Panasonic's GXLHM horizontal oscillator disable circuit service flowchart.

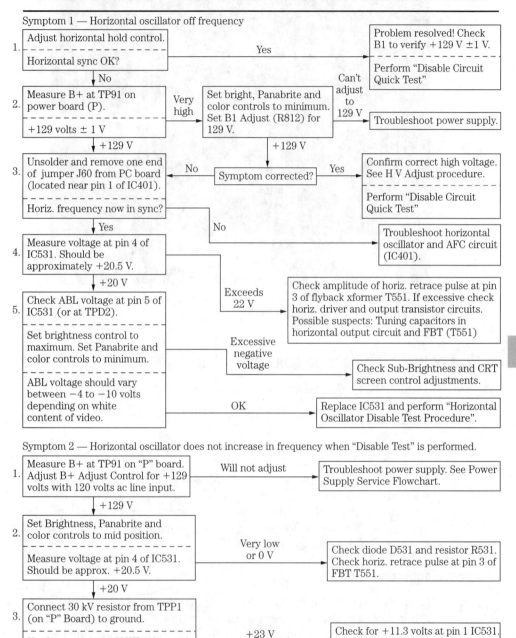

Symptom 1 — Horizontal oscillator off frequency

1. Adjust horizontal hold control.
   Horizontal sync OK?
   — Yes → Problem resolved! Check B1 to verify +129 V ±1 V.
   Perform "Disable Circuit Quick Test"
   — No ↓

2. Measure B+ at TP91 on power board (P).
   +129 volts ± 1 V
   — Very high → Set bright, Panabrite and color controls to minimum. Set B1 Adjust (R812) for 129 V.
   — Can't adjust to 129 V → Troubleshoot power supply.
   ↓ +129 V

3. Unsolder and remove one end of jumper J60 from PC board (located near pin 1 of IC401).
   Horiz. frequency now in sync?
   — No → Symptom corrected? — Yes → Confirm correct high voltage. See H V Adjust procedure.
   Perform "Disable Circuit Quick Test"
   — No → Troubleshoot horizontal oscillator and AFC circuit (IC401).
   ↓ Yes

4. Measure voltage at pin 4 of IC531. Should be approximately +20.5 V.
   ↓ +20 V

5. Check ABL voltage at pin 5 of IC531 (or at TPD2).
   — Exceeds 22 V → Check amplitude of horiz. retrace pulse at pin 3 of flyback xformer T551. If excessive check horiz. driver and output transistor circuits. Possible suspects: Tuning capacitors in horizontal output circuit and FBT (T551)
   Set brightness control to maximum. Set Panabrite and color controls to minimum.
   — Excessive negative voltage → Check Sub-Brightness and CRT screen control adjustments.
   ABL voltage should vary between −4 to −10 volts depending on white content of video.
   — OK → Replace IC531 and perform "Horizontal Oscillator Disable Test Procedure".

Symptom 2 — Horizontal oscillator does not increase in frequency when "Disable Test" is performed.

1. Measure B+ at TP91 on "P" board. Adjust B+ Adjust Control for +129 volts with 120 volts ac line input.
   — Will not adjust → Troubleshoot power supply. See Power Supply Service Flowchart.
   ↓ +129 V

2. Set Brightness, Panabrite and color controls to mid position.
   Measure voltage at pin 4 of IC531. Should be approx. +20.5 V.
   — Very low or 0 V → Check diode D531 and resistor R531. Check horiz. retrace pulse at pin 3 of FBT T551.
   ↓ +20 V

3. Connect 30 kV resistor from TPP1 (on "P" Board) to ground.
   If voltage at pin 4 of IC531 increases.
   — +23 V → Check for +11.3 volts at pin 1 IC531. If OK replace IC.

219

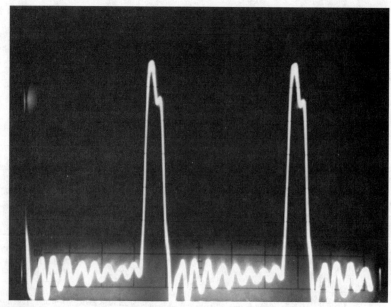

■ **4-29** *Scope the horizontal output transistor waveform when the shutdown circuits do not function.*

Five actual high-voltage case histories

The following are five actual problems and their solutions.

Poor focus—Zenith 20W60

When the focus control was varied, the picture would focus up at one point in the rotation. The picture tube was tested and all guns were fairly normal. Voltage measurement at pin 1 of CRT should be somewhere between 5 kV and 6 kV (Fig. 4-30). This voltage was quite low at the picture tube, and at the extreme end would come into focus. Replacing focus and screen part number RX3299A solved the focus problems.

Intermittent width—shutdown

In an RCA CTC145 chassis, the picture width would pull in some- times, and would remain normal at other times. Finally, the chas- sis shut down with a blown fuse. No dc voltage was found upon the collector terminal of the horizontal output. Sometimes resistor R4413 (680 ohms) would run red hot. In checking over the flyback winding and yoke circuits, capacitor C4415 (0.25 µF) was found to be intermittent and leaky (Fig. 4-31).

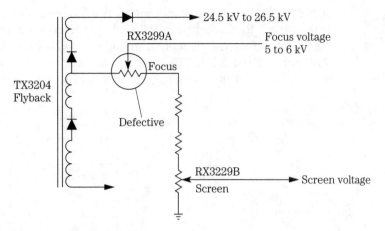

■ 4-30 *Poor focus in a Zenith 20W60 chassis was caused by a faulty focus divider network.*

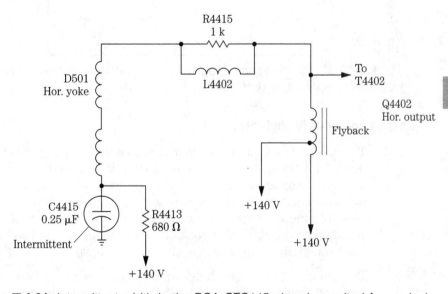

■ 4-31 *Intermittent width in the RCA CTC145 chassis resulted from a leaky C4415 (0.25 µF) capacitor.*

Low filament voltage—Panasonic CT-1320W

No brightness was indicated in this portable, with 22.3 kV of high voltage measured at the anode of the picture tube. After the picture tube was checked and found good, voltage measurements were fairly normal on all elements of the CRT. In fact, no filament light could be seen at the glass of picture tube.

A continuity check was made across the filament of the CRT and was normal. Removing the CRT socket and taking another measurement across the filament winding yielded a measurement of 210 ohms. Resistor R372 (2.4 ohms, 2 W) was found corroded and was replaced, solving the no-filament voltage problems (Fig. 4-32).

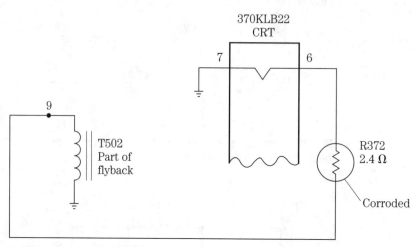

■ **4-32** *Very low filament voltage in a Panasonic CT-1320W was caused by a corroded resistor R372.*

Chassis normal—HV shutdown

The chassis in a Quasar NTS-989 chassis would come up and look normal, and then shut down. A quick HV test and scope waveform test at horizontal output transistor Q501 was made, and the voltage was monitored at the line-voltage regulator STR380 (IC801) to determine what section caused the shutdown. Slowly the line voltage was varied by the isolation line transformer, and when it was brought up to 90 volts, the chassis shut down.

To eliminate the voltage shutdown circuits, one end of D513 was removed from the circuit (Fig. 4-33). Again, the line voltage was raised, and the chassis operated at 120 Vac. This indicated problems within the high-voltage protection section. Q554 was tested and proved leaky. Q553 tested normal, but both Q554 and Q553 were replaced to cure the HV shutdown problem.

No HV—no picture—no raster

Although low voltage was found upon the collector terminal of output transistor Q602, no HV was measured at the picture tube. Q602 was found to be open and was replaced with a universal

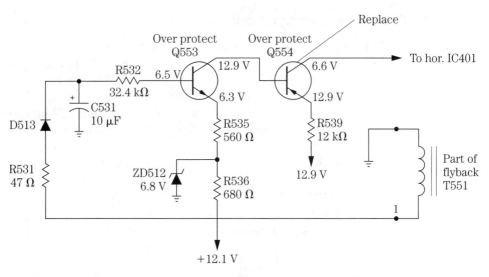

■ 4-33 *High-voltage shutdown was eliminated by replacing Q554 and Q553 in the overvoltage-protection circuits in a Quasar NTS-989.*

ECG89 transistor. Now the chassis blew the main fuse F701 (4 amp). All safety components were checked in the output collector circuits and tested good.

Q602 was replaced again, and slowly the line voltage was raised to 60 volts ac. Q602 began to get hot (Fig. 4-34). The drive waveform was fairly normal, indicating an overload in the flyback circuits. The red lead of yoke DY601 was removed from the circuit, and the results were the same. Replacing flyback T602 with the exact replacement in the Sharp 19F90 model solved the no-HV problem.

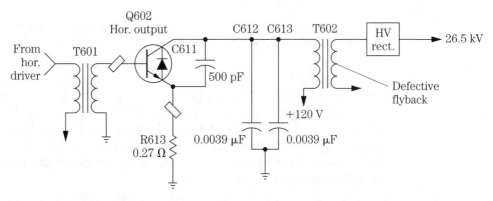

■ 4-34 *The no-high-voltage/no-raster symptom resulted from a leaky flyback T602 in a Sharp 19F90 model.*

RCA CTC166—High-voltage shutdown

After replacing the horizontal output transistor Q4401 in an RCA chassis, the high voltage shutdown circuits were tested by shorting 1 and 2 of J4901. The chassis would not shut down. In this circuit the HV is monitored by CR4901, which rectifies pulses from the horizontal output transformer T4401 (Fig. 4-35). When the high voltage increases, the rectified voltage at the cathode of CR4901 increases; with an increase of rectified voltages, zener diode CR4902 triggers to conduct, which shuts down the set.

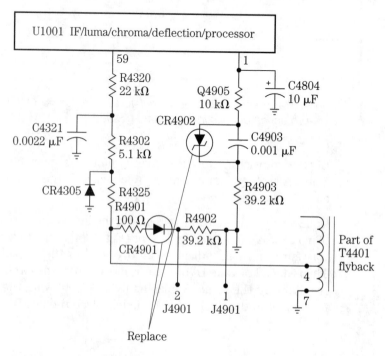

■ **4-35** *High-voltage shutdown in an RCA CTC166 chassis was caused by CR4901 and CR4902.*

To troubleshoot this shutdown circuit, remove one end of CR4901 from the circuit. Raise the line voltage with the variable isolation transformer and increase it to around 85 to 90 volts. Repair the shutdown circuits. Both CR4902 and CR4901 were replaced. Now raise the line voltage up to 125 to 130 volt ac, and see if the shutdown circuits are working. Test the shutdown circuits by shorting 1 and 2 terminals of J4911 together.

CR4902 is an 11-volt zener diode and should be replaced with original part number 159429 or NTE5019T1; SK9970 is the universal replacement. Replace CR4901 with a general-purpose 1-A silicon diode.

Overloaded secondary circuits

Remember, overloaded circuits within the secondary winding voltage sources can shut down the chassis or damage the horizontal output transistor. Any low resistance between a leaky IC, transistor, capacitors, or regulators found in the secondary circuits, such as sound, brightness, horizontal, color, IF, and video circuits, can cause chassis shutdown. Often, there will be a lag or delay in shutdown with either a sound or picture symptom.

For instance, if the picture went down to a white line before shutdown, check the vertical circuits overloading the flyback circuits. When the picture goes black with a delay in high voltage in the CRT, suspect a leaky video component. Likewise, in chassis that depends on a secondary voltage source to supply voltage to the deflection IC, check the IC for leakage leading to shutdown.

The sound was distorted before the chassis shut down in an Emerson MS250R portable (Fig. 4-36). Because the picture was normal for a few seconds with distorted sound, the sound voltage source was monitored. IC 1203 had only 0.75 volts at pin 2 when fired up, indicating a leaky IC (AN5836). The AN5836 IC was replaced with RCA SK9731, a universal replacement.

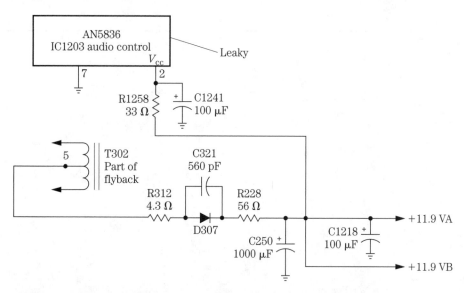

■ **4-36** *A leaky IC1203 overloaded the 11.9-V source of the secondary winding of flyback T302, shutting down the chassis on an Emerson MS250R model.*

Conclusion

Measure the high voltage at the CRT anode connection with the high-voltage probe on a set with a no-sound, no-raster symptom. Next, check the collector voltage at the horizontal output transistor. Remove the input lead to the tripler unit with low voltage at the output transistor to determine if the leaky tripler is overloading the horizontal output circuits. Draw a small arc from the input lead with a screwdriver blade to see if the flyback is normal. Replace the horizontal output transformer if the horizontal output transistor runs warm with low (or no) high voltage. Feel the molded body of the transformer and, if quite warm, replace the leaky transformer. Check Table 4-3 for a high-voltage troubleshooting chart.

■ **Table 4-3 For easy servicing, check this troubleshooting chart for high-voltage circuits.**

What to check	How to check it
Check HV at crt with high voltage probe.	Check flyback and horizontal circuits.
Check focus and grid voltages.	Check flyback circuits.
Check continuity of flyback primary winding.	Open winding or isolation resistor.
Disconnect red yoke wire or lead.	Check horizontal sweep circuits and voltages.
	Check waveform on base of output transistor. Repair horizontal circuits.

Troubleshooting the vertical circuits

5

AFTER THE HORIZONTAL SECTION, THE VERTICAL STAGES rate next as causing the most trouble in the TV chassis. A horizontal white line and insufficient and intermittent sweep are the most common vertical problems. Vertical foldover, crawling, and rolling of the picture can occur in the vertical circuits. Black and white bars in the raster can be a little more difficult to locate when they occur in the latest solid-state TV chassis.

The horizontal white line can be caused by many critical components in the vertical circuits and is the result of no vertical sweep. Insufficient vertical sweep can appear as from 1 to 8 inches of raster. The top and bottom portion of the raster cannot be adjusted with either the vertical height or linear controls to fill out the screen. Improper vertical sweep can be caused by a leaky output transistor, burned resistors or diodes, and improper voltage in the vertical output circuits. Intermittent vertical sweep can result from transistors, ICs, or poor board wiring connections. Check for leaky output transistors with improper bias resistance for vertical foldover. Vertical crawling and rolling can be caused by defective filter capacitors and improper sync signal.

The horizontal white line indicated by a no-vertical-sweep symptom is the easiest vertical trouble to locate. Intermittent vertical sweep and foldover are more difficult to find. Often, insufficient vertical sweep symptoms are caused by abnormal voltages and can be located with the DMM or VOM. Look for poor filtering in the vertical voltage source causing vertical crawling.

The most useful test instruments are the oscilloscope, DMM, and VOM (Fig. 5-1). The crosshatch generator is ideal to check vertical height and linearity. Critical waveforms taken by the scope can indicate where the waveform is missing. Low voltage and resistance measurements with the DMM can quickly locate the defec-

■ **5-1** *The DMM and scope can locate most problems in the vertical circuits.*

tive component. Testing of the transistors and diodes in the circuit can be done with the DMM. Of course, the correct schematic diagram is a must when servicing the vertical circuits.

Critical test points

Three critical test points within the vertical circuit can determine where and if the vertical section is functioning (Fig. 5-2). Take a scope waveform of the vertical oscillator and output stages to determine where the improper vertical sweep symptom occurs. Measure the supply voltage to the vertical circuits.

A voltage measurement on the heat sink or metal collector terminal of the vertical output transistor can indicate improper or no B+ power source with no vertical sweep. A lower-than-normal power source can produce insufficient or vertical foldover problems. No voltage at a given stage can indicate a leaky transistor or component.

The waveforms taken at the output of the oscillator transistor indicate the beginning of the sweep circuit. A waveform check at the vertical coupling capacitor shows that the vertical stages do not have yoke problems or corresponding component problems. If a waveform is found at the oscillator transistor and not at the output stages, look for the defective component between the two test points. A low-amplitude waveform can indicate insufficient verti-

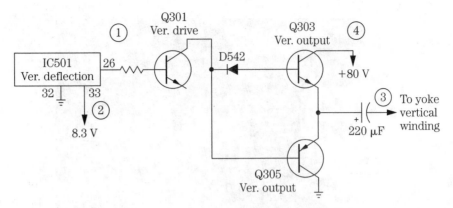

■ **5-2** *Troubleshoot the vertical section by the numbers; take waveforms at 1 and 3, and critical voltage measurements at 2 and 4.*

cal sweep. Scope waveforms taken throughout the vertical circuits, especially with feedback circuits, are useless in these type circuits.

Types of vertical circuits

The vertical blocking oscillator circuits are used extensively in the early tube circuits with capacitor coupling to the final output stage. Many of the early solid-state chassis used the single transformer-coupled output transistor (Fig. 5-3). Then came the multivibrator circuits. Today, many of the vertical circuits contain a relaxation oscillator circuit, using an npn and pnp transistor (Fig. 5-4). In the latest solid-state chassis, you might find the IC chips as both horizontal and vertical deflection in a frequency counter circuit (Fig. 5-5).

The relaxation oscillator and frequency counter circuits are directly coupled to the transformerless (OTL) output stages. No large output transformer is found in these vertical circuits. A pnp transistor is used as the top vertical output transistor, with an npn transistor as the bottom vertical output (Fig. 5-6). Direct or capacitor coupling to the vertical yoke section completes the vertical deflection operation.

Vertical frequency counter IC

The vertical output signal is developed inside an IC frequency counter chip (Fig. 5-7). The vertical drive signal is generated by charging C702 through the vertical height control (R705) to

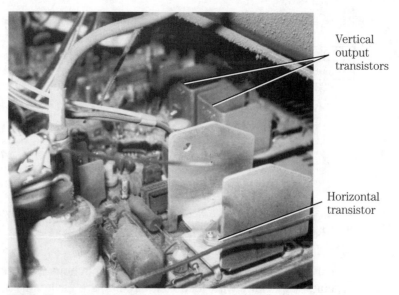

Vertical
output
transistors

Horizontal
transistor

■ **5-3** *The vertical output transistors are mounted on separate heat sinks in the vertical section.*

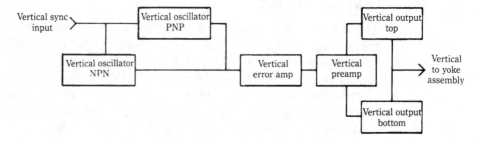

■ **5-4** *A block diagram of a transistorized vertical circuit, from the oscillator to the vertical output transistors.*

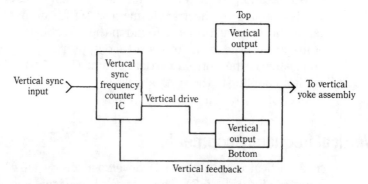

■ **5-5** *A frequency counter can provide signal and drive for the vertical and horizontal output transistors.*

Troubleshooting the vertical circuits

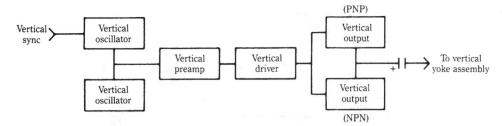

■ 5-6 *The vertical output circuits are coupled directly to the vertical yoke winding or through an electrolytic capacitor.*

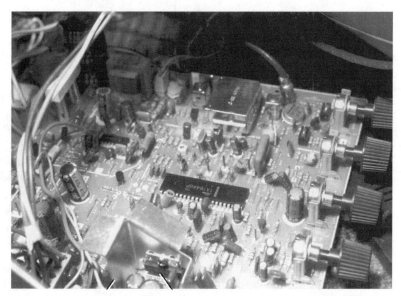

■ 5-7 *The latest TV chassis contains a vertical output IC on a heat sink instead of transistors.*

ground. The charging and discharging of Q702 generates a sawtooth voltage. The sawtooth voltage is amplified by a vertical amp in the IC, and is fed to the base terminal of the bottom vertical output transistor (Q703).

Check the sawtooth waveform at pin 18 of IC701 (Fig. 5-8). If there is no waveform at all, measure the dc voltage on pin 17. Go to the 24-V voltage source in the low-voltage power supply with no or improper voltage at terminal 17. Suspect IC701 of leakage if the 24-V source is fairly normal and low voltage exists at terminal 17. With normal B+ voltage and no sawtooth signal, check the vertical sync waveform at pin 24. If the sawtooth signal is fairly normal at pin 18 and there is insufficient vertical sweep, check the vertical output stages.

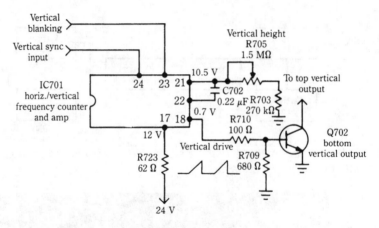

■ 5-8 *Check the vertical drive signal at pin 18 with the scope.*

Sylvania C9 vertical circuits

The video IF IC employs a synchronized countdown divider system for generating the vertical signal (Fig. 5-9). The divider system has an internal frequency-doubling circuit that doubles the horizontal oscillator signal to 31.5 kHz within the countdown divider. That means twice the horizontal frequency; in other words, two clock pulses are applied to the vertical countdown divider system during one horizontal scan line period. Due to this divider system, no vertical frequency adjustment is needed.

The divider system operates with two reset windows. One window is a wide window, active from 488 to a count of 722 clock pulses. The second window is active between 522 and 528 clock pulses. The wide or narrow window is activated by the up/down counter located within the countdown divider. The counter increases its count value by one each time the separated vertical sync pulse is within the search window. When the sync pulse is not within the window, the counter value is lowered by one.

There are three different working modes for the divider system. The first mode uses the wide window while looking for vertical sync and when changing channels, or while operating with a nonstandard TV signal that is outside the limits of the narrow window. The second mode (narrow window) is used when the up/down counter has reached its maximum count of 12 vertical sync pulses. The narrow-window mode is the normal operating mode for the system. When the divider operates in this mode and a vertical sync pulse is missing within the window, the divider is reset at the end of the window and the counter value is lowered by one (525 to 524).

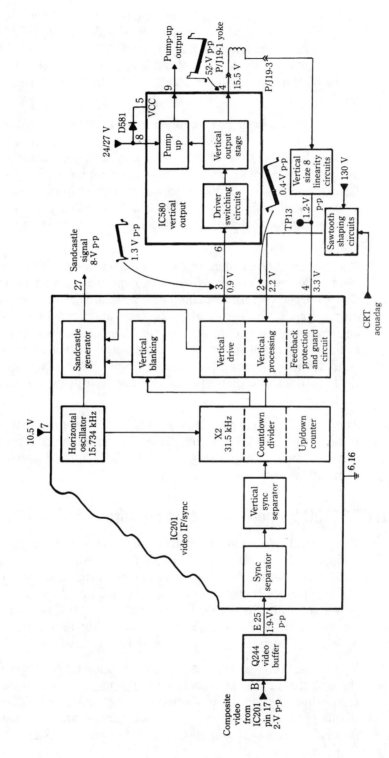

■ **5-9** *A block diagram of Sylvania's C9 vertical circuit.* Philips Consumer Electronics

233

At a counter value below 10, the divider system switches over to the large window mode in order to once again locate the vertical sync pulse. The last mode is used when no TV signal is found. With this condition, only noise is found and the divider is reset at a count of 528. This keeps a stable raster displayed with normal height.

Some VCRs operating in feature modes such as still picture or search generate a distorted vertical sync, which can cause a rolling picture. Usually VCRs reinsert vertical sync in the feature mode. In case the vertical sync is missing or does not meet proper specifications, whether it is VCR-related or created from some other source, the divider system within IC201 is designed to reset its counter automatically and generate a pulse that replaces vertical sync.

The divider system generates the vertical blanking pulse fed to the sandcastle output circuit. When the composite video sync signal is distorted, the vertical blanking pulse generated at the sandcastle output (pin 27) is developed by the vertical processing and sandcastle generation circuits. The vertical blanking pulses then start at the beginning of the first equalizing pulses instead of when the countdown divider is reset by proper count. Generation of the vertical sawtooth is accomplished by the sawtooth shaper circuit connected to pin 2.

The vertical output signal from pin 3 of IC201 is applied to the vertical output IC (IC580), which drives the vertical yoke. Feedback voltage from the output stage via the yoke is applied to pin 4 for gain and linearity correction.

RCA CTC130C vertical deflection circuits

The vertical output circuit receives a drive pulse from pin 18 of the deflection IC (U401). The vertical output transistor (bottom) (Q502) acts as the drive transistor for the output transistor (top) (Q501). Diodes provide switching and bias functions. The main power supply for the vertical output circuits is 26 V, which is supplied by R509. Additional B+ to aid in speeding up retrace time is supplied through R506 from the regulated B+ source (Fig. 5-10).

The vertical drive signal is generated by charging capacitor C502 through the vertical size control to ground. A vertical retrace is started by an internal switch between pins 21 and 22 of U401, which discharges C502. When capacitor C502 charges and discharges, it generates a sawtooth voltage coupled via C503 to pin 20 of U401.

Internally the signal is amplified to the deflection IC, and amplified and compared to the feedback voltage at pin 19. The feedback voltage is derived from the vertical yoke current through R508 and coupled to pin 19. This feedback maintains vertical linearity.

The vertical drive signal from pin 19 of U401 is applied to the base of bottom output transistor Q502. During the first half scan, Q502 amplifies this signal and controls the base drive of the top output transistor (Q501) through diode CR504. During the second half of the vertical scan, Q502 acts as an output device by pulling vertical yoke current through CR505. Pin 23 of IC401 provides a vertical blanking pulse that is applied to the sandcastle circuit to provide vertical blanking of the video signal during vertical retrace.

Vertical output retrace

Current will continue flowing from the coupling capacitor through the vertical yoke and Q502 to ground until the end of the scan, at which time Q502 will turn off due to initiation of retrace. At the time Q502 turns off, the current in the yoke is maximum.

Because it is impossible to immediately stop current flow through an inductor, the current in the yoke continues flowing. But because Q502 is turned off, the current now flows through diode CR503 into the base-collector junction of the top output transistor.

With this action Q501 conducts in the reverse mode (emitter to collector), storing up retrace energy in capacitor C506. Operation of the top vertical output transistor (Q501) in this mode is not detrimental to the performance or reliability of the device. C506 and the vertical yoke winding form an LCR series resonant circuit during retrace, while CR501 and CR502 prevent the retrace current from flowing into the 26-V supply.

As the current flowing through the yoke and into capacitor C506 decreases toward zero, the beam is moved back toward the center of the picture. When this occurs, the top output transistor turns on and saturates due to stored charge in C505. Q501 now conducts the energy stored in C506 back into the vertical yoke to move the beam to the top of the picture.

To shorten the retrace time, the voltage at the top output transistor is boosted by regulated B+ during the second half of the vertical trace. This action supplies extra voltage to force current into the vertical yoke and to move the beam back to the top of the picture tube.

235

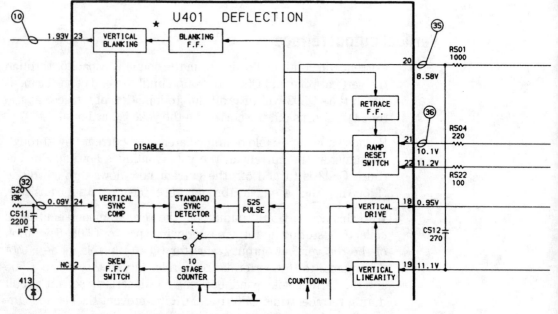

■ **5-10** *The RCA CTC130C vertical output circuits.* Thomson Consumer Electronics

VERTICAL

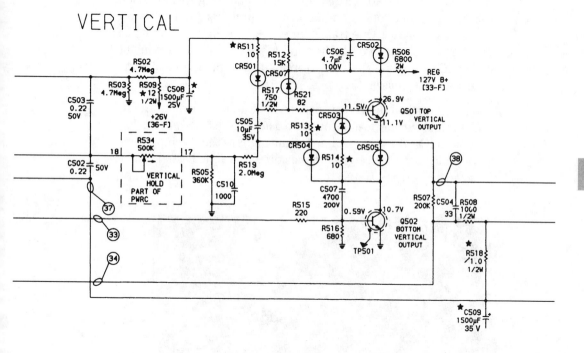

The current flowing through Q501 into the vertical yoke is now maximum at the end of the second half of vertical retrace. At this point, the bottom output transistor (Q502) is again turned on. Q502 begins shunting away base current from Q501 through diode CR504. This begins the first half of the vertical trace cycle.

Vertical output circuits

In the tube and early solid-state chassis, the vertical output circuits were transformer-coupled to the yoke assembly. In the 1970s, two transistors of different polarity were incorporated in a transformerless (OTL) vertical output circuit (Fig. 5-11). Each transistor conducts only one-half of the input signal and cuts off with reverse polarity. One transistor sweeps the bottom half while the other transistor sweeps the top half of the raster. You might find one IC containing all vertical input and output circuits in some of the latest TV chassis (Fig. 5-12).

A separate vertical driver transistor can be found between the IC oscillator and the vertical output transistors (Fig. 5-13). In the latest vertical ICs, the vertical amp is built right in the IC chip. Thus, the amplified sawtooth signal applies directly to the base terminal of the vertical output transistor.

238

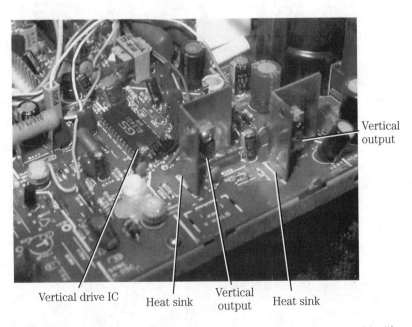

Vertical output

Vertical drive IC   Heat sink   Vertical output   Heat sink

■ **5-11** *The vertical output transistors, on separate heat sinks, with different voltages on the collector terminals.*

Large IC

■ **5-12** *The entire vertical circuit assembly may be included in one large IC in today's TV chassis.*

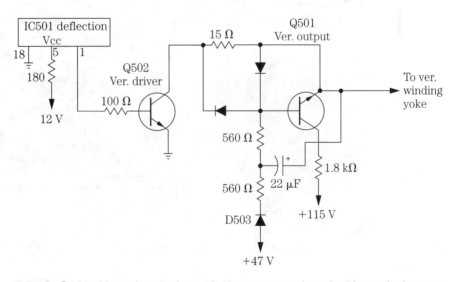

■ **5-13** *Q502 drives the single vertical output transistor in this vertical output circuit.*

Most vertical sweep problems are related to a defective component in the vertical output circuits. A horizontal white line, insufficient vertical sweep, intermittent sweep, and foldover can be caused by a defective output transistor or component. Usually, the vertical output transistors are mounted on a separate heat sink or metal TV chassis. You might find a top hat or flat mounting vertical output transistor (Fig. 5-14). The flat transistors can be insulated from the heat sink with small pieces of insulation. These flat-type transistors have a tendency to break down under load and still test normal when out of the circuit.

Troubleshooting the vertical output stages is fairly easy with a known oscillator sawtooth signal. Check the input waveform and output signal with the scope. With a fairly normal sawtooth voltage at the output of the vertical IC and no waveform at the vertical output transistors, suspect problems within the vertical output stages. Measure the voltage at the metal collector terminals of each output transistor. The collector terminal of one transistor is usually grounded, with the other having a high B+ voltage. Correct voltage measurements on each transistor can determine if the transistor or surrounding components are defective.

■ **5-14** *In older vertical circuits, you may find flat top and top hat transistors, while the new chassis have one IC power output transistor.*

Vertical output IC

Today, most vertical output circuits consist of one fairly large power IC (Fig. 5-15). The vertical output IC is mounted to a separate heat sink upon the TV chassis. The vertical IC receives a drive pulse from a vertical countdown or sweep IC and is coupled directly to the input terminal. The output terminal of vertical power output IC is coupled to the vertical winding of the deflection yoke.

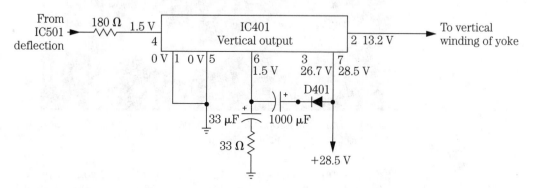

■ 5-15 *The vertical output circuit consists of a few capacitors, resistors, and diodes beside the vertical power output IC.*

241

The vertical IC may have one or two different voltage sources at voltage supply pins (VCC). When the supply pin voltage is low, suspect a leaky vertical output IC, or improper voltage from the supply source. Take voltage measurements upon all pin terminals and compare to the schematic. Check drive waveforms at input and output terminal of vertical output IC (Fig. 5-16). If no input waveform is found, check the vertical sweep IC circuits. Suspect a defective vertical IC with adequate input waveform and no output waveform. Check the yoke winding with a fairly normal output waveform from vertical output IC.

Servicing the Sylvania C9 vertical circuits

When troubleshooting the vertical circuit in the C9 chassis, leave the chassis connected to the picture tube and other components. The chassis can be removed to inject voltages into the vertical circuit if it's difficult to get to the vertical countdown IC terminals. Check the vertical drive waveform at pin 3 of IC201 and pin 6 of IC580 (Fig. 5-17). No vertical drive waveform can indicate a dead IC201.

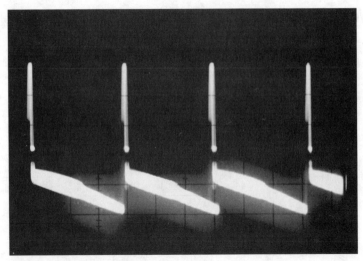

■ **5-16** *The vertical output pulse at the output IC and as it is applied to the vertical yoke winding.*

IC201 deflection circuits can be tested by injecting 10 volts at pin 7 (VCC) of IC201. Take a waveform at pin 3. Remember, the vertical waveforms are not very steady, even with the best scopes. If the vertical and horizontal countdown IC is functioning, a vertical waveform can be found at pin 3 and a horizontal waveform can be found at pin 26. Do not plug the set into the receptacle for this test.

If a sawtooth waveform is found on pin 6 and there is no vertical sweep, suspect IC580, the low-voltage source, or the vertical yoke. Scope pin 4 for a vertical output waveform. Measure the voltage at pins 8 and 4 of IC580. Low or no voltage indicates a leaky vertical output IC, or an improper voltage source. Disconnect pin 8 from the circuit, and if the voltage rises above 27 V, replace IC580. You can inject a 27-V source at pin 8 (IC580) and a 10-V source at pin 7 of IC201, and take vertical waveforms.

An open yoke winding or plug connections can produce a horizontal white line. Check the vertical yoke winding with the ohmmeter. Check TP13 for feedback voltage. Measure the resistance from each pin of IC580 to ground to determine if a component or IC580 is leaky. If there is insufficient vertical height, shunt capacitors C585 (10 µF) and C584 (3,300 µF) and look for dried-up or open conditions.

Troubleshooting RCA CTC130C vertical circuits

To determine if the vertical deflection IC (U401) is performing, scope terminal 18 for a vertical drive waveform. If the screen shows a horizontal line, you know the horizontal and vertical deflection IC is operating because both occur in the countdown circuits of U401. Although it's difficult to take vertical waveforms in many vertical output circuits, the input pulse (pin 18) and output pulse at the emitter of Q501 will indicate if the vertical output circuits are normal (Fig. 5-18).

Take critical voltage measurements on the collector terminals of both output transistors (Q501 and Q502). Higher-than-normal voltage at the collector of Q501 (top) can indicate that either Q501, CR505, or Q502 are open. Remove one end of each silicon diode for an accurate diode test with the DMM. Carefully check resistors R511, R513, and R154 for open conditions. Make sure the 26-V and 127-V B+ are present in the vertical circuits. Most problems can be eliminated in the vertical output circuit with transistor, diode, and resistance measurements.

Troubleshooting Realistic 16-410 vertical circuits

The 1-V negative drive pulse from pin 23 of internal vertical driver (IC9701) is fed to pin 6 of vertical output IC IC9501 (Fig. 5-19). The output pulse is shaped and amplified, providing a 10-V pulse to the vertical winding of the deflection yoke. The output pulse is found at pin 2 of output IC9501. Pin 1 is at ground potential, with pin 9 as a supply pin (VCC) of 23.5 volts. Often, the vertical output IC has a higher supply voltage than any other IC in the chassis.

Before replacing the output IC9501, check the input and output waveforms on pins 6 and 2. Suspect the vertical output circuits with a normal waveform on pin 6 and no or weak waveform on output pin 2. Check all voltages upon IC pin terminals and compare them to the schematic. Test each capacitor and the resistors tied to the pins of IC9501. If the supply pin voltage is real low at pin 9, suspect a leaky output IC or improper supply voltage. Check for open yoke windings and capacitors and resistors tied to common ground with a fairly normal output waveform at pin 2.

Horizontal white line

No vertical sweep is the easiest symptom to locate in the vertical circuits. Practically any defective component in the vertical circuit can produce a horizontal white line. First, measure the B+ voltage feeding the vertical circuits. You might find a different voltage supplying the vertical IC chip and the vertical output circuits. Improper voltage at any stage can cause no vertical sweep.

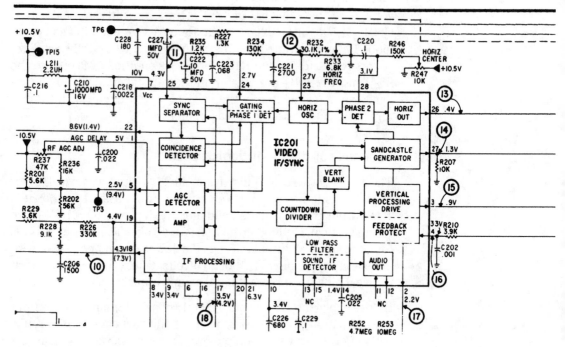

■ **5-17** *Sylvania C9 circuits are contained in IC201 and IC580.* Phillips Consumer Electronics.

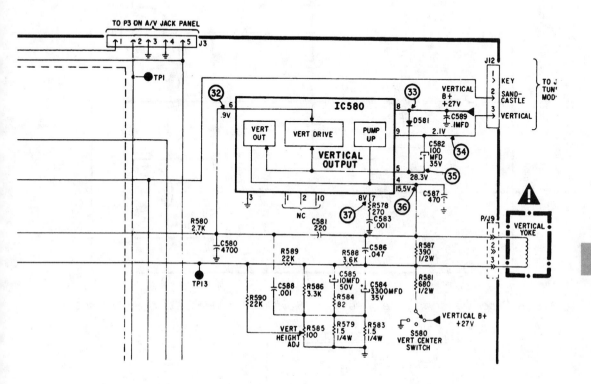

Horizontal white line

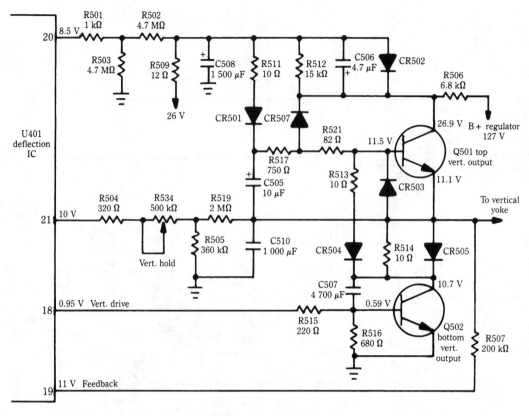

5-18 *Check the vertical drive pulse at pin 18 of ICU401 in the RCA CTC130C chassis.*

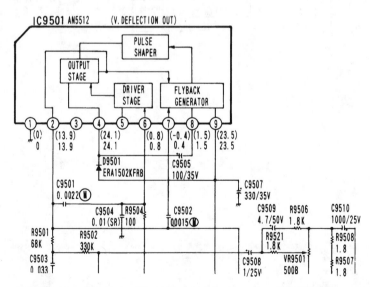

5-19 *The Realistic 16-410 portable has one IC (IC950) controlling vertical output, feeding the deflection winding.* Radio Shack

Only a horizontal white line was found in the raster of a Sharp 19D72 (Fig. 5-20). No voltage was found at either vertical output transistor. A voltage check at the 48-V voltage source indicated an overheated R520 (22-Ω) resistor. D503 was leaky when tested in the circuit with the diode test of the DMM. Replacing both R520 and D503 restored voltage and proper vertical sweep to the vertical circuits.

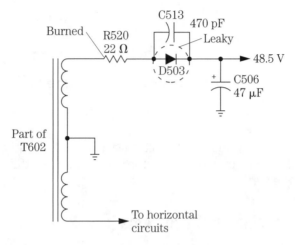

■ **5-20** *A leaky diode D503 and burned resistor R520 prevented dc voltage from reaching the vertical output circuits.*

Another common problem is no waveform at the vertical output transistors and a vertical sweep symptom. Check the oscillator waveform at the transistor or IC oscillator stage. Check for a waveform at the base and collector terminal of each transistor within the vertical circuit.

No waveform was found at the collector terminal of TR602 in a Montgomery Ward 12926 (Fig. 5-21). A normal signal was found on the base terminal. TR602 was checked in the circuit with the diode transistor test of a DMM. High internal leakage was found between the emitter and collector terminals. TR602 was replaced with an ECG159 universal transistor.

In directly coupled vertical circuits where vertical feedback controls the vertical oscillator circuits, the scope is useless. When no waveforms are found, check each transistor in the circuit with the DMM diode test. The vertical transistors can be checked within seconds with the DMM. If leakage is noted between two elements, check the schematic for resistors or diodes across the two elements causing the leakage measurement. If in doubt, remove the

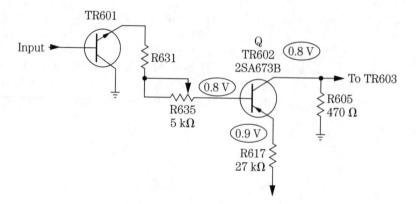

■ 5-21 *The circled voltages occurred with a leaky TR602 in a Wards 12926.*

emitter terminal with solder wick and take another measurement. The open or leaky transistors can be found within a few minutes in the vertical circuits.

Vertical oscillator IC problems

Check the vertical oscillator IC with a scope and voltage tests. Check the vertical sawtooth waveform fed to the vertical driver or output transistor. Improper or no waveform signal can indicate a defective IC, bad surrounding components, or no voltage input. Next, measure the B+ voltage at the IC terminal. If the voltage is low, suspect a defective IC, components, or voltage source. Remove the voltage source at the B+ pin. Suspect a leaky IC if the correct voltage returns.

Only a horizontal white line with no vertical waveform at pin 7 was found in a Sharp 19C81B (Fig. 5-22). Only 2.7 V was noted at the B+ voltage source (pin 11). Pin 11 was removed from the circuit with a solder wick. The supply voltage increased to 18 V, indicating that IC1501 was leaky. Extra care must be exercised when replacing the defective IC with a universal replacement. Sometimes the new replacement can be damaged during installation, or the new chip can be defective. If the replacement IC does not work, try another one. Replacing IC1501 with the original part number (RH-1X0094CEZZ) solved the no-vertical-sweep problem.

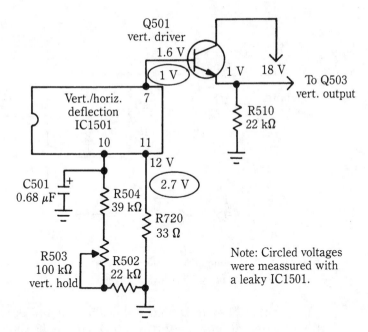

Q501
vert. driver
1.6 V

1 V   1 V   18 V
                    To Q503
                    vert. output

Vert./horiz.
deflection
IC1501   7

R510
22 kΩ

10   11

12 V

C501   +
0.68 μF

2.7 V

R504
39 kΩ

R720
33 Ω

R503
100 kΩ
vert. hold

R502
22 kΩ

Note: Circled voltages
were meassured with
a leaky IC1501.

■ 5-22 *No vertical waveform was found at pin 7 in a Sharp 19C81B.*

White line—output circuits

Check the vertical output circuits if the proper vertical oscillator or driver signal is feeding the output transistors. Measure the voltage at each collector terminal. Go to the power source if there is no voltage on both collector terminals of the transistor. Low collector voltage on both transistors can indicate a leaky transistor. Test each transistor in the circuit for an open or leaky condition with the DMM diode test.

Suspect burned resistors or leaky diodes used as bias components. Remove one end of each resistor or diode to check for correct resistance. Bridge critical electrolytic bypass and coupling capacitors in the output circuits. If trouble still exists, measure the resistance of each resistor in the vertical circuits, removing one terminal to check for correct resistance.

A horizontal white line was found in the raster of a K-Mart KMC1921G (Fig. 5-23). No voltage was found at the collector terminal of the driver transistor (Q303), and high voltage was mea-

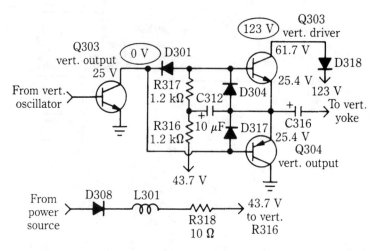

■ 5-23 *Check for open resistor (R318) providing voltage to output transistor in a K-Mart KMC1921G.*

sured at the collector terminal of Q305. The 43 Vdc was missing from the output circuits. R318 (10 Ω) in the 43-V power supply was open. Replacing the open resistor did not solve the no-sweep problem.

A quick in-circuit transistor test of both transistors revealed that Q305 was open and Q304 was leaky. While the transistors were out of the circuit, the bias diodes and resistors were checked for leakage and correct resistance. Replacing both transistors and R318 solved the no-vertical-sweep problem. Always replace both vertical output transistors when one is found defective.

Although most sweep problems in the vertical output circuits are caused by the output transistors, this was not the case in a Sylvania E08-1 chassis (Fig. 5-24). Both Q300 and Q302 tested normal in the circuit, with higher-than-normal voltage on both collector terminals. A sawtooth waveform was fed to the base of Q300. All resistors within the vertical circuits were normal. The full raster was restored when C342 (2,500 µF) was bridged with a good capacitor.

Critical vertical waveforms

Defective transistor output waveforms are difficult to obtain or read with a defective component in the vertical output circuit. You can check the output waveform and see (if the waveform is visible) what is feeding to the vertical yoke winding. While in vertical output IC components, critical waveforms can determine if the vertical input or output IC circuits are defective (Fig. 5-25).

250

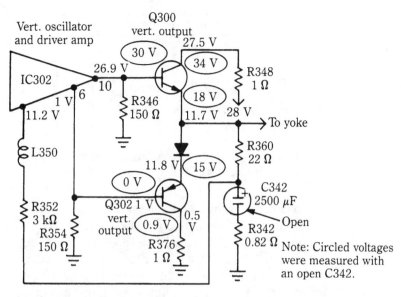

■ 5-24 *Open capacitor C342 (2500 µF) electrolyte produced a horizontal white line in a Sylvania E08-1 chassis.*

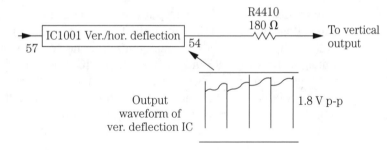

■ 5-25a *The output drive waveform from the deflection IC to the vertical output IC.*

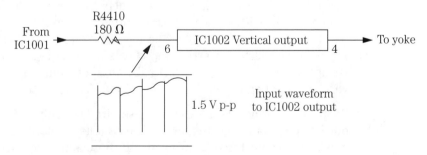

■ 5-25b *The input waveform at the output IC.*

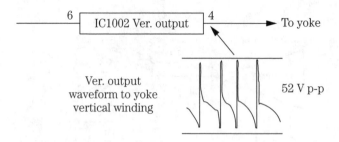

■ 5-25c *The vertical output waveform that drives the vertical yoke deflection winding.*

With no vertical sweep (a horizontal white line), scope the output terminal feeding to the deflection yoke. If there is no waveform or an improper waveform, attach the scope probe to the output IC's input terminal. Next, scope the vertical drive waveform from the sweep or countdown IC. Notice the peak-to-peak voltage at each waveform. Correct vertical output waveforms can determine what stage or circuit is defective.

Vertical voltage injection

The vertical sweep IC and output IC circuits can be serviced when the voltage source is taken from the flyback secondary winding by voltage injection. Check the schematic for the voltage supply terminal of the sweep IC and vertical output IC. Inject a voltage that is shown into these VCC terminals from an external universal power supply. Two different voltages must be injected at each IC. This means a dual variable power supply, or separate battery pack. Trace the scope waveform at the vertical sweep output terminal to the deflection yoke winding.

Servicing directly coupled vertical circuits

In many of the early TV chassis, directly coupled transistors formed the vertical circuits. The vertical oscillator might be a multivibrator type with vertical error and preamp circuits between the oscillator and output circuits. The feedback circuit from the output to the oscillator or error amp stages must be before or after vertical operation circuits. Practically any one component in the vertical circuit can keep the vertical circuit from functioning.

Scope waveforms are useless in these circuits. Only the vertical output check can indicate that vertical circuits are normal. No vertical sweep indicates yoke problems. If there is no vertical output signal, start with voltage tests. Check for positive voltage on both collectors of Q506 and Q507 (Fig. 5-26). No or low voltage on vertical output transistor Q506 (top) can indicate a leaky transistor or poor low-voltage power supply (116 V). If B+ voltage is present, suspect resistor R526 (15 kΩ).

Some technicians inject a vertical signal at the base of Q507 (bottom) and notice if the vertical sweep returns (Fig. 5-27). If the vertical sweep returns, suspect vertical error, preamp, or oscillator transistor circuits. When the vertical sweep does not return, check the output transistors. If the collector voltage of Q506 (top) is very high, suspect that either Q506, CR508, or Q507 are open. Only one of these components might be open. Test both output transistors (Q506 and Q507), CR503, CR510, CR506, and CR508 in the circuit. If any one does not test normal, remove the collector or cathode lead from the circuit and check it again.

Sometimes one of the output transistors will test normal, but it then breaks down under load. Also check the deflection yoke (CY100), CR505, C424, and pincushion transformer T403. Check the vertical windings of the yoke and pincushion transformer with the low-ohm scale of the ohmmeter (Fig. 5-28).

The vertical multivibrator oscillator circuits must oscillate before there is any vertical output sweep signal. Check Q501 and Q502 when no signal is applied to the base terminal of the error amp (Fig. 5-29). Remember, vertical waveforms are not very steady even with the best scopes. When the vertical oscillator is off frequency, check electrolytic capacitors C508 and C509, the vertical hold (R4206), and C424 (Fig. 5-30). Dried-up or open electrolytic capacitors can produce weak or no vertical oscillation.

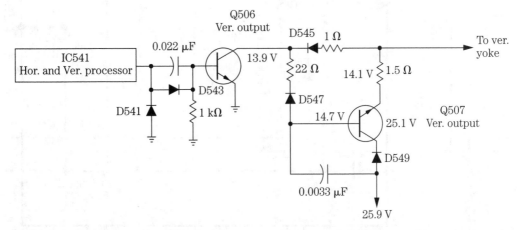

■ **5-26** *Take critical voltage measurements on Q506 and Q507 when a drive waveform is found at the input.*

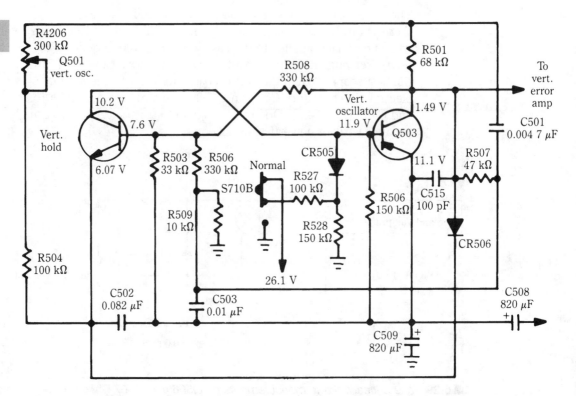

■ **5-27** *The correct vertical output waveform applied to the yoke winding.*

■ **5-28** *A multivibrator circuit with Q501 and Q503 as oscillators.*

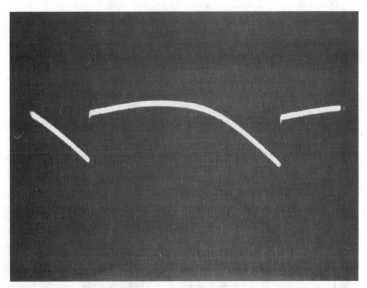

■ **5-29** *Vertical drive waveform from directly-coupled oscillator transistor to base error amp.*

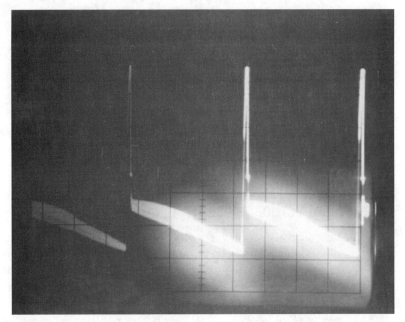

■ **5-30** *The vertical drive waveform at the output transistor to drive the vertical yoke winding.*

Critical voltage measurements on the transistor terminals can indicate a leaky or open transistor. Remember, in directly coupled circuits, the voltage on one transistor can directly affect another. Inject internal voltage to the vertical oscillator circuits from a separate power supply when scan-derived voltages are found (Fig. 5-31).

For vertical foldover or linearity problems check C506, C507, C508, C509, and C424. Check the feedback circuits and bias resistors in the vertical output transistors. Check R521, R533, CR510, CR506, and CR508 for foldover and linearity problems. Do not overlook an open or dirty service normal switch (S701B) when there is no vertical sweep and only a white horizontal line.

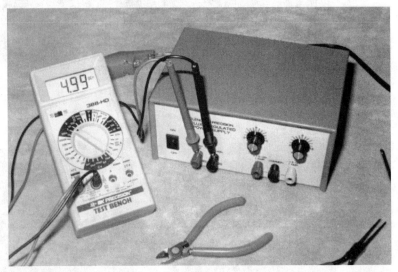

■ **5-31** *Inject external dc voltage from a dual-power supply when the vertical circuits are powered by the secondary circuits of the flyback.*

Insufficient vertical sweep

Often, insufficient vertical sweep and linearity problems are located within the vertical output circuits. Improper drive voltage can cause insufficient vertical sweep. Vertical sweep might be only 2 or 3 inches in height, or the raster might lack a fraction of filling out the top or bottom area of the picture tube. Poor vertical linearity (Fig. 5-32) at the top or bottom of the raster can be caused by leaky top and bottom transistors or a change in the bias resistors. A leaky vertical IC or components tied to the IC can produce insufficient vertical sweep. Check for the correct sawtooth waveform at the vertical IC output terminal.

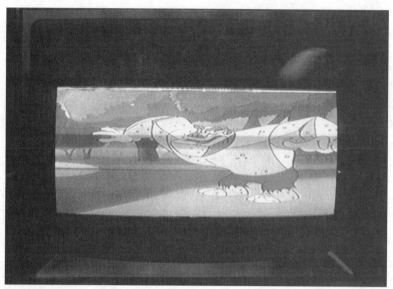

■ **5-32** *Poor vertical linearity can be caused by faulty output transistors, ICs, and an improper dc voltage source.*

Sometimes those flat-type vertical output transistors will test normal in the chassis and then break down under load. Replace both transistors if voltage tests are good. Only 1.5 inches of vertical sweep was found in a Sanyo 31C41N (Fig. 5-33). Both output transistors (Q455 and Q902) were replaced with GE-32 universal replacement transistors. Now the height lacked 2 inches at the top of the raster. A voltage measurement indicates only 0.57 V at the base of Q902, which should be around 46 V. Replacing a leaky diode (D454) solved the insufficient-vertical-raster problem.

Intermittent vertical sweep

Of all the vertical problems, intermittent vertical sweep symptoms are the most difficult to locate and fix. The intermittent symptom can be caused by transistors breaking down, poor board connections to a loose collector transistor mounting screw, or many other things. Suspect a poor board connection, loose component terminal, or poor eyelet or griplet connections if the raster collapses when the chassis is touched or moved. Try to narrow the intermittent to a certain section of the board. Intermittent transistor and IC components can be located by application of several coats of coolant. Monitoring the vertical circuits with the scope and DMM can help locate the intermittent stage and component.

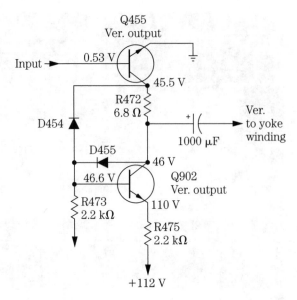

Q455
Ver. output

Input →

0.53 V

45.5 V

R472
6.8 Ω

D454

Ver.
to yoke
winding

+
1000 μF

D455

46 V

46.6 V

Q902
Ver. output

R473
2.2 kΩ

110 V

R475
2.2 kΩ

+112 V

■ **5-33** *D454 was found leaky in a Sanyo 31C41N vertical output circuit.*

Sometimes the raster would go to white line and at other times a white horizontal line would appear in the picture with normal sweep in an RCA CTC68AF chassis (Fig. 5-34). The vertical drive signal from the MAG001B module was fairly normal when the raster went to a white horizontal line. A voltage test with the collapsed raster found higher-than-normal voltage on the base and emitter terminals of Q101. Replacing leaky Q101 and Q102 solved the intermittent raster problem.

Both vertical output transistors should be replaced when one is found to be leaky or open. The collector voltage can be quickly checked by measuring the voltage at each heat sink (Fig. 5-35). Often, heat sinks mounted in the center of a board are not grounded directly to the metal chassis. Suspect a vertical top output transistor if the top part of the raster pops down with a normal bottom half.

Intermittent vertical sweep—Sylvania E21

The picture in a Sylvania E21 chassis would operate for several hours, and then collapse. Again the chassis might operate for days.

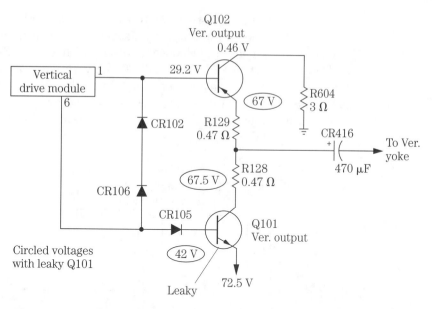

■ **5-34** *Check for leaky or open output transistors if there are improper voltages on all terminals.*

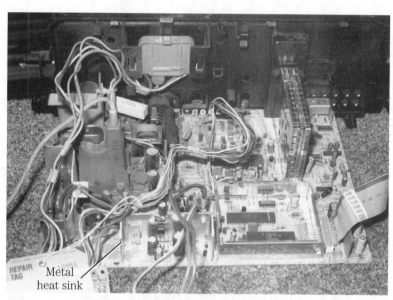

■ **5-35** *The metal heat sink of the vertical output IC is grounded.*

The vertical output transistors were suspected, so both were replaced. Sometimes these output transistors test normal, but can still open up (Fig. 5-36). Q300 was replaced with an RCA SK3083 and Q302 was replaced with RCA SK3054. Everything was normal for several days, and again the raster collapsed. When IC302 was in the intermittent state, several coats of cold spray were applied. The picture popped in. Replacing intermittent vertical driver IC302 restored the TV.

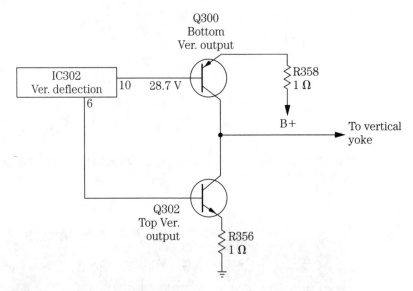

■ **5-36** *IC302 was found intermittent in a Sylvania E21 chassis.*

Vertical pincushion problems

Check the pincushion circuits when the sides of the picture are pulled inward. The pincushion circuits prevent TV picture distortion, in which each side of the raster can sag toward the center of the screen. Usually, pincushion problems do not cause too much trouble in small screen sizes. Special pincushion circuits are found in 27-inch or larger screens.

Pincushion problems are caused by a defective pincushion transformer, or poor solder joints. A narrow or bowed picture can result, with a shorted primary winding of the transformer. Replace any pincushion transformer that constantly has a high-pitched squeal noise.

Solder the terminal connections of the transformer in all pincushion circuits. Check the primary winding resistance and compare with the schematic. In transistorized pincushion circuits, test the pin output transistor for leakage and improper voltages on the collector terminal (Fig. 5-37). Test all other transistors in the circuit for leakage or open conditions. Often the pin output transistor becomes leaky when horizontal output transistor is found shorted.

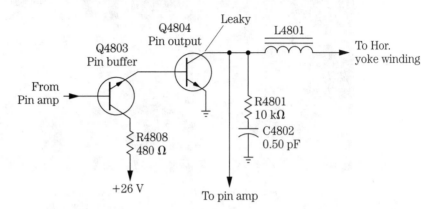

■ **5-37** *Vertical bowing was caused by a leaky pin output transistor in a CTC140 RCA chassis. Check all electrolytic capacitors in feedback circuits for vertical foldover and poor linearity.*

Vertical foldover

Check for leaky output transistors and improper bias resistors when vertical foldover occurs. Improper negative and positive supply voltages applied to the vertical circuits can produce vertical foldover. Most vertical foldover problems occur in the vertical output circuits. Vertical foldover can occur at the top or bottom of the raster. Usually, vertical foldover occurs with insufficient height problems.

The raster was down 5 inches at the top with vertical foldover in a J.C. Penney 685-2124. This particular chassis was made by General Electric (Fig. 5-38). The complete vertical chassis can be removed for transistor and component tests out of the circuit. The vertical output transistor (Q267) was found to be leaky. Q267 was replaced with an SK3054 and Q268 with an SK3083 universal transistor. Again, replace both vertical output transistors when one is found to be defective.

Vertical rolling

Insufficient vertical sync can cause the picture to roll vertically up or down. Scope the vertical input circuits for correct vertical sync

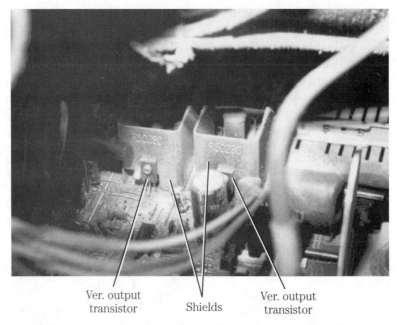

Ver. output
transistor

Shields

Ver. output
transistor

■ **5-38** *Vertical foldover can be caused by a defective output transistor, bad IC, or malfunctioning electrolytic capacitors in the output and feedback circuits.*

pulse. Check the low-voltage supply feeding the vertical circuits for excessive rolling and crawling in the picture. Suspect improper voltage or a change of resistance in the base and collector circuits when the picture rolls only one way. Remove one end of each resistor for accurate resistance measurements. If the raster collapses to a white line when adjusting the vertical hold control, suspect an open vertical hold control.

Excessive vertical rolling and pulling of the picture was noticed in an RCA CTC81C chassis (Fig. 5-39). A scope check of the vertical dc supply indicated poor filtering. The picture was restored when C305 was shunted with a 4,000-µF electrolytic capacitor.

Vertical crawling

Scanning lines slowly moving up the picture with a dark section is called vertical crawling. A dark bar might be at the top or bottom of the crawling area. Improper filtering in the low-voltage source feeding the vertical circuits produces vertical crawling. Lower dc voltage goes along with poor filtering. Check the speaker for additional hum in the sound.

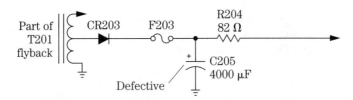

■ 5-39 *A dried-up filter capacitor C205 caused vertical pulling and rolling in this RCA CTC81C chassis.*

Scope the low-voltage power supply circuit to locate the defective electrolytic capacitor. Each filter capacitor can be shunted to locate the defective one. Always discharge the capacitor before clipping across the suspected one. You can damage transistors and IC circuits while shunting filter capacitors with voltage applied to the circuits. Some of these filter capacitors might be above 1,000 µF.

Vertical crawling was found in a Motorola D18TS chassis with the picture pulled down from the top of the screen (Fig. 5-40). The negative supply voltage feeding the vertical circuits was low. The picture returned to normal when another electrolytic capacitor was clipped across C809.

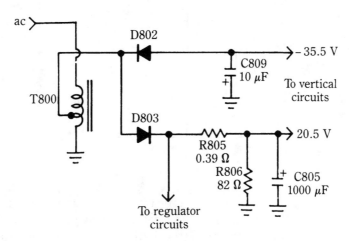

■ 5-40 *Vertical crawling was caused by a bad capacitor C809 in a Motorola D18TS chassis.*

Black top half

You might find the top or bottom half of the picture missing in the newer vertical circuits. Go directly to the top vertical output transistor when the top section of the picture is black. Likewise, check the bottom vertical output transistor when there is a black area at the bottom portion of the picture. With heavy retrace lines at the top of the raster in directly-coupled vertical circuits check the switch driver transistor.

The top half of the picture was black in an RCA CTC93E chassis (Fig. 5-41). The voltages of the top output transistors were way off. Q408 was removed from the chassis and was leaky. The transistor was replaced with one having the same part number (142691). The results were the same. Q407 was replaced with the same results, a black top section. An in-circuit diode test of CR416 revealed that it was leaky. Replacing both transistors and CR416 solved the black section at the top.

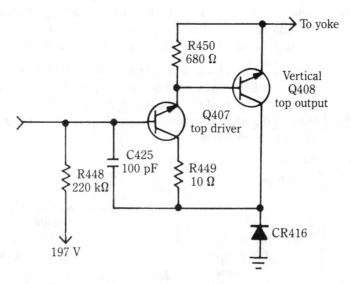

■ **5-41** *Leaky transistors Q408 and CR416 caused the top half of the picture to be black.*

After a General Electric 10JA chassis was on for 10 minutes, the top part of the screen appeared black with no vertical sweep. Each vertical output transistor was sprayed with coolant. When coolant was applied to Q268, the picture was normal. Heat was applied to the top vertical transistor and the top portion collapsed. Application of coolant and heat spray located the defective top-half output transistor (Fig. 5-42).

Shutdown after horizontal white line

The raster went into a horizontal white line and then the chassis shut down after being on for several minutes in a J.C. Penney

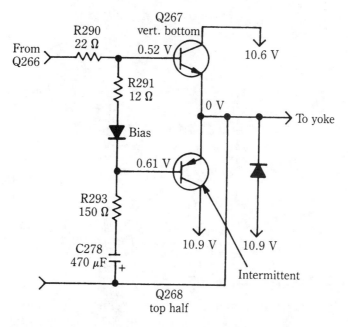

From Q266

R290
22 Ω

Q267
vert. bottom

0.52 V

10.6 V

R291
12 Ω

0 V

To yoke

Bias

0.61 V

R293
150 Ω

C278
470 μF

10.9 V   10.9 V

Intermittent

Q268
top half

■ **5-42** *Q268 was found intermittent with application of temperature extremes.*

685-2084. This chassis was manufactured by Samsung Corporation. After the set shut down, a variable line transformer raised the ac voltage to 80 Vac, and any higher voltage made the TV shut down. Before shutdown, the raster would go into a white horizontal line.

Voltage measurement on the deflection IC (IC501) (Fig. 5-43) indicated the 12-V source at pin 3 was very low (5.6 V). A resistance measurement from pin 3 to chassis ground was 71 Ω. To determine if the IC was leaky, pin 3 was removed from the circuit with solder wick and a soldering gun. The low resistance reading was still measured, indicating a leaky IC501.

Because IC components have a tendency to break down and become leaky, the 12-Vdc terminal (3) was removed from the circuit with the low-ohm measurement. These IC terminals can be easily removed with a mesh solder-removing material. Of course, C252 and R225 could have caused the low voltage at pin 3. The leaky vertical IC was placing a heavy load on the flyback winding of the horizontal output transformer causing shutdown. A low resistance measurement on any IC terminal that is not grounded can indicate a leaky IC component.

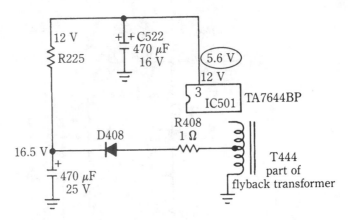

5-43 *Improper voltage at pin 3 in IC501 caused a horizontal white line in a J.C. Penney 685-2084.*

Bunching vertical lines

The symptom of heavy scanning lines was found at the top of the picture in a General Electric ECA chassis. After a 10-minute warmup, the bunching lines would appear. The picture was normal until the set had been operating for awhile. Adjustment of the vertical height control would take out the lines, but insufficient height would remain.

All voltages were normal with the driver and output transistors. Because the output transistors have a tendency to cause various lines in the raster, Q607 and Q609 were replaced (Fig. 5-44). The bunching lines still remained. Bias and emitter resistors were checked for correct resistance with one terminal removed from the circuit. The vertical output coupling capacitor (C619) was shunted with another and turned out to be the defective component.

Vertical-related problems

Sometimes insufficient or absent vertical sweep can happen outside the vertical circuits. An open yoke winding can produce a white horizontal line. A leaky or shorted yoke can look like a trapezoid pattern on the TV screen. Poor yoke socket wiring can develop into an intermittent vertical problem. Remember, a bad electrolytic capacitor in the convergence circuits can produce insufficient vertical sweep.

Only 1 inch of vertical sweep was found in an RCA CTC74AF chassis and it looked like a soft drink or liquor was accidentally spilled inside the chassis. Often, components or board arcing can develop

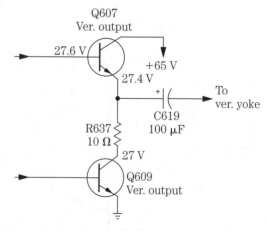

■ **5-44** *C619 (100 µF) produced bunching lines in the raster of a General Electric portable TV.*

when liquid is spilled inside the TV chassis. Cleaning up the chassis did not cure the vertical problem.

The vertical output transistor and voltages were normal. A fair size sawtooth output waveform was found on the scope. Troubleshooting the yoke and pincushion components located a burned control in the pincushion circuits (Fig. 5-45). Replacing the 500-Ω control solved the insufficient sweep problem.

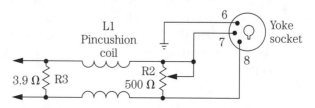

■ **5-45** *Check outside of the vertical circuits for vertical problems within the pincushion or yoke circuits.*

Troubleshooting the Sharp 19C81B vertical section

From the TV screen of a Sharp 19C81B insufficient vertical sweep was indicated, with only 2 inches of sweep (Fig. 5-46). Although the sawtooth waveform at the vertical amp terminal (7) of the deflection processor IC was fairly normal, very little waveform was found at the output coupling capacitor (C507). A waveform test at the driver transistor (Q501) was normal, pointing to possible trouble in the output transistor circuits.

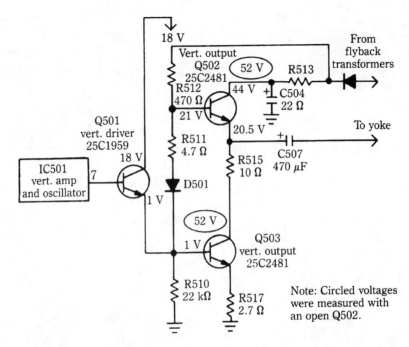

■ **5-46** *An open Q502 produced only a 2-inch vertical sweep in a Sharp 19C81B chassis.*

A voltage measurement at the collector terminal indicated higher than normal voltage. Both collector terminals had about the same voltage (52 V). The voltage on Q503 should be around 20 V. Q502 was tested in the circuit with the transistor test of the DMM and was found open. Q503 tested good in the circuit. Both transistors were replaced with a universal ECG373 replacement, restoring the full picture.

Unusual vertical problem

The customer complained the picture in an RCA CTC120 chassis would go down to a horizontal white line after several hours of operation. Sometimes the set would operate for days without any problems. When the TV chassis was on the bench, no vertical sweep was found on the TV screen. Pushing around on the chassis did not cause the raster to pop out. A normal vertical waveform was found at vertical IC401.

Both transistors tested normal in the circuit with the transistor test of the DMM. A dc voltage check indicated no 24-V source. R509 was found open (Fig. 5-47). R509 was replaced with a 12-Ω,

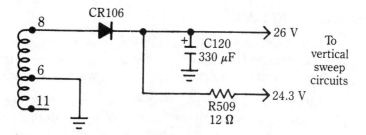

5-47 *R509 caused intermittent vertical sweep after several hours of operation.*

1-W resistor. Because one of the vertical output transistors might have temporarily broken down, opening R509, both transistors were replaced. The TV chassis operated for a week and was returned to the customer.

Three weeks later the set was returned to the shop with only 3 inches of vertical sweep. The chassis played in the shop for 5 days and finally collapsed again. R509 was running red hot (Fig. 5-48). Coolant was applied to IC401, Q502, and Q501 without any results. Next, each diode was sprayed and the raster went to a full screen. Heat was applied to CR504 and the sweep collapsed. Replacing CR504 solved the intermittent vertical sweep problem.

269

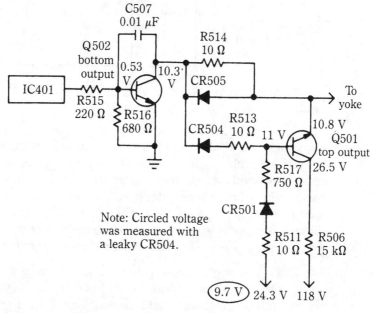

5-48 *Coolant sprayed on capacitor CR504 caused the picture to become intermittent, then collapse.*

Vertical "tough dog"

Only 4 or 5 inches of vertical sweep was noticed in a J.C. Penney 685-2020. In fact, the only vertical sweep was below the center of the vertical raster. The top half of the raster was missing.

Checking the schematic, the vertical sweep section has a top and bottom transistor output stage (Fig. 5-49). Both vertical output transistor Q22 and Q24 tested normal in and out of the circuit. Because the vertical transistor stages were directly coupled, all transistors were checked in the circuit with the diode transistor test of the DMM.

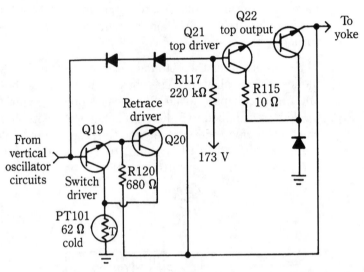

■ **5-49** *Replacing leaky transistors Q19, Q20, and R117 restored the missing top half of the vertical sweep.*

The switch driver (Q19) and retrace switch (Q20) were found to be leaky. After replacing both transistors with the original parts, the chassis operated for 2 minutes and went back to insufficient vertical sweep, with a bright horizontal line at the top. The retrace switch transistor (Q20) was operating red hot. Again Q20 was replaced, because it was possible it was a defective replacement. The results were the same. After a few seconds Q20 began to get very warm.

All voltage measurements were quite close, and some were off only a volt or two. Both bias diodes in the base circuit of Q21 checked normal with one terminal lead removed. Next PT101 (62-Ω cold) and R120 (680 Ω) were checked for correct resistance. R117 and R115 were checked in the circuit and were fairly close in resistance.

Actually, the "tough dog" culprit was located by removing each resistor lead from the circuit board. Each resistor was checked for correct resistance. When R117 (220 kΩ) was checked again, the resistor was found to be open. Sometimes we overlook a high resistance measurement that might be quite close to the resistor we are checking in the circuit. Always remove one end of a diode or resistor to check for correct resistance.

Troubleshooting vertical IC circuits

Vertical IC output circuits are much easier to service than in transistor output circuits. Just take a vertical drive waveform at vertical sweep IC and trace to the input of vertical output IC (Fig. 5-50). If no waveform at the sweep or countdown IC, repair this sweep circuit.

Next take the output waveform at the output terminal IC that feeds to the vertical yoke winding. Suspect a defective output IC, corresponding components, or improper voltage source with a normal input waveform and no or a weak output waveform. Measure the supply voltage (VCC) and voltage on all other pin termi-

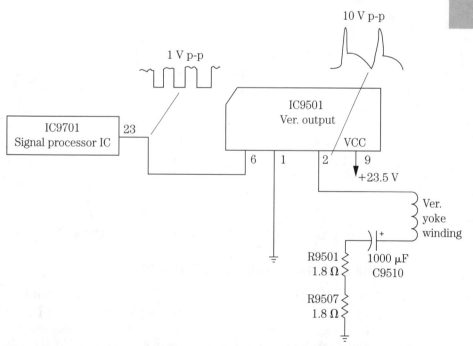

■ **5-50** *Check the output waveform of the vertical signal IC and the output of the vertical power IC for correct vertical waveforms.*

nals. Check all resistors and capacitors that connect to each pin terminal. Replace output IC with normal components and input vertical sweep signal. Make sure all electrolytic capacitors are normal on the pin terminals of output IC. Suspect yoke winding and return circuits components with normal output waveform and improper vertical sweep.

Five actual case histories

The following case histories are actual vertical problems that have occurred in different TV chassis. Each trouble has a different symptom with quick servicing methods.

No vertical sweep

Only a horizontal white line was found upon the screen of an Emerson EC10R TV. Voltage measurements were taken upon the output terminals of vertical output IC501 after a fairly normal waveform was fed into pin 4 (Fig. 5-51). The voltage in pin 3 was very low, indicating a leaky output IC. When pin 3 was removed from the foil, the voltage source measured 26.7 volts. IC501 was replaced with original part number 4152078500.

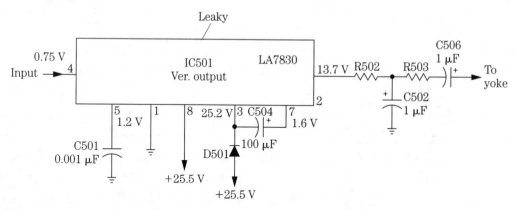

■ 5-51 *A leaky IC501 caused a lack of vertical sweep in an Emerson EC10R TV.*

Insufficient vertical sweep—Goldstar CMT 2612

The vertical drive output IC was suspected in this Goldstar chassis with insufficient sweep. Scope wave input waveform at pin 8 was fairly normal, indicating a defective output IC or corresponding components (Fig. 5-52). IC301 was replaced with an ECG1797 universal IC and the results were the same. D310 was found leaky and replaced with a general-purpose 1-amp silicon diode (SK3311).

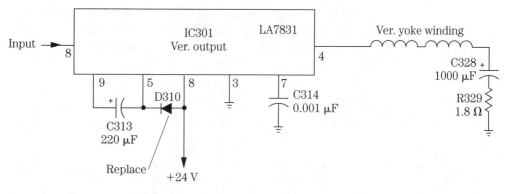

■ 5-52 *Replace D310 for insufficient sweep in the Goldstar CMT-2612 model.*

Vertical foldover—Sharp 19J65

At first, a Sharp portable had very little vertical sweep with low voltage upon pin 6 (1.7 V). IC501 was ordered out (X0238CE) since there was no universal replacement. The input waveform was fairly normal, with no waveform at output pin 2.

After IC501 was replaced, the vertical picture had a foldover problem. C512 was replaced, curing the foldover problem (Fig. 5-53). Check C505, C506, C509, C510, and C512 in the vertical circuits for possible foldover or poor vertical linearity.

273

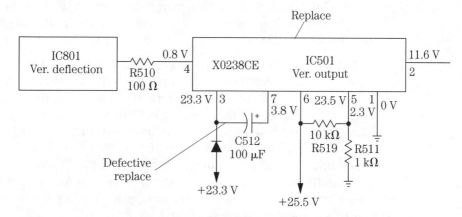

■ 5-53 *Vertical foldover in a Sharp 19J65 chassis was caused by C512. Also check C505, C506, C509, C510, and C512 if the unit displays vertical foldover and poor linearity.*

Vertical lines at top of picture

Several horizontal lines were found at the top of the picture in an Emerson MS250XA portable TV. All voltages were fairly normal.

One end of D305 and D304 were removed and tested good (Fig. 5-54). Q303 and Q302 were tested out of the circuit. Electrolytic capacitor C305 (4.7 µF) was almost open, causing lines at the top of the picture.

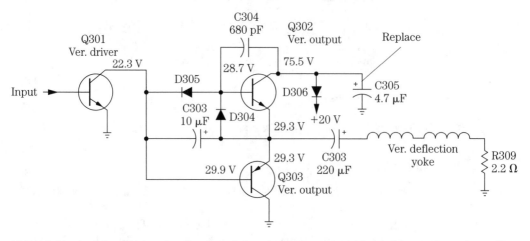

■ **5-54** *Vertical lines formed at the top of the picture because of a defective C305 (4.7 µF) in an Emerson MS250VA TV.*

Unusual vertical problem—RCA CTC145 chassis

The vertical sweep was intermittent, the picture would collapse to a white line, there was insufficient vertical sweep, and sometimes no sweep at all. This intermittent problem may occur after the set has run for a few hours.

U4501 was monitored with the scope at pin 4, and a voltmeter was clipped to pin 8. When the vertical section went into the intermittent state, very little output waveform was found, with practically no voltage change. Before replacing IC4501, all components were checked on each terminal. C454 is mounted quite close to the LA7831 IC, and this IC runs fairly warm (Fig. 3-55). When C454 was shunted, the picture returned. Several of these electrolytic capacitors have acted up in the recent RCA chassis and should be checked in the event of vertical problems. Replace C454 with longer leads and bend it away from U4501.

Conclusion

Scope the vertical oscillator and output transistors to isolate the defective stage. A voltage measurement on the collector (metal) terminals of the vertical output transistors will indicate if the tran-

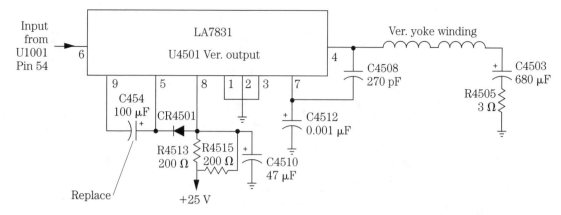

■ 5-55 *Replace electrolytic capacitor C454 (100 µF) across pins 9 and 5, leave longer leads, and bend the component away from U4501.*

sistors are defective or if there is no dc voltage source. Remember, the vertical output transistors can break down under load. Automatically replace them when all other components appear normal. Check outside of the vertical circuits for a defective vertical component. Do not overlook the convergence circuit with insufficient vertical sweep or a horizontal white-line symptom. Check for easy methods of servicing the vertical circuits in troubleshooting chart Table 5-1.

■ Table 5-1 Use this vertical troubleshooting chart for easy servicing.

What to check	How to check it
No vertical sweep.	Check low voltage source to vertical circuits.
Check waveform on base of output transistor or IC.	Check drive waveform on vertical sweep IC.
Check waveform at output of transistors or IC.	Test output transistors. Check voltages on output IC.
Check for open yoke winding.	Check for open ground return capacitor or resistor.
Vertical foldover.	Sub all electrolytic capacitors in output and feedback circuits.
Insufficient vertical sweep.	Power supply source. Check output transistors. Measure bias resistors. Check capacitors and diodes in output circuits.

Servicing IF video circuits

THE BIGGEST PROBLEM WHEN SERVICING THE INTERMEDI-
ate frequency (IF) stages is getting to these circuits. Often, the IF
circuits are found directly under the bell of the picture tube and
are enclosed in metal shields (Fig. 6-1). After separating the chassis from the picture tube mounting braces in a portable TV chassis, the IF stages are fairly easy to service. Sometimes the chassis
can be serviced from the bottom. In console models, the chassis
can be pulled back or removed. Accurate in-circuit voltage and
transistor tests can solve most IF repair problems.

Low voltages found in the IF transistor stages can be checked with
the DMM. Often, collector voltages can change or collectors can be
difficult to get at in the IF circuits. Accurate voltage measurements on the emitter terminal can indicate an open or leaky transistor (Fig. 6-2). These voltages are very low (1 to 3 V) across the
emitter resistor. No voltage can indicate an open transistor or an
improper dc voltage applied to the collector terminal. Check the
emitter resistor for an open or correct resistance with the low-ohm
scale. Often, the emitter resistance can vary from 1 kΩ to 47 kΩ.

Besides the DMM, the scope with demodulator probe, sweep
marker, and color-bar generator are useful test instruments in servicing IF circuits. The oscilloscope with a detector or demodulator
probe can be used to signal trace each IF stage with the color
broadcast signal or from the color-bar generator. The tuner-subber
can be used to inject signal at the IF input and to the base of the
second IF amp transistor with the picture tube as monitor to locate a defective first or second IF stage. The sweep-marker generator can be used for complete IF alignment.

IF circuits

The IF circuits can consist of three or four transistor stages or
one IC component (Fig. 6-3). The conventional IF circuits can
contain tuned output or some form of tuned input with transis-

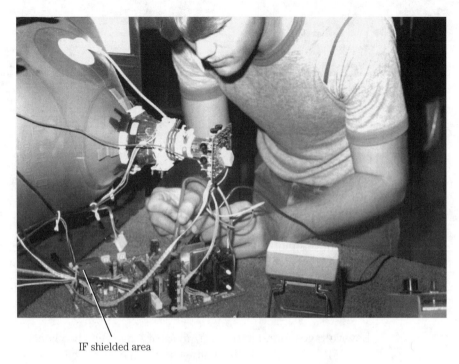

IF shielded area

■ **6-1** *Most IF stages are shielded, with components inside the shielded area.*

■ **6-2** *A block diagram of the IF amplifiers and order buffer circuits.*

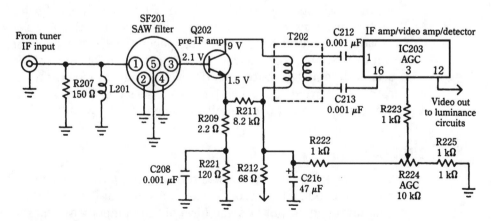

■ **6-3** *The SAW filter network is connected between the tuner and IF stages.*

tor RC coupling methods. In earlier solid-state TVs, stacked IF circuits were quite common. Today, the IC component has taken over the IF circuits.

The output signal of the tuner is fed through several tuned traps and connected to the base circuit of the first IF video transistor (Fig. 6-4). The amplified IF signal is transformer-coupled to the second and third IF video stages. Notice the second IF video stage is controlled by AGC voltage from 5.5 V to 7 V. The video detector is a fixed diode transformer coupled to the first video amp. The sound take-off coil can be tapped at the collector terminal of the third IF video or video amp.

Although universal transistors can be used as replacements in the IF section, it is best to use parts with the original part numbers. In some RCA chassis, universal transistors cannot function as IF amplifiers. You will save time by using the right part, because original transistors will work every time. The universal transistor terminal leads must be cut to the exact length of the original and spaced for mounting. Do not mount the IF transistors underneath the chassis. Replace each defective IF transistor inside the shielded area. They are easier to replace under the chassis, but they must be shielded. Always replace shields and solder banded areas to prevent stray signals from entering the IF section (Fig. 6-5). IF alignment is not required with direct transistor replacement.

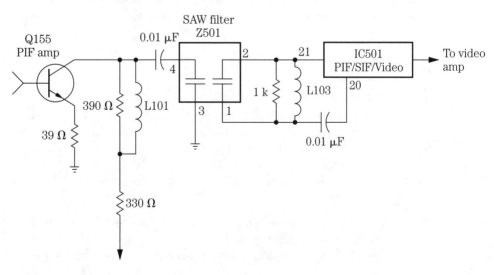

■ **6-4** *Today's PIF amp, SAW filter, and IC501 PIF/SIF/video IC.*

Varactor tuner   IF

■ 6-5 *Notice the shielded IF with varactor tuner nearby.*

IF ICs

In the early solid-state IF stages, the first and second IF video circuits were contained in one IC component (Fig. 6-6). Later on, the tuner was coupled to the IC IF stages with a single input transistor. Today, the IF input amp transistor is located between the tuner, SAW filter network, and IF AFT-AGC IC circuits (Fig. 6-7). The RF and IF AGC circuits are developed inside the IC component.

You can use a complete plug-in IF module with the modular TV chassis. Simply remove and plug in a new IF module when there are IF trouble symptoms. These same IF modules can be found in

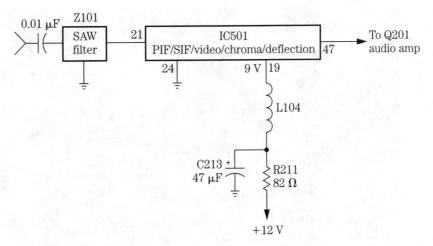

■ 6-6 *The IF stages may consist of a picture IF amp, SAW filter, and IC501 (which contains PIF/SIF/video), and a deflection IC.*

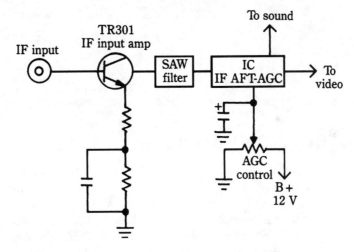

■ 6-7 *The SAW filter does not require alignment.*

the newer one-piece chassis (Fig. 6-8). Make sure all IF terminals plug into and are firm on the TV chassis. Fasten the module to the chassis with mounting screws to prevent future intermittent IF problems. Check for a voltage adjustment on some models when replacing the IF module.

Varactor
tuner

■ **6-8** *The varactor tuner has replaced the tuner module in many TV front-end circuits.*

RCA CTC130C IF video circuits

A surface acoustical wave (SAW) filter and an IF processor IC are found in the CTC130 chassis. The SAW filter eliminates the IF interstage tuned circuits. The SAW filter device generates the proper IF responses without the interstage transformer found in the conventional tube chassis. The SAW filter makes it possible to consolidate the rest of the IF stages within one IC. The single IF IC contains three stages of IF amplification, with AGC, AFT, and synchronous video detection (Fig. 6-9).

The SAW filter is made up of a piezoelectric material onto which are plated two pairs of transducer electrodes. One is the input transducer and the other the output transducer. The SAW filter frequency response is like that of a conventional discrete IF system.

Because the IF response is determined during SAW filter manufacturing, no field alignment is necessary. In fact, there are no adjustment screws. The SAW filter characteristics cannot be changed.

The IF signal from the tuner is amplified by Q301. The 47.25-MHz trap in the base circuit of Q301 eliminates the adjacent channel sound carrier. This extra amplification is necessary to make up for the SAW filter insertion loss. The output of the SAW filter is an IF signal with proper frequency response. It is applied differently to the IF IC. The signal passes through three stages within the IC. The IF signal is then applied to the synchronous video detector. The lowest possible distortion in the video output signal is provided by synchronous detection. After detection, the video signal is amplified and passed through a noise inverter. The composite video output signal is found at pin 12 of IC U301.

Sylvania C9 IF processing

The C9 chassis has separate processing for the video and audio. The video is processed within video IF sync IC (IC201). The sound is processed by the sound IF detector (IC202). The purpose of the split sound system is to achieve a better signal-to-noise ratio for the stereo signal (Fig. 6-10).

The 45.75-MHz IF signal from the tuner is coupled to the IF pre-amp (Q240), where it is amplified before entering the SAW filters (Y200 and Y201). Adjacent channel sound (47.25 MHz) is trapped within the tuners before the IF signal is applied to the IF preamp. The IF band-pass signal from SAW filter (Y200) is applied to pins 8 and 9 of the video IF sync IC for video processing. The IF band-pass signal from the SAW filter (Y201) is applied to pins 1 and 18 of the sound IF detector IC for sound processing.

IF amplification occurs inside IC201 and the signal is supplied to the IF video detector block. The reference signal needed for detection is tuned at pins 20 and 21. This signal is adjustable by L205 (45.75-MHz video detector AFT adjust) and is adjusted to the picture carrier. After detection, the composite video signal is amplified, receives noise correction, and exits the IC at pin 17 with a level of 2 V p-p.

SAW filter network

Today a new circuit found in the IF stages is called the surface acoustic wave SAW network. The SAW filter component is made up of a piezoelectric material with two pairs of transducer electrodes. One is the input and the other the output transducer. Voltage applied across the positive and negative terminals causes distortion and mechanical waves. The SAW filter establishes the proper IF frequency, which in turn eliminates IF alignment. Here the SAW filter is located between the IF preamp and the IF processor IC (Fig. 6-11).

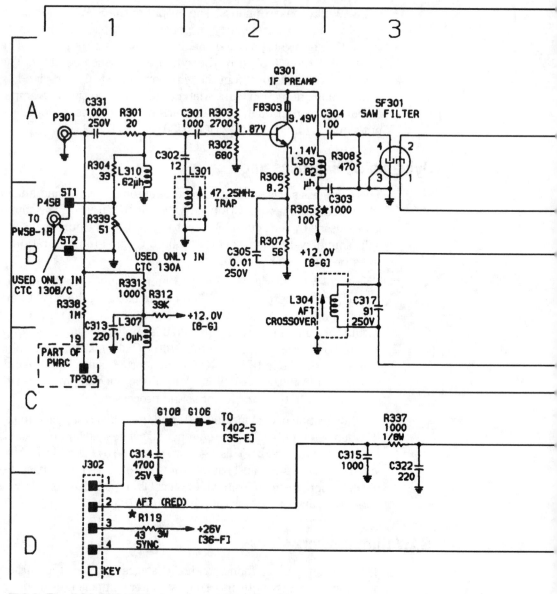

■ 6-9 *The RCA CTC130C chassis consists of an IF preamp transistor (Q301), SAW filter, and IF processor IC (U301).* Thomson Consumer Electronics

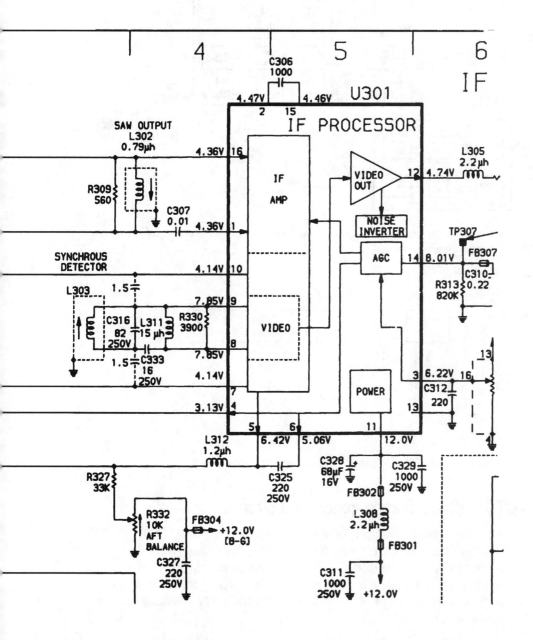

285

SAW filter network

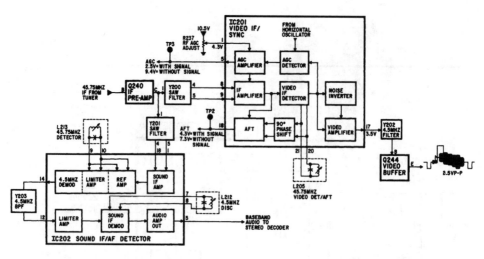

6-10 *In the Sylvania C9 chassis, the IF preamp (Q240), SAW filter, and IF video/sync IC (IC201).* Phillips Consumer Electronics

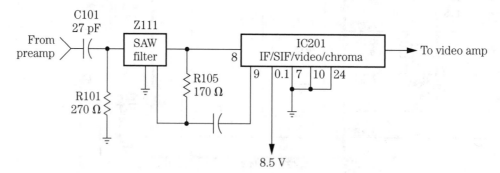

6-11 *The SAW filter (SF301) establishes the correct frequency.*

RCA CTC130C sandcastle generation circuits

The sandcastle waveform must be present at pin 7 of IC U701. If the complete sandcastle waveform is not present there will be no output from the IC (Fig. 6-12). The sandcastle waveform is developed by combining the horizontal blanking, vertical blanking, and burst keying pulses prior to application to the chroma/luminance processing IC.

The burst/clamp keying pulse from Q801 is coupled by CR705 and R710 to the anode terminal of CR706 and CR707. The horizontal blanking pulse at pin 10 of T402 is coupled by R705 and CR707. The vertical banking pulse from pin 23 of U401 is coupled by R704 to the cathode of CR706.

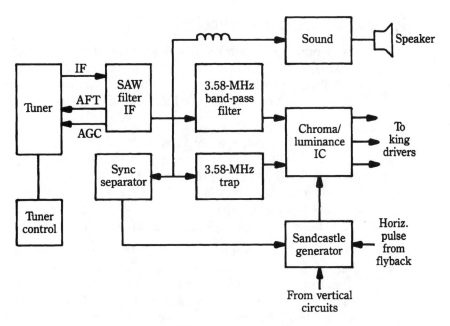

■ **6-12** *Block diagram of the RCA CTC130C sandcastle generation circuits.*

During the horizontal and vertical blanking period these three separate pulses are combined at the cathode of CR706 to form the sandcastle waveform. This sandcastle waveform is coupled by R744 to pin 7 of the chroma/luminance IC (U701). The normal sandcastle waveform can be scoped at the test point (TP806) (Fig. 6-13). The sandcastle generation circuits are found in many recent RCA chassis.

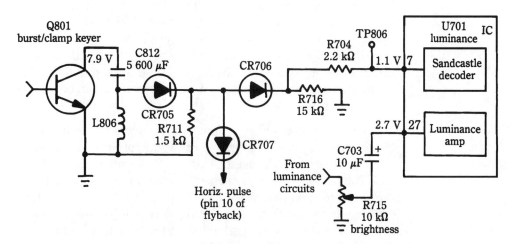

■ **6-13** *The RCA CTC130C chassis sandcastle IC test point (TP806) and luminance IC (U701).*

Sylvania C9 comb filter circuits

The comb filter circuit interfaces with the chroma/luminance IC (IC640). The features of these circuits include the use of a glass delay line to provide improved chroma and luminance signals, luminance delay, and a peaking circuit for the luminance signal (Fig. 6-14).

The composite video signal is applied to the delay line driver (Q600) where the signal is inverted and sent to pin 1 of the delay line (Y600). The in-phase signal is fed through R607 and R600 to be combined with the delay line at pin 4. The 1-H delayed signal has out-of-phase chroma (color) at the wiper of R601, which cancels. Thus, the luminance signals are in phase and added.

The luminance inverter (Q605) and the chroma signal coming from the input at the base of Q600 across R605 is allowed to pass to pin 3 of the chroma/luminance IC (IC640). With the chroma signal canceled at R601, the luminance is passed to the luminance buffer (Q610) for further processing. The high-frequency components of the luminance signal are developed by the circuitry of C606, C602, C607, and the luminance equalizer (Q615). The peaking signal is applied to pin 10 of IC640. Now the luminance signal is passed through delay circuitry to provide the proper timing relationship between the luminance and chroma signals. The delay circuit components are C609, L609, C610, L610, and C611.

IF trouble symptoms

The most common troubles in the IF section are no picture and no sound with white raster; a faint picture and garbled sound, or a smeary picture with multiple ghosts; and a snowy picture with fair sound. Weak or no color can result from a defective IF stage. In earlier tube and solid-state chassis, a "birdie" type sound could be heard with a defective IF stage. Most IF symptoms can be seen on the picture tube.

The same symptoms can be found with defects in the tuner, detector, video, and AGC circuits. A snowy picture can be caused by the first or second IF stage or by a defective tuner and AGC circuit. The faint and weak picture symptoms can occur in the IF picture detector and first video amp (Fig. 6-15). Similar symptoms can be found in the IF circuits. Improper IF AGC voltage can cause a snowy picture or weak garbled sound with a poor picture indicating a defective IF circuit.

288

Before attempting to repair the IF section, make sure the tuner and IF AGC circuits are normal. Substitute the suspected tuner with a tuner-subber. Measure the IF AGC voltage at the IF test points or in the AGC circuits. Clamp the IF AGC voltage with the internal power supply to determine if the AGC circuits are normal. After determining that the IF circuits are defective, locate the correct schematic and voltage chart. The defective IF stage can be located by scope waveforms and voltage measurements.

Troubleshooting the IF stages

A dead or weak IF stage can be located with a strong local station tuned in and scope waveforms taken at each IF stage. The oscilloscope must have a demodulator or detector probe attached. The color-bar generator is clipped to the antenna terminals and the signals are traced in the IF subber signal at the input terminal of the IC, noticing the picture on the screen (Fig. 6-16). Check the scope waveform at the output terminals of the IF section with a strong station or the color-bar generator signal. Often, IF IC terminals are easily located under the chassis with the IF section shielded and covered in a metal box. Accurate voltage measurements on the IC terminals or on the emitter terminal of the IF transistors can locate the open or dead stage.

Troubleshooting SAW filter circuits

Although the SAW filter network does not cause too much trouble, you can check it with the VOM and crystal checker. It's best to remove the SAW filter from the circuit because low-ohm resistors are found in the input and output circuits (Fig. 6-17). The primary leakage can be checked with ohmmeter probes across terminals 3 and 4. Likewise, the output circuit can be checked across terminals 1 and 2. No significant resistance should be measurable across these terminals.

A resistance check does not necessarily indicate the SAW filter is functioning. Check the SAW filter in or out of the circuit with a crystal checker. The crystal checker will determine if the SAW filter is oscillating (Fig. 6-18). Check the input and output terminals in the same manner. The crystal meter will provide a high reading when tested out of the circuit and a lower reading in the circuit. You can build your own crystal checker by following the instructions found in *Build Your Own Test Equipment* (from TAB Books, an imprint of McGraw-Hill).

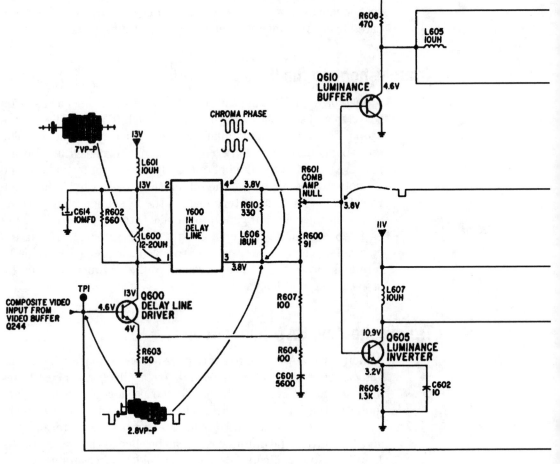

■ 6-14 *The Sylvania C9 comb filter circuits with waveforms.* Phillips Consumer Electronics

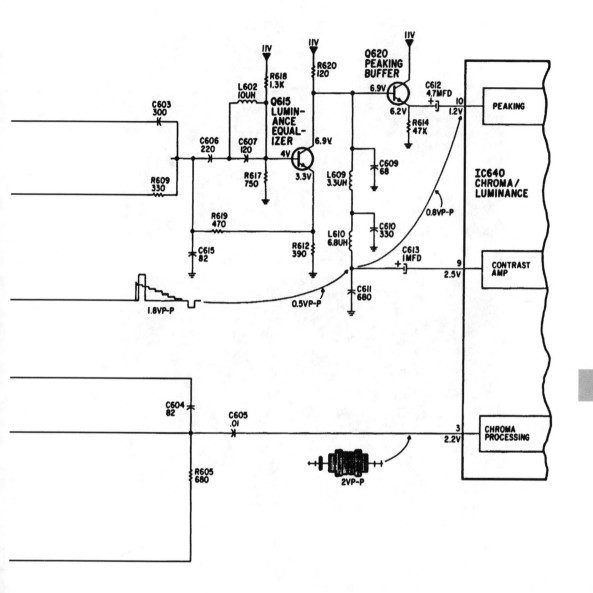

■ **6-15** *The SAW filter network is located close to the IF section.*

■ **6-16** *A tuner-subber can be used to locate a dead or open first or second IF transistor stage.*

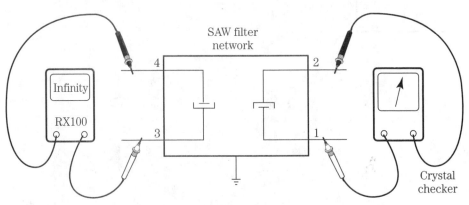

■ **6-17** *Test the SAW filter with resistance and a crystal checker.*

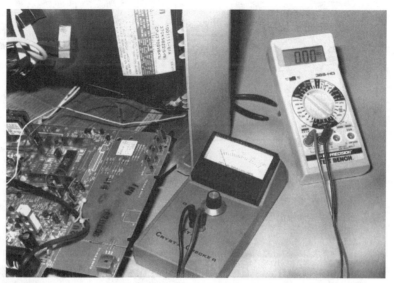

■ 6-18 *The crystal tester can check both input and output terminals of SAW filter.*

Servicing RCA CTC130C IF circuits

To determine if the tuner section or IF stages are defective, connect a tuner-subber to the IF cable. Plug the IF cable into the subber and tune in a local station. If there is no sound or picture with normal raster, suspect problems within the IF stages. When a normal picture is found at the CRT screen, repair the tuning system.

Inject an IF signal at the IF input cable and check for a picture signal at the CRT. If normal, check the tuner, AFT, AGC, and tuner AGC. Scope the video waveform at pin 12 of U301 (Fig. 6-19). If there is no signal, check the waveform at pin 16 of the IC. If there is still no signal, proceed to the collector of the IF preamp (Q301). Although the signal will be quite weak, you can still determine if the IF preamp and tuner circuits are normal.

Improper IF AGC voltage can shut down the IF picture stages. Measure the AGC voltage at TP307. This voltage can vary between 7 and 11 Vdc. If there is no video waveform at pin 12 (U301) or at the base of the video amp, inject AGC bias voltage at TP307 or pin 14 of U301. If the video waveform appears with external bias voltage, suspect a defective AGC system. When the picture does not return with proper AGC voltage, check the voltages in preamp transistor Q301, and look for a shorted IF cable or an open L309, L302, R306, or R307.

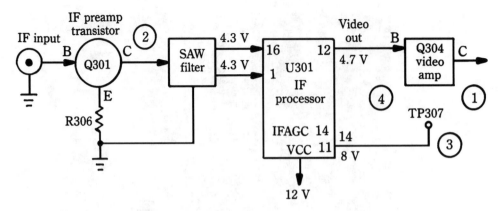

■ 6-19 *Service the RCA CTC130C IF stages by the numbers.*

Check all voltages on pins 1, 3, 11, 12, 14, and 16 of the IF processor (U301). If the IF signal is entering the IF processor and not coming out, suspect the IC. Regular TV broadcast signals can also be used to scope the preamp and U301 circuits with a demodulator probe and scope. Check the SAW filter (SF301) with a crystal checker if in doubt.

RCA CTC157 IF circuits

The IF output signal is taken from the tuner to the IF preamplifier (Q2301) and the output is applied to a SAW filter network (SF2301). The IF signal from the SAW filter is applied directly to the CTV processing IC (U1001). Inside U1001, the signal is amplified and passed on to a video detector circuit (Fig. 6-20). Both AFT and AGC voltages are developed internally by U1001. The AFT voltage is routed back to the A1U control (U3300) in the system control circuit. This voltage determines if the tuner frequency is correct for a given channel by the A1U system. The AGC voltage is applied to the tuner to control tuner amplifier gain.

Troubleshooting Sylvania C9 comb filter circuits

When troubleshooting the C9 comb filter circuits with the scope, check the composite video input at TP1 (Refer to Fig. 6-14). Scope pins 1, 3, and 4 of the delay line (Y600). Scope the waveform at the base of Q610 and at pin 10 of chroma/luminance IC640. Measure the voltages on the comb filter transistor and IC640. Two different voltages are required to operate the comb filter circuits (11 and 13 Vdc). These voltages are derived from the switched-mode power supply (SMPS). If they are missing, check and repair the SMPS.

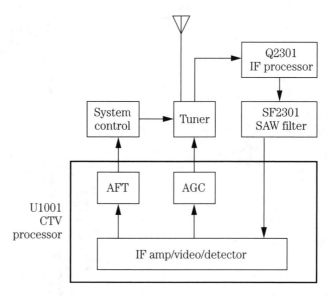

■ **6-20** *A block diagram of RCA CTC157 IF circuits.*

RCA CTC130C luminance circuits

Most present-day TV chassis incorporate the luminance and chroma processing in one large IC component. The CTC130 chassis uses a combined chroma/luminance processing IC (U701) that provides all the processing for both luminance (brightness) and chroma (color). Combining the two signals in one processing IC provides proper color and contrast tracking in one IC component (Fig. 6-21).

Luminance signal is applied to pin 27 of IC U701. The signal is amplified and applied to the chroma/luminance matrix amplifiers in the IC. The contrast is controlled with a variable voltage at pin 26 of U701. The output of the picture control amp is applied to the luminance amplifier providing color and contrast tracking. The picture control amp is also controlled by the beam limiting circuit to provide contrast tracking with beam limiting. The beam limiter circuit is controlled by the high-voltage resupply line to the IHVT and the brightness (black level) control.

Brightness is set by comparing the setting of the brightness control at pin 24 to the level of the blue blanking output signal. During horizontal retrace, the brightness dc voltage and the luster of the blue output blanking signal are compared with a resultant voltage that controls the brightness of the luminance amplifier. This maintains a consistent brightness that depends on the incoming blanking levels.

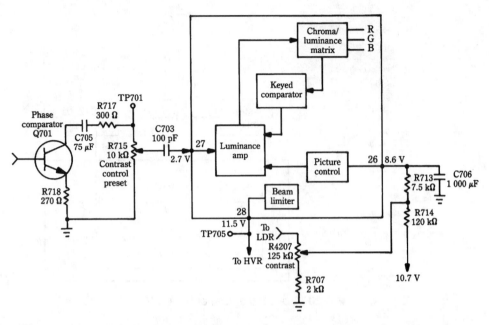

■ 6-21 *Take critical voltage and waveform tests at TP701, TP705, and pins 26 and 27 of luminance IC U701 in the RCA CTC130C chassis.*

Brightness limiter operation is accomplished by sampling the dc beam current to the high-voltage windings of the IHVT through a resistor that is connected to pin 28 of IC U701. The voltage at pin 28 is approximately 12 V with a normal brightness picture. When the beam current increases, the voltage at pin 28 drops. When the dc voltage drops below 12 V, the beam limiter circuit conducts, reducing the beam current by acting on the luminance amplifier in the IC.

IF or tuner?

The no-picture, no-sound symptom can occur if the tuner or IF video components are defective. Connect a tuner-subber to the IF cable. If the picture is normal with the subber, suspect the tuning system. If there is still no picture, troubleshoot the IF and video circuits. The tuner-subber can be used for signal injection to the IF IC.

Tuner-subbers were used constantly years ago, and there were several different manufacturers. Today, it's difficult to locate a tuner-subber. You can make your own out of a transistorized VHF tuner and power source. How to build a tuner-subber is described in *Build Your Own Test Equipment*, (from TAB Books, an imprint of McGraw-Hill). The tuner-subber can save a lot of valuable service time, which means money!

Snowy pictures

Besides the tuner, check the first and second IF stages for a snowy picture. Measure the IF AGC voltage without signal to the tuner. Improper AGC voltage can produce a snowy picture. Take accurate voltages on the emitter and collector terminals. Check the dc voltage source fed to the IF section for low voltage. If possible, make an in-circuit test of each transistor with the DMM. The snowy picture can be caused by an open or leaky IF transistor.

In an RCA CTC44W chassis, all VHF stations were snowy. The local UHF station picture contained a little snow. A quick in-circuit test of each transistor located an open second IF transistor. Besides transistor tests, accurate voltage measurement can locate that leaky or open IF transistor.

Brightness, no picture, garbled sound

A smooth white raster with not a ripple of snow or noise is caused by the IF stages. Usually, a rushing noise in the sound and some channel noise is seen without a station tuned in. Garbled sound indicates poor sound takeoff at the IF stages. The dead IF stage can be caused by an open or leaky IF transistor or IC.

The sound was garbled, with a white raster in a J.C. Penney 685-2124. The IF stages were signal-traced with the scope and demodulator probe. Video signal was found at the base and none at the collector terminal of the third IF video transistor (Fig. 6-22). Zero voltage was found at the emitter terminal, indicating that Q103 or R136 was open. The collector voltage was high, at 20 V. Q103 was found to be open and was replaced with an SK3018 universal transistor replacement.

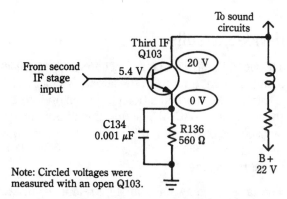

■ 6-22 *Q103 produced a white raster with garbled sound in a J.C. Penney 685-2124.*

AGC or video

A pulling, washed-out picture can be caused by a defective AGC or video IF circuit. Often the picture is out-of-sync, and the AGC control has no effect on the picture. First, check the AGC voltage at the base of the first and second IF transistors. If it is normal, the AGC circuits are working.

Voltage measurements in the IF stages of a J.C. Penney 2857 found low voltage at the collector terminal of the second IF transistor (Fig. 6-23). A forward-bias voltage between the emitter and base terminals indicated the transistor was normal, but the transistor was double-checked with an in-circuit transistor test. A resistance measurement from the collector to ground was 939 Ω. Capacitor C114 was found to be leaky to T104. Removing and replacing C114 solved the AGC and video look-alike picture.

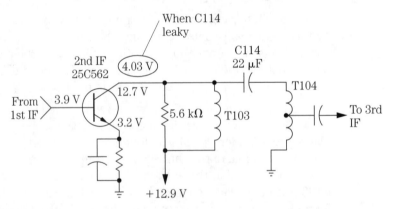

■ **6-23** *Leaky C114 caused a pulling, washed-out picture in a J.C. Penney 2857.*

Latest IF video circuits

The video IF circuit may be included in a single large IC that covers many different circuits in the latest TV chassis. IC1001 contains the IF/LUMA/chroma/deflection circuits in one large processor. Besides these circuits, AGC/AFT/sound IF and preamp circuits are found in one section of the 64-pin IC. This processor IC, when found defective, can be replaced with an ECG1790, SK9850, or NTE1790 universal replacement.

The large IC may have an IF preamp transistor and SAW filter within the input circuits, between tuner and IC1001 (Fig. 6-24). RF and IF AGC circuits are included with picture IF input pin 21,

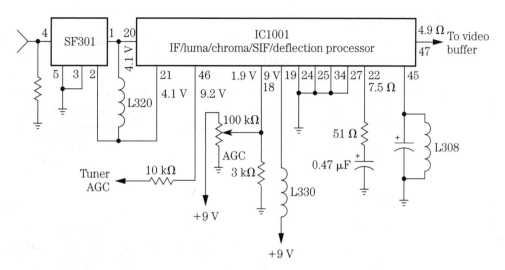

■ 6-24 *The latest IF circuits are included in one large IC with luma, chroma, SIF, deflection, horizontal, and vertical circuits.*

RF AGC out pin 46, RF AGC delay pin 18, picture AGC pin 22, RF supply voltage pin of 9 volts (19), PIF AGC TC pin 23, video detector tank pin 45, AFT output 42, and pin 47 of video output.

To troubleshoot, inject a video signal at the IF input and check for video on the face of the picture tube. If there is no video at the CRT, check the video waveform at the base of the video buffer transistor. When video is missing at the base, inject AGC bias to pin 22 of IC1001. Check all voltages, waveforms, and components upon IC1001, pins 18 through 47. Connect a bar sweep generator to the antenna terminals and check for correct waveforms in IF video circuits.

IF alignment

IF alignment should be checked after all circuits are found to be normal in the IF stages when there are faint, smeary, fluttering, or oscillating pictures. Do not try to adjust IF cores or capacitors to improve a weak or faint picture. These adjustments do not change unless someone turns them. IF alignment should be done with a sweep-marker generator. If you do not have the correct alignment test equipment, it is best to take the chassis to the manufacturer or to a qualified TV alignment technician. Most alignment generators collect dust on the shelf because they are seldom used in TV servicing. IF alignment is not needed in the latest TV chassis with a SAW filter network.

Video test equipment

The video stages can be serviced with signal tracing of IF stages with the oscilloscope. Connect a color-dot-bar generator to the TV set antenna terminals for signal tracing. Take in-circuit transistor tests of IF and video circuits. Check critical voltages with a DMM. A video or audio square-wave generator can be used to locate the stages causing a smear or oscillating picture. All IF and video signals are taken with the color dot-bar generator and scope, according to Howard Sams schematics.

Video problems

A weak or a washed-out picture can be caused by a defective video stage. Very little or no control of brightness can occur in the video circuits. Heavy scanning or retrace lines in the picture are caused by the video circuits or by improper voltages at the picture tube. Most smeary picture symptoms are caused by open peaking coils and leaky video transistors in the video circuits. A defective video output stage can cause a no-picture, no-brightness symptom.

Some video symptoms can be caused by the picture tube circuits. Determine if the picture tube circuits are at fault by taking critical voltage measurements on the CRT terminals or corresponding circuits. With a low or insufficient brightness symptom, the service/normal switch can be switched to service to determine if the CRT is weak. Rotate each screen grid control to form a white line. If a bright line is noted, assume the CRT and circuits are normal. The same method can be used on TV chassis without the service switch. Simply turn up the screen or sub brightness control and notice if the picture tube responds.

Video IC circuits

With the early solid-state chassis, most video circuits contained only transistors. Today, video circuits can be incorporated in one IC with AGC sync and color circuits (Fig. 6-25). Because fewer components are found in the IC video circuits, they are easier to service. Simply scope the signal in and out of the IC component. Take critical voltage measurements if the video signal goes in and not out of the IC terminals. Check each component tied to the IC video terminals before replacing the suspected IC.

■ **6-25** *Check for one large IC with most all the TV functions in one component.*

Comb filter circuits

A new video circuit called the comb filter is found between the IF and luminance circuits in the latest color TV chassis (Fig. 6-26). Its purpose is to separate the luminance and chroma video information to eliminate color bleeding or cross color from the picture. There are many additional circuits provided with the comb filter. The comb filter circuits are developed with an IC processor.

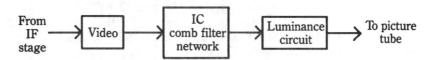

■ **6-26** *The comb filter is located between video and luminance circuits in the latest TV chassis.*

No video, normal sound

A normal sound and white raster symptom without any picture occurs with a video circuit problem. The video circuits can be signal traced with a tuned station signal and scope or color-bar generator and scope. Start at the first video amp with the scope probe and check the signal at the base and collector terminal of each video transistor until the signal is lost. Take critical voltage measurements on the suspected transistor or IC component.

Sharp 9B12

The sound was normal with no video in a Sharp 9B12 (Fig. 6-27). Because the sound was normal, the video signal was scoped after the sound takeoff point. The scoped signal stopped at the first video amp (Q202). A quick voltage check showed no positive voltage at the base and collector circuits. Although the 12-V supply source was normal, voltage was not applied to the vertical circuits. On closer inspection, the PC board was found to be cracked behind the balance control. No doubt the TV set had been dropped on the knobs, cracking the board wiring. Solid hookup wire soldered across the cracked wiring brought back the picture.

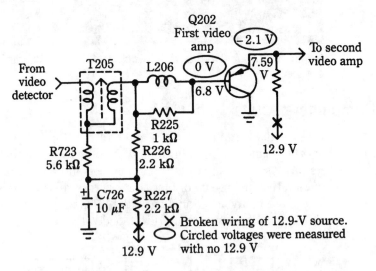

6-27 *A broken foil wire in a Sharp 9B12 chassis caused improper voltages on Q202.*

No video, no raster

Check for high voltage at the CRT anode connection with a no-video, no-raster symptom. If the high voltage is normal, suspect video problems at the last video stage or the CRT and circuits. Next check all voltages on the CRT terminals. Normal CRT voltages indicate a defective video output circuit.

The high voltage was normal with no video or raster in an RCA CTC86D chassis (Fig. 6-28). When the CRT voltages were measured, very low screen voltage was found. R435 and R441 were running red hot. A quick resistance measurement found 546-Ω leakage on the 200-V source. After disconnecting the screen circuits from the 200-V source, other circuits tied to the same voltage source

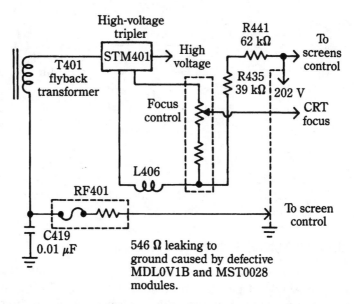

■ **6-28** *Check voltages at the picture tube socket to correct a no-video/no-raster symptom.*

were checked. Both the luminance/sync module (MDL001B) and tuner module (MST002B) were replaced to bring back the screen grid voltage source. Remember, more than one module can be defective in a modular chassis.

Weak or washed-out picture

A weak and/or negative picture can be caused by an open peaking coil, leaky video transistor, or defective picture tube. Check the picture tube with a good CRT tester. Very light and weak pictures with normal sound can be caused by leaky or open video transistors and IC components. Suspect a leaky delay line for a washed-out picture with poor contrast. A washed-out picture with retrace lines can be caused by an open delay line.

Weak picture, red outline

No picture was noticed until the color control was turned up in a J.C. Penney 19YC chassis (Fig. 6-29). The picture would disappear when the color was turned down. The video signal was scoped to the base terminal of the third video amp, and no signal was found at the collector. Q325 tested normal in the circuit with a DMM. When the fourth video amp was tested in the circuit, Q330 was found to be open. Replacing Q330 with a GE-18 universal transistor solved the very-weak-picture symptom.

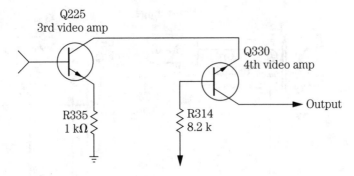

■ **6-29** *An open Q330 caused a weak picture in this J.C. Penney 19YC chassis.*

Washed-out picture, retrace lines

Video problems were suspected in a Panasonic CT-994 with a washed-out picture and retrace lines (Fig. 6-30). The video signal was found at the collector terminal of the first video amp (TR301) and not at the base of the second video amp (TR302). Continuity resistance measurements of L301, L302, and L303 indicated an open in the delay line. Repairing the broken lead connection at one end of the delay line cured the washed-out picture symptom.

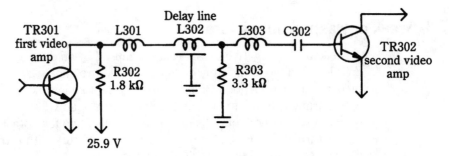

■ **6-30** *An open delay line in a Panasonic CT-994 caused a washed-out picture with heavy retrace lines.*

Loss of picture and sound

Sometimes the picture would go, and then the sound. At other times, the color would be missing in a CTC111 RCA chassis. Again, all of the above symptoms would occur or only one of them (Fig. 6-31). Suspect a leaky coupling capacitor (C611) from the video buffer transistor to the comb processor IC (U600).

304

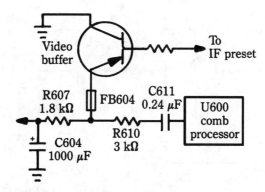

■ 6-31 *A leaky C611 caused a loss of color and picture in an RCA CTC111 chassis.*

Intermittent video

The sound was normal with an intermittent picture in a Montgomery Ward WG-17196AB (Fig. 6-32). Sometimes just touching the chassis would cause the raster to go black. The picture would really act up when the video output transistor (Q312) was touched. Replacement of transistor Q312 with a GE-251 universal replacement and replacement of resistor R370 solved the intermittent problem.

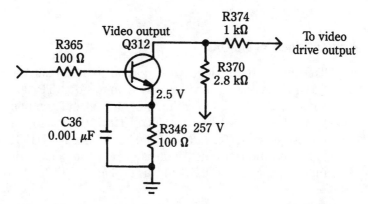

■ 6-32 *An intermittent Q312 in a Wards 17196AB chassis caused an intermittent picture with normal sound.*

Intermittent black screen

The picture would "pop" in and out of a J.C. Penney 685-2014 just about every minute. The video was traced to the emitter terminal of Q225 (Fig. 6-33) and not at the video driver transistor (Q404).

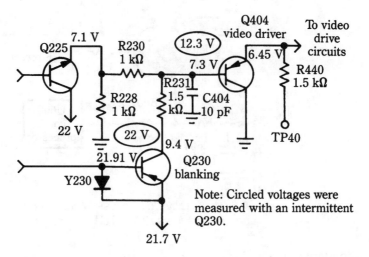

6-33 *High base voltage at Q404 was caused by a leaky and intermittent picture (Q230) in a J.C. Penney 685-2014 chassis.*

In the intermittent mode, the base voltage of Q404 would increase. At first, Q404 was suspected of open conditions. A quick in-circuit test indicated Q404 was normal. Very high voltage (22 V) was found at the collector terminal of the blanking transistor (Q230). The leaky blanking transistor was replaced with an ECGI59 universal replacement.

Intermittent video, audio OK

The intermittent video with normal audio occurred in an RCA CTC146 chassis. The video was monitored at TP2307. The video signal was normal, indicating the intermittent component was beyond the contrast preset control (R2716). When the video was monitored at pin 52 and 53 of the luminance IC (U1001), the video appeared intermittent. Upon checking the schematic, the intermittent delay line (DL2701) was replaced (Fig. 6-34).

Fuzzy pictures

You could tap around the tuner or IF amp transistor (Q2300) in an RCA CTC140 chassis, and the picture would intermittently appear fuzzy. In one model, poor pin connections were found at the IF amp transistor (Q2300). In another CTC140 chassis, poor soldering of the pins of U2300 caused the same symptom. It's wise to solder all the pins on U2300 and Q2300 for intermittent fuzzy pictures.

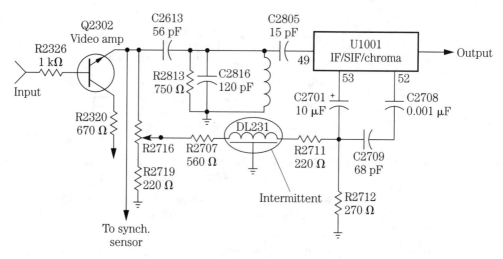

■ 6-34 *Check the delay line (DL231) for intermittent video symptom.*

Very little brightness

Check the video output stage, CRT, or picture tube circuits when the symptom of very little brightness occurs. Turn up the screen controls and service switch to determine if the picture tube circuits are normal. Suspect a defective boost voltage or CRT when the brightness cannot be turned up with the screen or color output controls. If the bias or screen controls raise the brightness with retrace lines, check the video output circuits. Check the automatic limiter brightness transistors and circuits. Do not overlook an open brightness control.

No brightness, normal high voltage

Lack of brightness with retrace lines was noted in a K-Mart SKC1970. These occurred with brightness and screen controls wide open (Fig. 6-35). Although the CRT tested weak, the no-brightness problem must have originated in the video circuits. A quick signal and voltage check of the second video amp (Q300) indicated an open transistor. Although Q300 tested normal out of the circuit, the transistor was replaced with an SK3114 universal transistor. The voltage and brightness conditions were the same. Resistance measurements found R308 was open.

Can't turn down the brightness

When the brightness cannot be turned down, suspect trouble in the picture tube circuits. The picture tube is conducting too much and

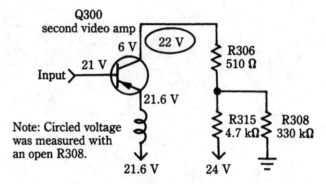

6-35 *An open R308 in a K-Mart portable produced a black screen with normal high voltage.*

cannot be cut off with the controls. Check for missing boost, screen, and drive voltage at the CRT. Heavy retrace lines with no control of brightness can develop with a defective normal/service switch. Suspect a defective automatic brightness limiter (ABL) when the brightness cannot be turned down. If the picture tube circuits are okay, check the video output circuits where the brightness control is located. Look for a sub-brightness control adjustment.

The picture tube circuits can be double-checked with the service switch and screen controls. With many brightness control problems, the screen controls cannot be lowered far enough to turn out the brightness. Measure the boost voltage at the screen and picture tube circuits for missing or high boost voltage.

Brightness with retrace lines

The brightness could not be turned down in a Goldstar CR401 with retrace lines in the raster (Fig. 6-36). Only 48 V were found on the cathode elements of the CRT (pins 3, 8, and 12). The normal voltage is around 144 V. R521 in the flyback boost voltage source was open. After testing D410 for leakage, R521 was replaced, restoring the boost voltage.

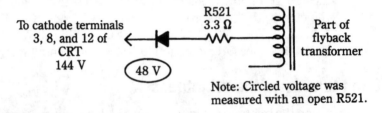

6-36 *An open R521 caused the brightness to be fixed at maximum in a Goldstar CR401.*

In another similar symptom, the brightness or screen control had no effect in an RCA CTC92A chassis (Fig. 6-37). With the service switch at service position, a very bright-white vertical line could not be turned down. Voltage and resistance measurements found R102 was open in the ground leg of the screen control.

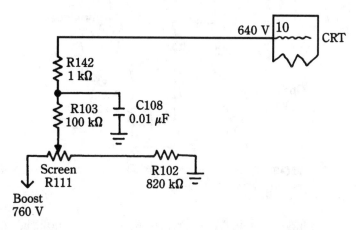

■ 6-37 *No brightness or screen control was found in this RCA CTC92A chassis.*

Cannot turn down brightness

Heavy retrace lines and no brightness control were found in an RCA CTC90D chassis. The video and other modules were replaced, with no change in the symptom. All voltages were normal on the picture tube socket. Correct waveforms were entering the luminance/sync module MDL002A. Voltage checks on the luminance/sync module indicated improper voltage at pin 17 (Fig. 6-38). The voltage remained at pin 17 when R4202 was rotated. Resistance measurements of the brightness control and R310 were normal. A continuity check between R309 and the service switch indicated S301 was open. An open or corroded service switch can cause many different brightness and picture problems.

No brightness control

Check the brightness and sub-brightness control adjustments when no brightness control is found in the raster. Improper brightness control can be located with a defective component in the automatic brightness limiter (ABL) circuits. A broken or open raster service switch can produce uncontrollable brightness. A very bright raster with chassis shutdown can be caused in the lumi-

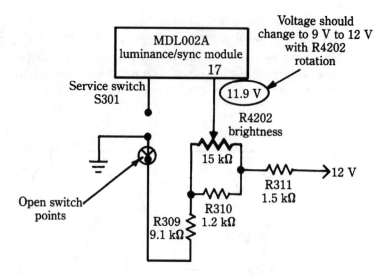

6-38 *Poor condition of the service switch terminals resulted in no brightness control.*

nance IC. A leaky luminance IC can cause no brightness control. Check for leaky transistors or capacitors in the video circuits.

No control of the luminance IC

The brightness and contrast controls had no effect on the raster in an RCA CTC109 chassis. Critical voltage measurements on U701 turned up low voltage at pin 26 (Fig. 6-39). All resistors checked normal. A resistance of 768 Ω was found between pin 26 and the chassis ground. One lead of C710 was removed from the board. No leakage was found. When pin 26 was unsoldered from the PC wiring, 768 Ω of resistance was measured from IC pin 26 to ground. Replacing leaky U701 with original part number 146858 cured the no-brightness control problem.

Video IC replacement

After locating the defective video or luminance IC, extreme care must be exercised in removing and replacing the IC component. Excessive solder can be removed from the wiring side with a solder gun and solder wick material. Be careful not to loosen tie points or component leads next to the IC terminals. If they are loosened, resolder them back into the circuit. Too much heat can pop or buckle the PC wiring.

Use a low-wattage or battery-operated soldering iron to solder each terminal of the new IC chip (Fig. 6-40). Small-diameter solder is

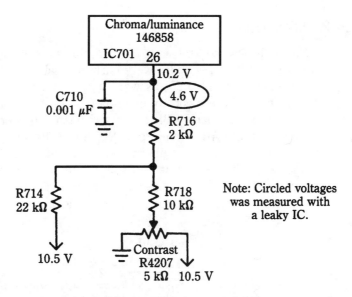

■ **6-39** *A leaky IC701 caused poor brightness control in this RCA CTC109 chassis.*

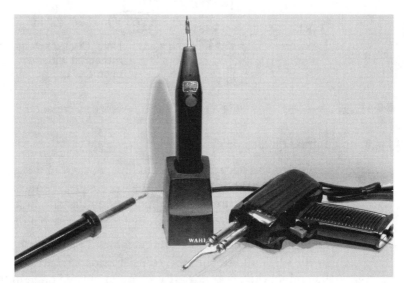

■ **6-40** *Choose a battery-operated soldering iron for transistor or IC connections.*

ideal so that excess solder does not flow between the pin terminals. Make a good contact, but do not leave the iron on one pin too long. Place solder on both sides of the pin area. Clean out excess solder or paste between pin contacts with the back edge of a pocket knife to prevent excess solder from touching the adjoining pin terminal.

Very low brightness

No brightness control with very low brightness was found in a Sharp 19A63. All voltages were fairly normal on the picture-tube elements. The video stages were normal, except the brightness and sub-brightness control had no effect on the raster. A voltage measurement in the ABL circuits indicated higher-than-normal voltage on the ABL transistors (Fig. 6-41). Because the voltages were about the same on all transistor terminals, the ABL transistor was tested in the circuit and was found to be good. R431 was found to be open in the base circuits, causing high voltage at the collector terminal with no brightness control.

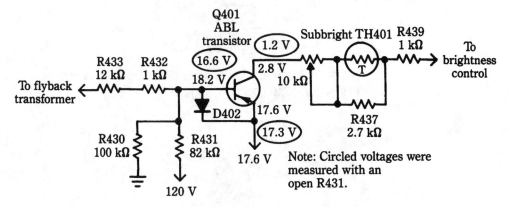

■ **6-41** *Very low brightness in a Sharp 19A63 chassis was caused by open R431.*

Faint, no contrast picture

The picture was very faint, without any contrast in a Montgomery Ward GEN-12907A portable. There were some overscan lines at the top of the picture without any function of the brightness or contrast controls. No video was noticed at the video amp.

After carefully checking over the video and picture tube circuits, high voltage found on the ABL transistor focused attention on the ABL circuits (Fig. 6-42). R431 was found burned open in the base circuit of the brightness limiter transistor (Q401). Still, the picture and brightness was very faint. All resistors were checked in the brightness circuits. A short was found between R433 and ground. C614 in the flyback circuit was directly shorted, producing higher-than-normal voltage on Q401, which in turn left a faint and uncontrollable picture. The loss of brightness was not directly in the video circuits, but in the circuits controlling it.

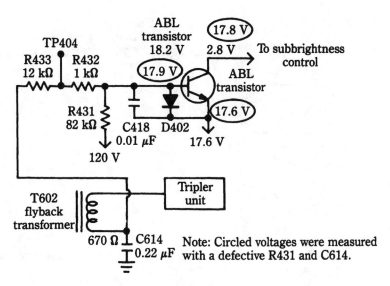

■ **6-42** *An open R431 in a Wards GEN-12907A portable caused a faint picture with little contrast.*

Intermittent brightness

Intermittent brightness can occur in the video circuits, ABL, picture tube, and with the voltage source feeding these circuits. Monitor the dc and boost voltage source for intermittent voltage. Check the brightness, sub-brightness, and screen controls for erratic operation. Scope the video circuits when the chassis is in the intermittent mode. Voltage and coolant tests can locate an intermittent transistor or IC circuit. Check the delay line for an open or intermittent connection. Do not overlook possible intermittent peaking coils between the video transistors or delay lines.

No brightness after one hour

The intermittent brightness problem would appear after the chassis was on at least one hour in a K-Mart 1320 portable TV. Video signals were monitored at the input terminal of video IC601 (Fig. 6-43). The video would disappear after the TV was on for awhile, and it would stay out. A signal check at the collector terminal of the video amp (Q301) was normal. The delay line (TD301) was found to be open between Q301 and IC601. Video scope waveform tests with low resistance measurements located the intermittent delay line.

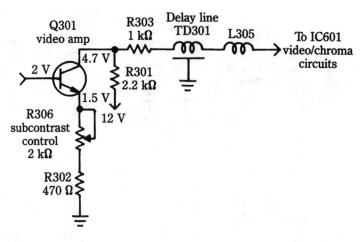

6-43 *An open TD301 delay line caused intermittent brightness in a K-Mart portable.*

Very bright screen and shutdown

A defective picture tube or shorted spark-gap assembly can cause the raster to come on very bright without a picture, and shut down after a few seconds. Often the brightness control has no effect on the raster. Remove the high-voltage anode connection. Isolate the high-voltage lead away from the TV chassis. If the chassis comes up and does not shut down, replace the shorted picture tube.

Sometimes the chassis will shut down before any voltages can be measured. Remove the picture tube socket and anode connection to determine if the video circuits are causing the shutdown. Check the video circuit if the chassis does not shut down with the CRT socket removed. Take voltage measurements on the video output transistor or IC luminance components.

Brightness shutdown in an RCA CTC109 chassis

The brightness shutdown problem was isolated to the chroma/luminance IC700 circuit. Critical voltage measurements on pin 26 of U700 were fairly low (Fig. 6-44). Either U700 or some component tied to pin 26 was leaky. A resistance measurement of 1.5 kΩ was found between pin 26 and chassis ground. Pin 26 was removed from the PC wiring with no resistance between the pin and chassis. This indicated IC700 might be good. A 1.2-kΩ leakage was found across the terminal of CR705. Replacing CR705 solved the very bright raster shutdown problem.

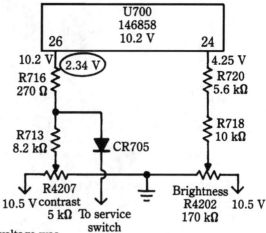

Note: Circled voltage was
measured with a leaky CR705.

■ **6-44** *A leaky diode (CR705) caused chassis shutdown in an RCA CTC109 chassis.*

Smeary pictures

Leaky video transistors and open peaking coils in the video circuits cause most smeary picture symptoms. Improper universal video transistor replacement can produce a smeary picture. Substitute a video transistor of another brand, and see if it clears up the video problem. If not, try to locate the original. A defective video IC luminance component can cause a smeary picture.

An unusual video problem

Heavy retrace lines with no control over the brightness occurred in a J.C. Penney 2039. Sometimes small firing lines could be seen across the raster, with no sound. The tuner control system was dead. The tuner control module was replaced, with no change. Because the tuner would operate with no sound, no picture, and only retrace lines on the raster, the defective component must be tied to all circuits (such as a dc power source).

Voltage measurements were taken on the tuner terminals, indicating low voltage and no negative voltage or waveform to the tuner memory module. Tracing these connections from the tuner module to the flyback transformer indicated a low 24-V source (Fig. 6-45). Only 13.6 V was found at this point. The low-voltage rectifier and isolation resistor (R415) were normal.

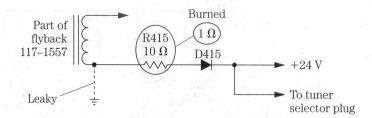

6-45 *Heavy retrace lines were caused by a shorted winding to core in a J.C. Penney 2039 chassis.*

Either some component was loading down the 24-V source, or the flyback winding was defective. The voltage came up a few volts with the load removed from the 24-V source, but it still was not up to normal. A resistance measurement of the flyback winding is difficult to take, but continuity was good. After several hours of frustration and going over tuner voltages once again, the trouble was located in the flyback winding. The continuity ohm measurement of the flyback winding was normal, but inside the transformer, the winding was grounded to the core of the transformer. Replacing the flyback transformer (117–1557) solved the unusual video problem.

Five actual video case histories

The following are five real problems related to the video raster.

Intermittent video—normal sound

The picture would pop in and out of the screen in a Toshiba CT37C portable. Sometimes the TV chassis would operate properly for hours. A dot-bar generator was connected to the antenna terminals and the video stages were monitored with the scope at TP12. Very little video waveform was found at the emitter terminal of video amp Q201 when the set was intermittent. Coolant and heat applied to Q201 pointed out an intermittent transistor. Q201 (2SC1815Y) was replaced with an ECG85 universal replacement (Fig. 6-46).

Smeary picture—normal sound

The picture was smeary on all channels in a Goldstar NC-07X1 chassis. Because the sound was good, no doubt the defective component was in the video output circuit. Waveform tests were made in the comb filter and video amp stages. These waveforms were fairly normal up to Q205. A voltage check on collector of Q205 was

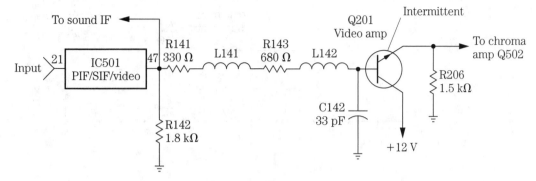

■ **6-46** *Q201 in a Toshiba CT37C portable caused intermittent video.*

low, and should equal the voltage source of 11.2 V (Fig. 6-47). Q205 was tested for leakage and was good. L202 was found to be open, lowering the dc collector voltage.

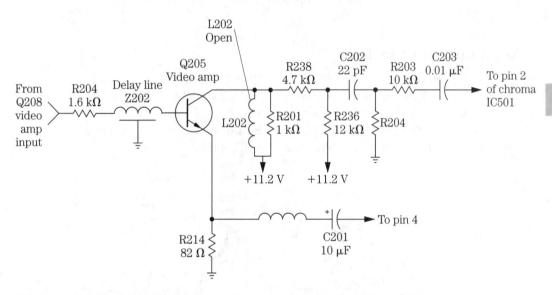

■ **6-47** *An open coil L202 produced a smeary picture in a Goldstar NC-07X1 chassis.*

Black lines across screen

The black lines streaking across the screen would come and go in a Sanyo AVM255 TV. The dot-bar-color generator was connected to the antenna terminals and waveforms were taken of the video circuits. The signal was normal up to the buffer amp Q313 and very weak on the base of video amp Q315. The delay line (L304) was checked and seemed to be open. L304 was replaced with exact replacement part number LG0005KH (Fig. 6-48).

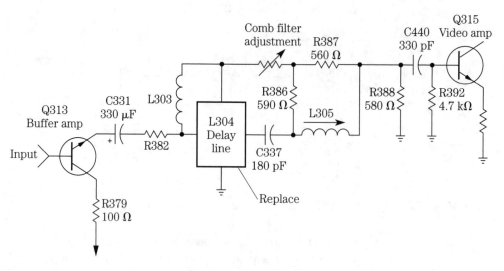

■ **6-48** *A defective delay line (L304) in a Sanyo AVM255 TV caused streaking black lines across the screen.*

Negative picture, retrace lines

The picture had a green tint to it with a faint, negative picture in the background and heavy retrace lines in an RCA CTC93E chassis (Fig. 6-49). Both brightness and contrast controls were wide open, with very little effect on the picture. When the color control was turned down, the faint picture went out.

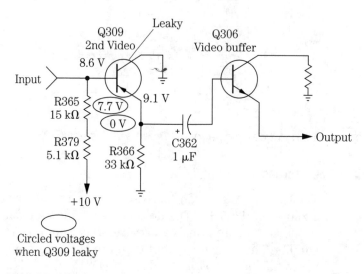

■ **6-49** *A leaky transistor Q309 caused a negative picture with retrace lines.*

Video signal was traced with the scope to the base of the second video amp (Q309). No signal was found on the emitter terminal. Voltage measurements indicated low base voltage and zero emitter voltage. The 27.5-V source feeding R366 was normal. Q309 indicated leakage when tested in the circuit with the diode transistor test of the DMM. After Q309 was removed from the circuit, a 78-Ω leakage was found between the collector and emitter terminals. Q309 was replaced with an original replacement part.

No brightness control

The brightness control had no effect on the raster in an RCA CTC111L chassis with normal sound (Fig. 6-50). The brightness reference transistor (Q703) voltage is used to control the gain of luminance IC701 to maintain a consistent brightness level. The voltages on Q703 were changing, and the low voltages were difficult to measure accurately. R702K was found to be open from base terminal to ground. The resistor network of R702 was replaced with the original replacement part because it contains many resistors in one component.

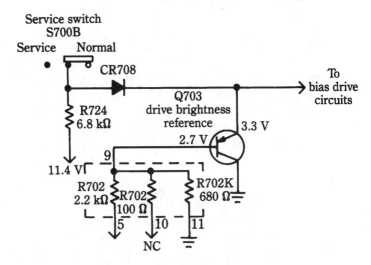

■ 6-50 *Open R702 in the base of Q703 resulted in no brightness control in an RCA CTC111L chassis.*

Picture-in-a-picture

You might be able to see more than one picture on the CRT at the same time with the special feature of picture-in-a-picture. The picture process allows you to have the large picture with several small pictures within the larger picture. You can zoom the small picture size, freeze, swap, or move it in some TV chassis.

When the picture is selected, the small picture appears in the lower right-hand corner of the screen (Fig. 6-51). The screen displays the big picture, with the small picture to the right. The small picture can be moved to just about any position on the screen by using four different arrow keys or buttons. The small picture can move until the arrow button is released. Of course, the sound is transmitted for the larger, original picture on the screen.

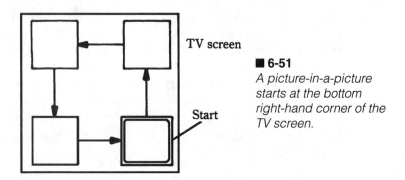

TV screen

■ 6-51
A picture-in-a-picture starts at the bottom right-hand corner of the TV screen.

Start

There are many different functions found with the picture-in-a-picture process. The picture can be moved with the move button and exchanged with the big picture by pressing the swap key. The freeze button can still either the small or large picture. Zoom in increases the size of the small picture each time the zoom-in button is pressed. The zoom-out button decreases the size of the small picture. Pan, strobe, multichannel, and special-effects modes can be found in some picture-in-a-picture models.

The picture-in-a-picture module is found alongside the TV set in the RCA CTC140 chassis. The picture module plugs directly into the TV chassis and uses voltage from the power supply and flyback circuit for operation. Both horizontal and vertical sync is applied to the picture-in-a-picture module from the TV chassis.

The TV and CAV video is selected with electronic switching to the picture-in-a-picture processor. Also, the luminance and color (Y and C) are selected to the input of an R-Y and B-Y decoder (Fig. 6-52). The R-Y and B-Y signal is applied to the picture-in-a-picture processor IC. A burst oscillator provides a continuous 3.58-MHz signal to the encoder and picture-in-a-picture processor. This burst oscillator is locked to the big picture chroma signal. The picture-in-a-picture

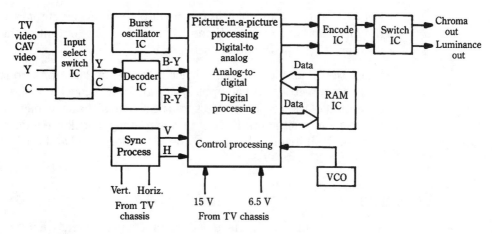

6-52 *A block diagram of the picture-in-a-picture processing circuits.*

processor uses the 3.58-MHz signal during multichannel mode to phase-lock its internal oscillator.

The horizontal and vertical sync outputs are applied to the picture-in-a-picture processor. The horizontal and vertical sync is locked to the composite sync output of the decoder stage. The picture-in-a-picture processor IC has a voltage-controlled 20-MHz oscillator (VCO) to synchronize internal timing with external horizontal sync signals.

Analog-to-digital conversion occurs in the picture-in-a-picture processor and converts analog luminance, B-Y, and R-Y signals into digital information. The digital video information is stored in the RAM IC. This digital information can be operated by the digital processing circuits.

The R-Y and B-Y and Y/C signal are fed through an encoder IC to a fast switch IC and then applied to the TV chassis. The signal information is connected to the on-screen display or line drive circuits of the picture tube.

To troubleshoot the picture-in-a-picture circuit, check for correct voltage applied to the picture-in-a-picture module. Some chassis have an internal power supply and regulator circuits. For a no-big-*or*-small picture symptom, check the input TV and CAV input circuit signal. Check for luminance (Y) and chroma (C) signals at the input stage. Signal trace the composite and component video signals throughout the input for proper switching.

For a symptom of no color in the small picture, check chroma at the input and check the B-Y and R-Y signals at the output of the decoder IC. Scope the burst oscillator IC.

When the small picture rolls vertically while the large picture is normal, scope the vertical input sync signal. Also check the vertical output of the vertical processor sync signal. If the small picture takes some time before locking in horizontally, check the adjustment of the horizontal frequency coil of the horizontal and vertical processor IC.

Conclusion

The oscilloscope can be the most useful test instrument when servicing the IF and video circuits. Locate the defective circuit with waveforms taken with the scope. Check signal waveforms at the input and output of the IC video component. Take voltage and resistance measurements within the suspected stage. Check each transistor in the circuit.

A tuner-subber can be used to check the first two IF stages with signal from the station injected at the base of each transistor. Use the audio square-wave generator in the video circuits to locate a smeary or oscillating symptom. Rounded or clipped waves will indicate the defective stage.

Always replace IF shields above and below the circuit board after repairing the IF or video sections. Replace IF transistors with the original when available. Most universal transistors and IC components work very well in the video circuits. Check troubleshooting chart Table 6-1 of IC and video circuits.

■ **Table 6-1 Troubleshooting chart for IF video circuits.**

What to check	How to check it
No IF or video output.	Test transistors in circuit.
Sub tuner to determine if tuner is okay.	Check IF IC voltages.
Check IF AGC voltages.	Check IF AGC circuits.
Suspect IF-video IC with normal voltages.	Check voltages upon each IC terminal.
Normal sound—no video.	Check video transistor.
Signal trace with scope with normal voltages.	Connect color-bar generator to antenna terminals.
Good sound—faint picture.	Check video circuits and IC.

AGC and sync circuit tests

7

SERVICING THE AGC AND SYNC CIRCUITS IS NOT AS COMPLI-cated as it was in the past. The automatic gain control (AGC) cir-cuit basically automatically controls the incoming signal level at the IF and RF amplifier stages. These stages are controlled with a small dc bias voltage that changes the gain of these stages. If the incoming signal becomes greater, the bias voltage rises, reducing the gain of the IF and RF stages. Likewise, if the incoming signal is less, the bias voltage is lower, increasing the gain of these stages (Fig. 7-1). Some of the early TV chassis had an RF AGC circuit separate from the IF AGC stage.

Today, in the keyed AGC and delayed AGC circuits, a flyback pulse is fed to the AGC circuits for a more accurate and dependable sys-tem. The incoming video signal is fed from the video stage to the AGC keyer circuit. A dc-controlled amplifier feeds the bias voltage to the first and second IF amplifier and AGC delay circuit (Fig. 7-2). The AGC delay circuit delays the bias voltage to the RF tuner stage until a fixed level of gain is obtained in the IF stage for max-imum performance.

In the latest TV receivers, you might find an IC component that in-ternally develops the AGC action and varies the gain of the first and second IF stages. A comparison voltage is developed to con-trol the RF AGC voltage. This type of AGC circuit is no longer keyed with a pulse from the flyback transformer circuit.

The AGC circuit can be tested by scoping the waveforms of the video input signal and flyback pulse to the AGC circuit. Critical voltage tests on the AGC transistor or IC circuits should be made, with no signal at the tuner. Turn the manual tuner between chan-nels or on a dead channel for accurate voltage measurement. All dc voltages shown on any schematic are without any signal applied to the antenna. When signal is applied to the TV chassis, these voltages will rapidly change. If not, the AGC circuits are not func-tioning. One good test is to notice the change in the AGC voltage

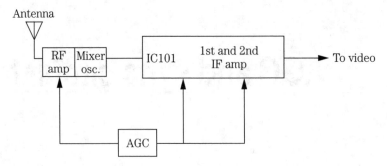

■ 7-1 *A block diagram of the AGC circuits feeding the RF amp, mixer, oscillator, and the first and second IF amps.*

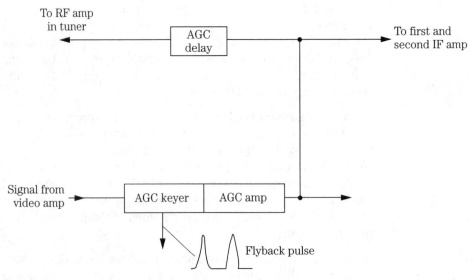

■ 7-2 *The delayed AGC is taken from the AGC amp and keyer.*

when a station is tuned in. The AGC circuits are working if there is a change in voltage.

Abnormal voltages taken on the AGC, transistor, or IC components can indicate the AGC stages are defective. High voltage fed to the base circuits of the IF stages can cut off the transistor, producing a white screen. Low voltage can cause overloading and pulling of the picture. Carefully check the IF and RF bias voltage listed on the schematic for normal AGC operating conditions (Fig. 7-3).

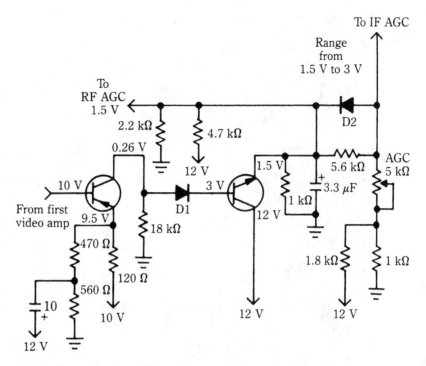

■ 7-3 *Compare the AGC voltage measurements with the schematic on the first and second If amplifiers.*

RCA CTC130C AGC and AFT circuits

When AGC is developed internally in the IF processor IC, it is used to vary the gain of the first, second, and third IF stages. RF AGC voltage is developed when the IF AGC is compared against the setting of the RF AGC control. The AGC delay control is found on pin 3 of the IF processor IC (U301).

In this AGC circuit, the AGC voltage is no longer keyed by a horizontal pulse. This makes the horizontal oscillator pull-in more effective, eliminating phasing problems between the horizontal keying pulse and the transmitted horizontal sync. The noise-limiting circuit eliminates the need for more conventional AGC circuits (Fig. 7-4).

The IF processor IC (U301) is powered from an 11.4-Vdc source. Recent RCA chassis use practically the same AGC circuits. The supply voltage (VCC) feeding the IF processor might vary a few volts. The AFT voltage supplied to the tuner is derived from pins 5 and 6 of the IC (U301). Normal AFT voltage is about 6.5 V with no input signal.

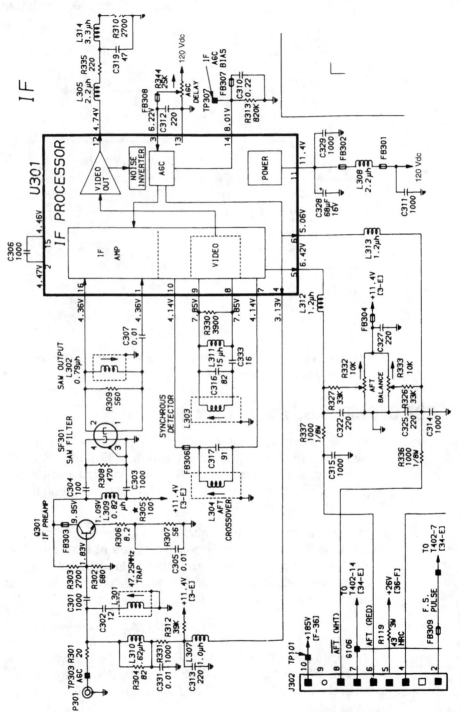

■ 7-4 *The IF and AGC circuits found in the RCA CTC130C chassis.* Thomson Consumer Electronics

In RCA's latest TV chassis (CTC156/157/158/ and 159), the AGC circuits are found in the CTV processing IC (U1001). Here the IF AGC voltage is used to vary the gain of the IF amplifier. It is also used by the RF AGC circuit (Fig. 7-5). The video detector stage passes the video to the IF AGC comparator circuit. The IF defeat signal from the tuning system computer IC is input at pin 22 of U1001. This input is only used in the monitor models, and is used to turn off the IF circuits in U1001 when using video inputs.

The RF AGC circuit compares the level of the IF AGC voltage to the level of the RF AGC in order to develop the RF AGC voltage. The RF AGC voltage to the tuner is found at pin 46. RF AGC adjustment can be made with R2314. The RF AGC voltage can be checked at pin TP2311 and the IF AGC voltage at TP2305. The RF AGC voltage applied to the tuner can be monitored at TP2310.

Service the AGC with a video input waveform at pin 16 and output at pin 12 of the IF processor (U301). Take AGC voltage measurements at pins 3 (6.22 V), 5 (6.42 V), and 6 (6.42 V), and at TP307 (8 V). Improper AGC voltage and VCC voltage at pin 11 (12 V) can indicate a defective IF IC or power-supply source.

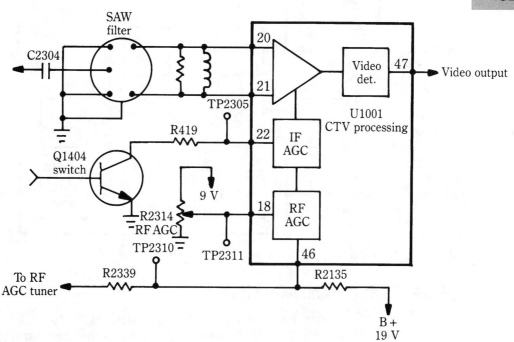

■ **7-5** *The AGC IC (U1001) circuits in the RCA CTC156 chassis.*

Sylvania C9 AGC and AFT circuits

The detected video is applied to the AGC detector which is keyed by the horizontal signal. The purpose of the AGC circuit is to supply a dc voltage to the tuning system (RF AGC) to control the gain of the RF/IF stages (IF AGC). RF is adjusted by R237 of the AGC circuit. Check the AGC voltage at TP3 (Fig. 7-6). It should be approximately 2.5 V with signal and about 9 V without signal.

The AFT circuit receives a phase-shifted input as a reference that leads the input signal by 90°. The product of these inputs develops a dc voltage to correct any drift or mistuning of the tuning system. Check the AFT voltage at TP2 (4.3 V) when tuned to a station. The voltage will be about 7.3 V with no station tuned in.

Service the AGC circuits with critical voltages on TP3 and TP2. The RF adjustment (R737) should vary the voltage at pins 1 and 5 of IC201. Make sure the horizontal keyed pulse is fed to the AGC circuit. Check the supply voltage pin of IC201 before attempting to replace the IC.

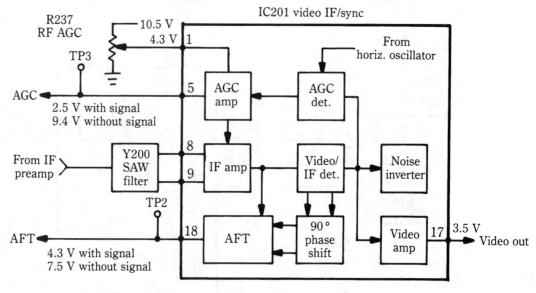

■ **7-6** *A block diagram of the Sylvania C9 chassis AGC circuits.*

RCA CTC157 IF and AGC circuits

The video signal from the video detector passes a signal to the IF AGC circuit (Fig. 7-7). An IF defect signal from the system control IC (U3100) is placed at input terminal pin 22. This input is used to

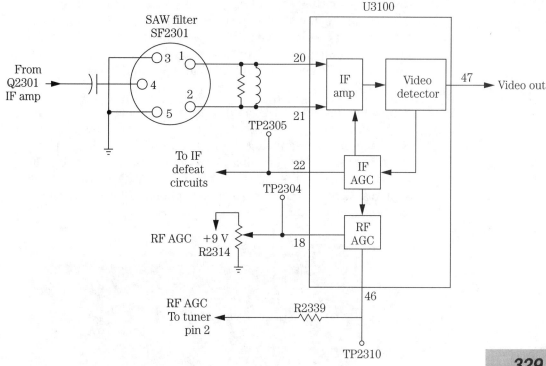

■ 7-7 *The AGC circuits (U3100) found in the RCA CTC157 chassis.*

turn off IF circuits in U1001 when using the video circuits. The IF AGC voltage is used to vary the gain of the IF amplifier and provide RE AGC. The RF AGC circuit compares the level of the IF AGC voltage level of the RF AGC. RF AGC control connects to pin 18, and the output RF AGC voltage connects at pin 46.

For a weak or snowy picture, check the AGC voltage at the test point (TP2310). This voltage should be around +8 volts with a weak signal, and +4.25 volts with a strong signal. Check components at pins 18, 22, and 46 of U1001 for weak or snowy picture. AGC operation can be checked by injecting a variable voltage at tuner pin 2 (RF AGC) with a varying voltage from +4 to 8.5 volts. If the picture returns to normal as voltage is varied, check components at pins 18, 22, and 46, the AGC alignment, or U1001.

AGC controls

In some TV chassis, you might find one AGC control, a separate RF delayed and AGC control, or no control at all. Improper adjustment of either control can cause picture rolling. The AGC control can be adjusted to lower the gain of the TV so local stations will not overload the picture. Weaker TV stations can be received by readjusting the control to provide more gain in a given area. Some TV chassis might list the AGC control as the noise control (Fig. 7-8).

You can assume the AGC circuits are functioning if rotating the AGC control makes the picture darker or appear as overloading. No reaction from adjusting the AGC control can indicate problems within the AGC circuits. Improper adjustment of the AGC control can cause the set to overload. The AGC control setting should always be done after repairs or when the TV is returned to the home.

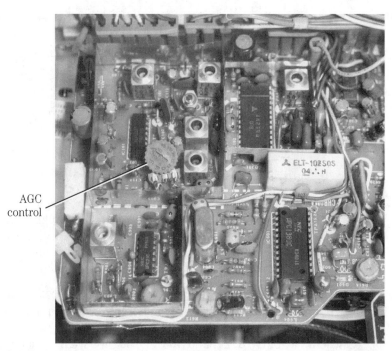

AGC control

■ **7-8** *Adjust the AGC control for a normal picture without snow or overload.*

AGC problems

AGC problems can include a very dark and unstable picture. A white raster can be caused by a defective AGC tuner or video circuit. Check the AGC circuit for a "flagwaving" picture with a buzz in the sound. Low contrast and erratic pulling in the picture can be caused by improper AGC voltages. Excessive snow in the picture can be caused by a defective AGC, tuner, or IF stages. Besides the AGC circuits, these same symptoms can be caused by defects in the tuner, IF, or video stages.

AGC clamping tests

To determine if the AGC circuits or tuner are defective, a variable positive voltage fed to the AGC terminal on the tuner can bring back a normal picture. If the picture is good, you can assume the tuner is normal with a defective AGC system (Fig. 7-9). This is done with an external power supply. A tuner-subber can be plugged into the IF cable to help prove the tuner is operating. Make sure correct dc voltages are found on the tuner.

Although clamping tests have limitations in some circuits, check for accurate AGC voltage applied to the tuner and IF stages according to the schematic. Excessive positive voltages at the tuner AGC terminal can cause snow, but when applied to the IF stages can produce a white screen. All AGC voltages should be taken with no signal at the TV. Simply rotate the tuner to a point between channels.

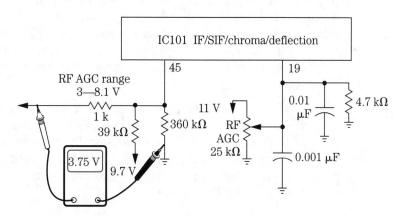

■ **7-9** Check the AGC voltage to the tuner to determine if the AGC circuits are operating.

AGC and tuner

In many cases, the tuner is removed and replaced with no improvement in the picture. This can be caused by AGC defects. The tuner-subber can indicate if the tuner is defective. If the tuner-subber brings back a normal picture, you can assume the original tuner is defective. Go a step further and measure the AGC and B+ voltages at the tuner with the tuner locked between channels or on an unused channel. Abnormal voltage at the tuner AGC terminal can indicate a defective AGC circuit. Replace the tuner if there are normal AGC and B+ voltages.

Poor AGC action

Most AGC problems are caused by leaky or open transistors and IC components. Besides voltage and transistor tests, the defective component can be located with coolant or heat application. Sometimes, spraying each transistor or IC will help find the intermittent AGC symptom.

Dried-up or leaky electrolytic capacitors within the AGC circuits can produce AGC problems. Sometimes shunting a good capacitor across the suspected one can solve the AGC condition. Remove one end of the capacitor for accurate leakage tests. A good in-circuit capacitor tester is ideal in locating defective capacitors in AGC circuits.

Take resistance measurements within the circuit when all voltages and in-circuit tests are normal. Remove one end of all resistors above 50 kΩ to make accurate resistance measurements. In many cases, an in-circuit resistor measurement can be very close to the resistor value and is passed over as good. Emitter bias and collector resistors should be checked for close resistance tolerance in AGC circuits.

Besides defective AGC components, poor IF cable and board connections can cause intermittent AGC look-alike problems. Flex the IF cable to find possible loose connections. Push up and down at various points on the chassis. Broken or cracked wiring can produce AGC symptoms (Fig. 7-10). Localize the intermittent to one section of the board (Fig. 7-11).

AGC circuits

The input signal of any AGC circuit is usually taken from the emitter circuit of the first video amp or from a separate cathode follower stage. Some of the earlier AGC circuits with transistors included four or more transistors. Today, practically all TV chassis use a section of an IC component for AGC control. The AGC IC is much easier to service than the transistor stages.

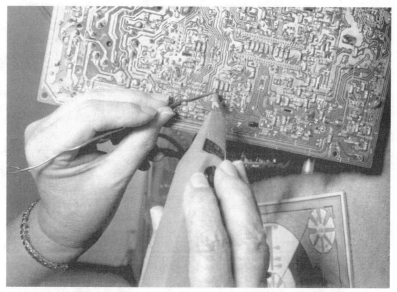

■ **7-10** *Repair AGC PC wiring on the foil with bare hookup wire.*

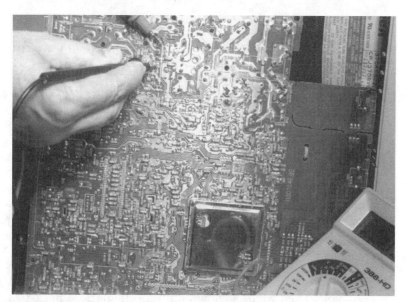

■ **7-11** *Try to localize the intermittent to one section of the board.*

In Fig. 7-12, the input signal is taken from a noise inverter stage, which gets the video signal from the first video amp. The gain of the AGC keying transistor is varied with AGC control R221. The keyed pulse from the flyback transformer winding is found at the collector circuit of Q201. The AGC amp output voltage feeds both

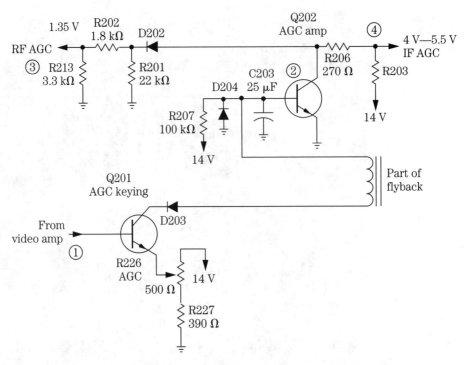

■ **7-12** *Check the AGC circuits by the numbers, with waveforms and voltage measurements.*

the RF AGC tuner and IF AGC voltage from the collector terminal of Q202. Critical voltage measurements at both RF and IF transistor base circuits indicate if the AGC circuits are working.

Any AGC circuit can be quickly checked by taking scope waveforms and voltage measurements. Check the video waveform at the input AGC transistor (Fig. 7-13). Scope the gated waveform at Q303 from the flyback transformer winding. Go from the base to the collector terminals of each transistor and compare the waveforms to those on the schematic.

Measure the AGC voltage output applied to the tuner and IF stages with no signal. Most AGC voltages will only vary a few volts from what is shown on the schematic. Check to see if the AGC control will vary the voltages at these different points. Often, the RF AGC post on the VHF tuner is connected with a white or green wire. If a test point is not provided for IF AGC voltage tests, measure the voltage at the base of the first and second IF video transistors. Abnormal voltage measurements at the RF tuner or IF stages indicate a defective AGC circuit. Take

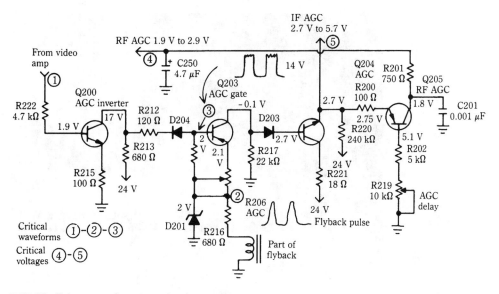

7-13 *Take waveforms and voltage measurements to locate the defective component in AGC circuits.*

accurate voltage measurements on each transistor with no signal at the tuner.

Most AGC problems found in the transistorized circuits are caused by leaky or open transistors, dried-up electrolytic capacitors, and a change in resistance. Transistors can be checked in and out of the circuit. Check for leaky or open diodes with the diode test of the DMM. Remove one end of all resistors for accurate resistance measurements.

ICs and AGC

Most color and black-and-white TVs manufactured today use IC components in the AGC circuits (Fig. 7-14). Locating the defective part is much easier with fewer components. Adjust the RF AGC control, if one is found in the circuit. Suspect a defective AGC circuit if no action is noted in the picture.

Take a voltage test at the RF tuner terminal (Fig. 7-15) and IF test points (1 and 2). Scope the waveforms at 3 and 4. Check the flyback winding or wiring board connections with no keyed waveform at point 4. Check all voltages on the IC terminals related to the AGC circuits without a signal at the tuner. Low voltages can indicate a defective IC. Shunt C111 with another electrolytic capacitor. Check all resistors for accurate resistance before replacing the suspected IC.

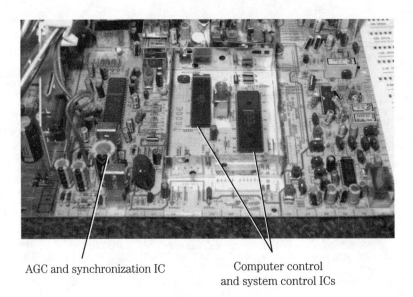

AGC and synchronization IC

Computer control
and system control ICs

■ **7-14** *One large IC may contain the AGC and sync circuits, along with many others.*

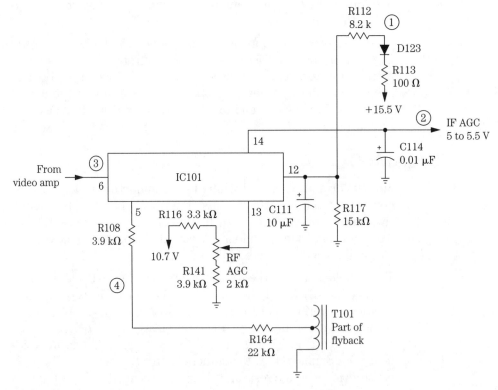

■ **7-15** *Low supply or pin voltages can indicate a leaky AGC IC.*

AGC and sync circuit tests

Electrolytic capacitors and AGC

Electrolytic capacitors in the power source feeding the AGC circuits can cause some very unusual pictures. Automatically shunt each filter capacitor. Clip into the circuit with the power off. Small electrolytic bypass capacitors within the AGC circuits can cause excessive pulling and tearing of the picture. Temporarily tack another capacitor across the suspected one. Remove one end of the capacitor and check for leakage.

Unusual firing lines

The picture in a Zenith 12AC10C15 chassis would pull and tear with small firing lines. Sometimes the lines were eliminated when the AGC delay control was turned down. Replacing the small polarized capacitor in the base circuit of the AGC delay transistor solved the problem (Fig. 7-16).

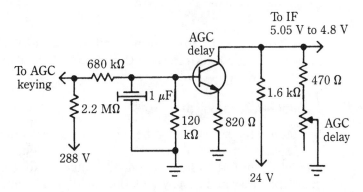

■ **7-16** *A twisting, pulling picture was caused by a defective nonpolarized capacitor (1 μF).*

Five actual AGC problems

The following items are examples of real AGC-related problems in common TV chassis.

Intermittent AGC and negative picture

The picture was intermittent and sometimes appears negative indicating AGC problems. Voltages were checked on pins 5, 6, and 7 of the VIF/SIF/AFT/Det. IC101. The voltages were way off in the intermittent mode, but would not stay off long enough for a complete voltage measurement. When the probe touched pin contacts, the set would act up (Fig. 7-17). All connections were resoldered

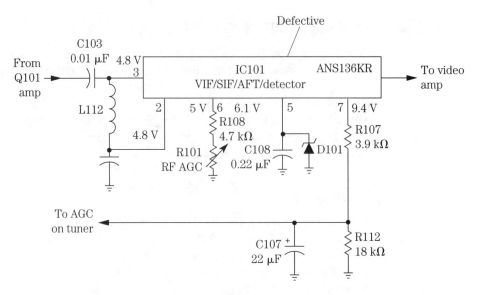

7-17 *Intermittent AGC and a negative picture was caused by IC101 in this J.C. Penney 685-2520 chassis.*

upon IC101 and the problem was still the same. IC101 was replaced with AN5136KR in the J.C. Penney 685-2520 chassis, thus fixing the problem.

Poor connection—AGC problem

Sometimes the set would be perfect, and other times the picture would snow up in a General Electric 19PC-J chassis. Pushing around upon IC101 and components would make it act up. The snowy picture was cured by soldering the RF AGC control contacts of R116 (Fig. 7-18). Remember, the voltage measurement on pins 12 and 16 will change as the IF and RF AGC controls are adjusted.

White screen and sometimes snow

The screen was white and sometimes snow was found on an unused channel, indicating an overloaded AGC problem in a Philco 20ST30B chassis. The tuner was subbed and AGC voltage was clamped at the tuner, indicating problems within the AGC circuit. The picture would return with the tuner subbed or a clamped AGC voltage.

All filter capacitors were shunted, because a defective electrolytic capacitor can cause problems with the AGC or sync circuits. Voltages within the AGC circuits were fairly normal. A scope check of the keyed AGC pulse from the flyback transformer was normal (Fig. 7-19). Bypass capacitors and transistors within the AGC cir-

338

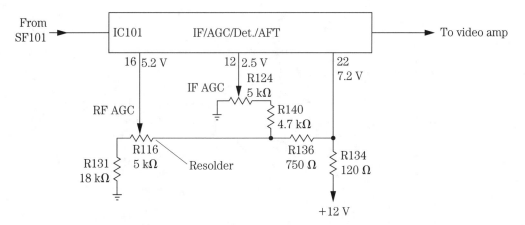

■ 7-18 *A poor contact on control R116 resulted in a snowy picture in a General Electric 19PC-J chassis.*

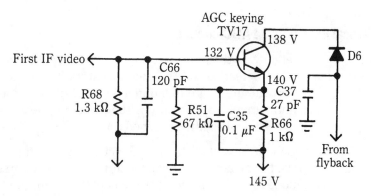

■ 7-19 *The screen would turn white in a Philco 20ST30B TV; resistor R51 was found open.*

cuits seemed normal. Each resistor of those over 50 kΩ was measured with one terminal removed. R51 was found to be open in the AGC keying emitter circuit to ground.

AGC or sync?

The picture in a J.C. Penney 286 had tearing sync or possible AGC problems. All transistors were checked in the AGC circuit with the diode transistor test of the DMM (Fig. 7-20). A scope test indicated a normal pulse from the flyback circuits. Voltage measurements were fairly normal. When C403 was shunted with another 10-μF capacitor, the picture returned to normal. Dried-up electrolytic bypass capacitors in the AGC circuit can cause the picture to appear as sync or AGC problems.

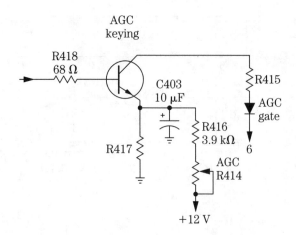

AGC
keying

R418
68 Ω

C403
10 μF

R415

AGC
gate

R416
3.9 kΩ

6

R417

AGC
R414

+12 V

■ **7-20** *Excessive tearing of the picture in a J.C. Penney TV was caused by a faulty capacitor C403.*

Vertical lines outside AGC circuits

Vertical lines at the left side of the picture have been noted to be caused by defective electrolytic capacitors in the AGC circuits. Four vertical lines were found in a Zenith 20Y1C48 chassis with a very normal picture. Sometimes they would appear in the background with the brightness turned up.

Although all capacitors in the AGC circuit were shunted without any results, the actual component turned out to be three blanking gate diodes found in the emitter circuit of the video amp (Fig. 7-21). Replace these diodes with 1N34A types. Replacing all three leaky diodes solved the vertical line problems.

Conclusion/AGC circuits

Always check the AGC control before attempting to service the AGC circuits. Momentarily short the RF AGC terminal at the tuner with a test clip or screwdriver and notice the change in the picture. If a big change is noticed, perhaps the AGC is working. There will be no AGC action if the IF, detector, and video amp sections are defective.

Measure the voltage at the RF AGC tuner terminal. Turn the TV off-channel, or in case of a push-button or remote-control set, select a dead channel. If a big voltage change occurs, the AGC circuits are probably working. Locate the IF test point. In case of no test points, measure the dc voltage at the base terminals of the first and second IF transistors. Compare voltage measurements

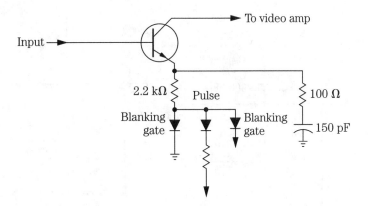

■ 7-21 *Firing lines in the picture were caused by leaky diodes in the emitter of the video amp.*

for the RF tuner and IF without a signal to those on the schematic. Suspect defective AGC circuits if there are lower- or higher-than-normal voltage measurements.

Clamp the RF AGC voltage with a variable dc voltage supply and notice if the picture returns. A fairly normal picture indicates AGC problems. Inject a variable dc voltage at the IF test point to determine if the IF AGC circuits are defective.

Scope the video input waveform at the input of the AGC circuit. Go from stage to stage and notice where the waveform is missing. Check the key waveform pulse from the flyback transformer winding. An open winding or connection can prevent the keyed AGC circuit from operating.

RCA CTC130C sync separator circuit

The 4.5-V p-p negative sync video signal from the video amplifier stage (Q303 and Q304) is applied to the sync separator transistor (Q305). The output of the sync separator stage is a positive-going composite sync (Fig. 7-22). This sync pulse is used to trigger the burst/clamp keys (Q801) and supplies sync to the deflection IC (U401).

To determine if the trouble is the sync or AGC, scope the collector terminal of Q305 where the sync splits to the vertical and horizontal sync circuits (Fig. 7-23). If there is no signal, check the sync amp or video composite signal in the base area of the sync separator (Fig. 7-24). With a normal sync signal coming from the sync separator transistor (Q305), scope the vertical input sync of U401 at pin 24 if there is no vertical sync. If there is no horizontal sync, scope the input terminal (1) of U401.

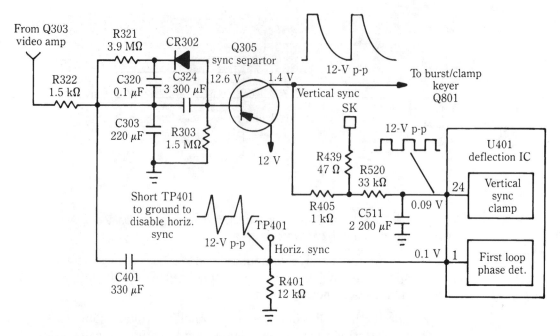

■ 7-22 *The vertical and horizontal sync circuits in the RCA CTC130 chassis.*

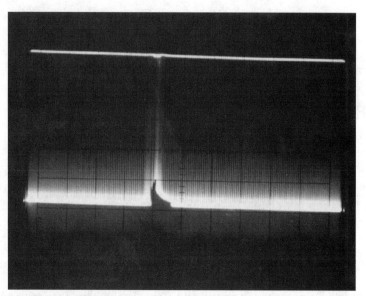

■ 7-23 *The waveform at the collector terminal of the sync separator transistor.*

■ **7-24** *The video waveform at the base of the sync separator.*

Suspect a defective deflection IC (U401) when either vertical or horizontal sync is found on the input terminals with poor sync in the picture. After locating possible stages for poor or no sync, take critical voltage measurements. The horizontal sync can be checked by shorting out TP401 to chassis ground. Notice if the picture starts to run sideways.

Servicing sync circuits

Poor vertical or horizontal sync problems can cause the picture to roll and pull horizontally. If the picture flips up and down and the horizontal is stationary, suspect sync problems within the vertical circuits. When the picture goes from side to side, check the horizontal sync circuits. Check the sync circuits with movement in both vertical and horizontal sweep of the raster.

The sync pulses are transmitted at the TV station to lock the pictures in at the TV set. Sometimes poor or snowy reception contains weak sync pulses, letting the picture intermittently roll or flip. In extreme fringe area reception, you might find the picture will roll or flip sideways with normal sync circuits. In most cases, the AGC control must be adjusted.

Because the two sync pulses are fed from the video circuits to the separate sync stages, the scope is the ideal test instrument to signal trace the sync waveforms (Fig. 7-25). You might find a single transistor or IC component in the sync circuits. Today, the sync circuits are found in one large IC with AGC, luminance, and color circuits (Fig. 7-26). Look for the sync and AGC circuits within the luminance module of the modular chassis.

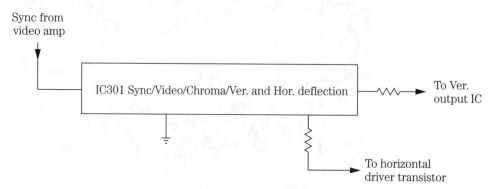

■ **7-25** *The sync pulses are fed from the video amp to the input of IC301 to synchronize the vertical and horizontal circuits.*

344

Sync/AGC/Vertical deflection IC

■ **7-26** *Today, the sync and AGC circuits are found in one large IC.*

The sync separator

The transistor sync separator is a very simple circuit with the video and sync pulses fed into the base circuits (Fig. 7-27). Both vertical and sync pulses are separated from the video at the collector circuits. The vertical sync pulse goes to the vertical input circuits, while the horizontal sync feeds the AGC and horizontal circuits. Both sync circuits can be affected if trouble exists in the sync separator stage. You might find a sync amp transistor preceding the sync separator in some TV sync circuits.

Very poor signal can cause insufficient vertical and horizontal sync, including AGC problems. A defective tuner, IF, and first video amp results in weak sync signal. Scope the video waveform at the emitter terminal of the first video amp and signal trace to the base of Q101. A weak or insufficient-height waveform pulse from the separator stage can result in vertical rolling. The 60-Hz sync pulse is fed through C1 to the vertical oscillator stage. C11 couples the horizontal sync pulse to the horizontal AGC circuits.

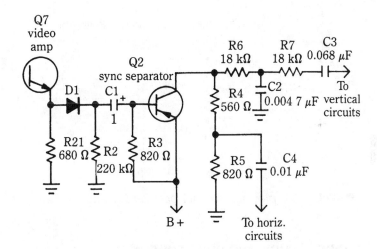

■ **7-27** *In a transistor stage, C3 couples sync to the vertical circuits and C4 to the horizontal circuits.*

The latest IC sync circuits

The latest sync circuits have not changed too much with the sync taken off of the video amp (Q208) emitter terminal and fed to the sync separator stages in one large IC801. The sync signal can be checked with the scope at pin 27. This sync separator provides sync for the deflection and chroma circuits inside IC801. Pin 25 is the vertical sync separator or terminal, and the horizontal sync is provided inside the IC. Check waveforms and voltages upon pins 27 and 25; take VCC voltage measurements on pin 18 (Fig. 7-28).

IC801 provides video, contrast, pedestal clamp, brightness, R-G, y-y and B-y output, color killer, tint, VCO, ACC, ground (gnd), sync separator, burst gate pulse, vertical sync separator, ramp generator, vertical driver, AFC, horizontal, horizontal oscillator, x-ray protect, horizontal pre-driver, and voltage supply pin 16 operations.

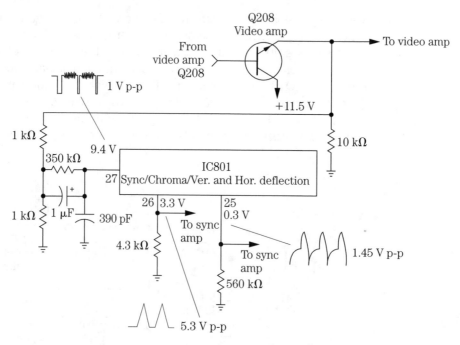

■ **7-28** *IC801 provides sync for both vertical and horizontal circuits.*

Troubleshooting the sync separator

Check the video composite waveform at the input of the sync separator or sync amp transistor. The TV receiver must be connected to the antenna with a normal picture tuned in, because the sync pulses are transmitted with the TV signal. Compare the video waveform with those on the schematic. Make sure the video signal has the correct amplitude or height.

Check the waveform at the collector terminal of the sync separator. A clear-cut narrow waveform should be found here. If the height of the vertical sync pulse is low, the picture might be rolling on the screen. Horizontal and vertical jitter can be caused by a "dirty" sync pulse. Excessive hash or noise in the sync waveform can produce vertical jitters. A bending or pulling of the sync pulse can indicate that hum or 60-Hz signals are getting into the signal at the sync separator or from the vertical or horizontal circuits. If the sync pulse contains video signal, the sync separator stage is not working properly.

Voltage measurements on the sync separator transistor can indicate a defective transistor or connecting component. Check the transistor in the circuit with the transistor test of the DMM. If not satisfied with transistor tests, remove the emitter terminal and take another test. Sometimes low emitter and bias resistors within the circuits can indicate false transistor leakage. Check each resistor with one terminal removed from the circuit.

The vertical picture would keep rolling with normal horizontal sync in a Packard Bell 98C38 chassis (Fig. 7-29). Video sync was fairly normal at the base terminal of the sync separator (Q305), with very low pips at the collector terminal. Voltage measurements were high on all transistor terminals. A careful inspection showed the collector load resistor was burned. Replacing R348 helped some, but the picture wanted to roll. R344 was found to be open in the emitter circuit of the sync separator transistor. Undoubtedly Q305 had shorted and took out both resistors. In this circuit, the horizontal sync is taken from the emitter with the vertical sync at the collector terminal.

347

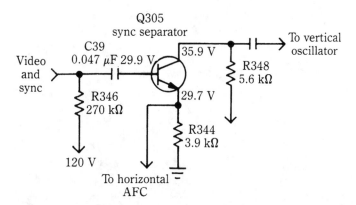

■ 7-29 *The picture kept rolling in a Packard Bell 98C38 chassis; the problem was burned resistors.*

Poor video and sync were noticed in a Montgomery Ward GGY-12949A portable TV. Because both video and sync were bad, the video signal was traced through IC400 (Fig. 7-30). Pin 4 of IC400 was fairly weak. IC400 was replaced with no change. Although voltage measurements did not show much, resistance measurements did. A video peaking coil was found to be open. Replacing L408 solved the poor sync and video picture.

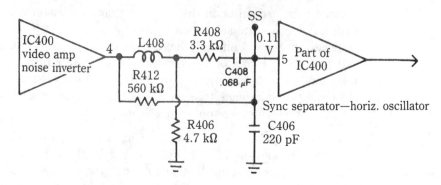

■ **7-30** *An open L408 caused poor video and sync in a Wards GGY-12949A portable.*

Six actual sync problems

The following accounts describe six actual problems that occurred in connection with sync circuits.

Poor horizontal and vertical sync in a Panasonic PC11T30R

Both horizontal and vertical sync were very unstable in a Panasonic PC11T3OR chassis. The sync signal was normal at pin 18 of the video IF (IC101) and poor at pin 17 of the sync separator. This waveform should have a 1.4- to 1.5-V p-p sync signal. C401 (3.3 μF) was found opened (Fig. 7-31).

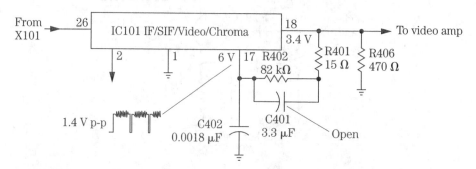

■ **7-31** *Poor horizontal and vertical sync were caused by an open C401 in a Panasonic PC11T30R chassis.*

Poor horizontal and vertical sync in a Goldstar CMT 2612

A normal sync waveform was found at the video amp in a Goldstar CMT-2612 portable, and poor sync at pin 27 of IC301. The voltage on pin 27 was very low. A resistance measurement from pin 27 to common ground showed no signs of leakage. R302 (360 kΩ) resistor was found open (Fig. 7-32).

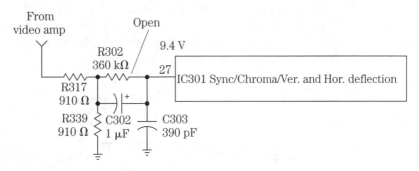

■ **7-32** *Poor sync was found in a Goldstar CMT-2612 portable with open R302.*

Poor picture and sync

The picture in a Zenith 4B25Cl9 chassis was very poor with a faint outline. The picture had poor sync and color, which could be video trouble (Fig. 7-33). The 25-V power source measured only 5.7 V. Checking over the 25-V power supply circuits located a leaky zener diode (X26). Replacing the 25-V zener diode with a 1-W type solved the poor picture and sync problem.

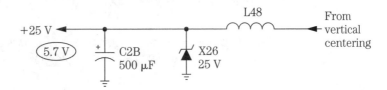

Circled voltage with
   leaky Zener X26

■ **7-33** *Check the power voltage source for poor sync in a Zenith 4B25C19 chassis.*

No vertical or horizontal sync

The picture would roll and slide sideways in a Sharp 19D72. After locating the sync circuits within IC501, voltage measurements were taken on each terminal. All voltage measurements were normal except 12 and 14. Higher voltage was found on pins 12 and 14 (Fig. 7-34). To determine if the voltage was cut down by the video amp or within IC501, pin 14 was unsoldered from the circuit. Actually, IC501 was leaky, which was causing sync problems.

The suspected IC component leakage could be identified by voltage and resistance measurement. After locating the terminal with abnormal voltage, remove that terminal from the PC board. Most IC terminals are numbered or can be identified from each end (Fig. 7-35). Here the large video AGC sync/luminance IC is numbered in several places on the wiring side of the PC board.

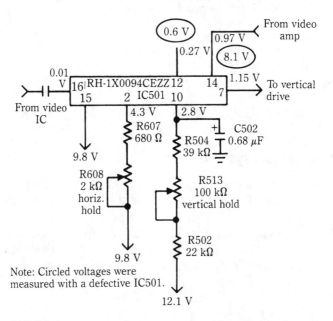

7-34 *Improper voltages on pins 12 and 4 indicated that IC501 was defective in a Sharp 19D72 TV.*

Poor horizontal sync

An abnormal sync waveform at the base terminal of a J.C. Penney CTC97 chassis indicated pulling of the picture sideways (Fig. 7-36). The vertical and horizontal waveform at the sync separator were quite normal. The keyed sync waveform from the flyback transformer was normal up to R132. The signal waveform was very low at C132. A quick resistance measurement of R132 was higher than normal. One end of R132 was removed from the circuit and it was found to be open. Waveform checks with the scope and resistance measurements found the defective component.

■ 7-35 *To check a certain pin on an IC, unsolder the pin with solder wick and an iron, then measure the pin to ground for leakage.*

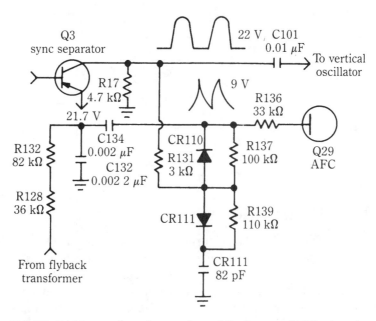

■ 7-36 *R132 was found open in a J.C. Penney CT97 chassis, producing poor horizontal sync.*

No vertical or horizontal sync in a Sylvania 20B1 chassis

Although the horizontal and vertical drive circuits were normal in the Sylvania B1 chassis, both pictures were moving up, down, and sideways. IC250 (TV signal processor) contains the drive pulse at output pin 24. The vertical drive pulse to the yoke at pin 21 was normal (Fig. 7-37).

The horizontal sync input waveform at pin 42 was normal and the vertical sync input at pin 44 was quite normal, but there was poor mixing inside IC250. The voltages on pin 42 and 44 were off a little, but with normal input sync. This indicated the TV signal processor (IC250) was defective. Although the sync input waveforms were fairly normal with normal sweep circuits, IC250 was replaced, solving both sync problems.

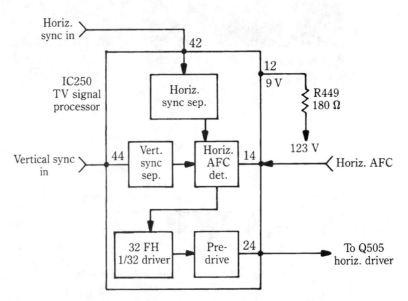

■ **7-37** *A defective signal processor (IC250) caused sync problems in a Sylvania 20B1 chassis.*

RCA CTC130C AGC delay

The AGC delay control (R334) has been preset at the factory. Readjustment of this control is only needed when the tuner has been replaced, when the IF circuit has been repaired, when the AGC control has been misadjusted, or when unusual local signal conditions exists. Adjusting the control to each extreme of rotation will usually provide poor signal-to-noise ratio. Adjust the AGC control so that all channels are good and free of color, beat, or cochannel interference. After the AGC adjustment, check all strong local TV channels.

Conclusion

Sync circuits are much easier to service than AGC circuits. The sync waveforms are distinct and clear compared to those in the AGC circuits. Check the input waveform at the base of the sync separator transistor and the output at the collector terminals. In ICs, check the input video signal and output unless the IC contains both vertical and horizontal oscillator circuits. Correct waveforms with accurate voltage and transistor measurements will locate most sync problems.

■ **Table 7-1 For easy troubleshooting,**
check this troubleshooting table for AGC and sync circuits.

What to check	How to check it
White raster or snowy picture.	Sub tuner with tuner-subber.
Determine if tuner or AGC.	Measure AGC voltage to tuner and IF stages.
Check AGC circuits.	Replace AGC IC.
Poor horizontal and vertical sync.	Check sync separator waveforms on IC.
Check sync waveform from video circuits.	Check video circuits.
Replace sync IC with normal voltages.	Make sure sync input is normal.

353

Tuner problems and cures

8

MOST TV TECHNICIANS ADMIT THE TUNER, IF, AND AGC circuits are the most difficult stages to test and repair in the whole TV chassis. The tuner can be the most confusing, because parts are difficult to get to. New varactor tuners with control boards have many different ICs, synthesizers, and processor components. Servicing the TV tuner can be made easy by taking each unit one step at a time.

There have been new names added to the tuner assembly since the old mechanical wafer tuner was first invented (Fig. 8-1). The varactor tuner can be controlled with a tuning system, control processor, initializer, presetter board, tuner control, memory, and frequency synthesizer control units (Fig. 8-2). A tuner control assembly can have soft-touch, keyboard, pushbutton, selector board, auto board, and remote-control tuning. They all do the same thing by controlling the tuner assembly, which in turn selects the correct channel (Fig. 8-3).

Tuner repair can be made easy with isolation and signal injection techniques. A defective tuner can produce a snowy, weak, or erratic picture. The defective tuner control unit can cause a dead, intermittent, or drifting varactor tuner with improper tuning voltage. Each unit can be isolated with voltage measurement and signal injection methods with a tuner-subber or tuner replacement.

No picture, no sound, white raster

A defective tuner, IF section, AGC circuit, or shorted IF cable assembly can cause a white raster without picture or sound (Fig. 8-4). Because there is no snow in the picture, check the mixer section and B+ voltage at the tuner terminals. An IF squeal or a birdie sound can indicate a defective IF section. The all-white screen can be caused by improper AGC voltage applied to the VHF tuner terminals. A low-ohm continuity check of the IF cable between the tuner and chassis may locate a leaky IF cable assembly.

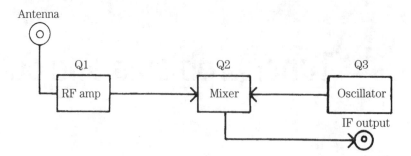

■ 8-1 *A typical block diagram of a mechanical VHF tuner.*

■ 8-2 *An RCA tuner and memory module mounted together in several different chassis.*

In an RCA CTC120 chassis (Fig. 8-5), the symptoms were no picture, no sound, and normal high voltage. A B+ voltage measurement at the tuner indicated no voltage at the tuner control. R119 was open in the 26-V source, which supplies B+ voltage to the tuner control unit. Replacing the burned resistor restored the picture and sound.

Snowy picture

The snowy raster can have a very faint picture with poor sound, or just a plain snowy raster. A snowy picture can result from improper signal at the antenna terminals, a defective tuner, improper AGC

356

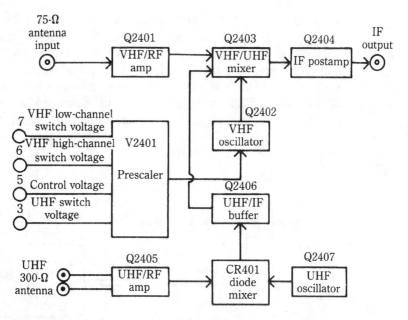

■ **8-3** *A block diagram of the RCA CTC120 tuner control module MST-005A.*

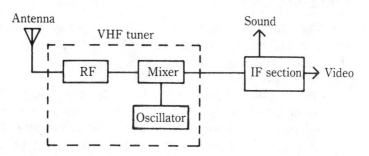

■ **8-4** *A defective tuner and IF section can cause a snowy picture, no picture and no sound, or only a white screen.*

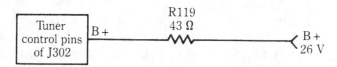

■ **8-5** *Absent picture and sound was caused by a lack of B+ voltage at J302 in this RCA CTC120 chassis.*

voltage, or a defective IF system. The poor antenna or cable system can be checked with another portable TV receiver (Fig. 8-6). A damaged balun circuit in the front end of the tuner can cause a snowy picture. Often, the oscillator and mixer stages of a solid-state tuner are normal, with snow in the picture indicating a leaky or open RF or FET transistor. A dirty tuner can produce a snowy picture. Check the AGC voltages at the tuner for excessively high or low AGC voltage. The leaky or open first and second IF stages can cause snow in the raster.

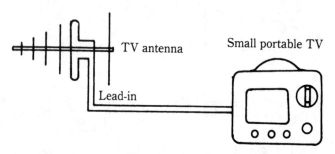

■ **8-6** *Check the antenna or cable system with a portable TV set.*

Intermittent picture

The intermittent picture can result from a broken antenna lead-in or a bad connection right at the TV. A dirty tuner can cause the picture and sound to drop out. Slightly move the tuner knob and see if the picture returns. Often, the intermittent picture with snow is caused by a problem in the front end of the TV receiver. An intermittent transistor within the tuner can produce the same symptom. Check for an intermittent AGC component causing improper AGC voltage at the tuner. Practically any IF or video circuit can cause the intermittent picture symptom without snow in the raster.

Cleaning the tuner

A dirty mechanical tuner can cause the station to drop off-channel, or intermittently produce a snowy and noisy picture. The customer might complain of no color in the picture, which relates to a dirty tuner. Slightly move the tuner knob, noticing if the picture acts up. Remove the tuner to properly clean it (Fig. 8-7).

Remove the bottom metal cover to get to the various contacts on the rotating wafers or drum area. The old drum tuner can be

■ **8-7** *Clean the mechanical tuner when it will not stay on stations. Use tuner wash and spray.*

cleaned by applying tuner cleaner to a cloth and holding it against the drum as the assembly is rotated. Clean the tuner contact springs with a silicone spray cleaner.

The wafer switch-type tuner is a little more difficult to clean. It does not matter if the switch contacts are of gold or silver; they still corrode. Some TV tuners have excess grease applied to the contacts when they're manufactured. The grease collects dirt and dust. Dislodge excess grease with a tuner wash spray. Select a silicone contact spray that will not destroy or damage plastic components within the tuner.

A good tuner cleanup consists of spraying the tuner, a tuner washout, and cleaning of each individual contact. Some technicians use a pencil eraser, the end of a pocket knife, or a small, stiff wire brush on each contact. Always remember to replace the metal cover to prevent stray RF signals from entering the tuner or upsetting tuner alignment. Send the tuner into a tuner repair depot for a supersonic wash when it will not clean up or appears to have excessively worn switch contacts.

Three quick tuner tests

Any tuner can be checked with voltage measurements on the tuner terminals, signal injection with a tuner-subber, or tuner substitution and correct AGC voltage (Fig. 8-8). Check for B+ voltage at the tuner to chassis ground. This voltage can vary from 10 V to 50 V on most solid-state tuners. The B+ voltage on varactor tuners can vary from 10 V to 30 V. Besides a B+ operating voltage, all varactor tuners have a dc tuning voltage between 1.5 V and 30 V. This voltage will change each time the channel is switched. AGC voltage can vary between 1.5 V and 8 V on a solid-state tuner. Remember this AGC voltage is always positive, while a negative AGC voltage is found on the tube-type tuner.

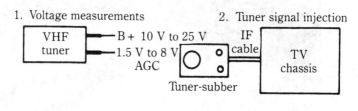

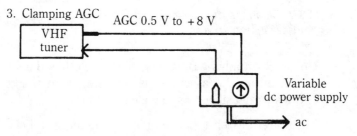

■ **8-8** *Make three different tuner tests with voltage, using an external voltage source and a tuner-subber.*

Signal injection with a TV signal generator or tuner-subber at the IF cable will quickly determine if the tuner is defective or if the problem is in the TV chassis. After a voltage measurement, plug the IF cable into the tuner-subber. If there is still no picture and sound, you can assume the tuner is normal and the signal problem lies within the TV chassis. The tuner should be repaired or sent in for repair if the picture and sound returns with the subber connected. Simply substitute another tuner module in the modular chassis to determine if the tuner is defective.

Clamping the AGC terminal at the tuner with the proper dc voltage can determine if the AGC circuits are normal. We must remember the tuner-subber does not have the AGC voltage connected to it when it is subbed into the circuit. The tuner-subber can show normal picture and sound, but the AGC voltage source might be defective within the TV chassis. Double-check the AGC voltage at the tuner by injecting a variable dc voltage at the AGC terminal or with accurate AGC voltage measurement (1.5 V to 7.5 V). The picture and sound might return with some snow with external AGC voltage. In this case, you can assume the AGC circuit is defective within the TV chassis. If not, you might send the tuner in for repairs when nothing is wrong with it. Besides additional tuner repair cost, you still have the very same signal problem.

The tuner-subber

The tuner-subber is nothing more than a regular solid-state tuner with self-contained batteries in a separate container (Fig. 8-9). Most subbers have an IF cable jack or plug-in cable with alligator clips so the unit can be easily attached to the TV chassis. An RF gain control should be adjusted to eliminate snow in the picture.

To operate, simply remove the IF cable from the TV tuner and plug it into the jack at the back of the tuner-subber. Attach the external antenna wire or cable to the antenna input terminals of the subber. Turn the unit on with the RF gain control full-clockwise. Tune in a station with the tuning selector and fine tune for the best color picture. The tuner-subber will eliminate a lot of guesswork, especially in servicing controlled varactor tuners.

You can make your own tuner-subber with a solid-state tuner out of an old TV set. (See *Build Your Own Test Equipment*, published by TAB Books, an imprint of McGraw-Hill.)

■ **8-9** *Inject a signal at the IF cable from the tuner-subber to determine if the tuner or the first IF stage are defective.*

Defective antenna balun coils

Although blocking capacitors are placed in each leg of a 300-Ω balanced antenna, a direct lightning strike can damage the entire antenna assembly. You might find small capacitors blown apart, with small wires burned off the antenna connectors (Fig. 8-10). A very snowy picture will be found on the TV screen.

The antenna coils must be repaired or replaced. Often, the small wire ends can be unwound a turn to get enough wire to solder back to the various components. Order a replacement for the entire antenna assembly if the unit is damaged beyond repair.

An RF balun transformer that changes the 300-Ω antenna input terminals to 75 Ω can be used by connecting the balun backwards. These balun transformers are used to connect the TV receiver to a 75-Ω cable system or VCR. Sometimes these antenna balun assemblies are difficult to obtain.

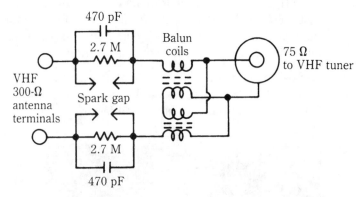

■ **8-10** *Broken wires at antenna terminal or lightning damage can cause a snowy picture.*

Conventional mechanical tuner

The mechanical tuner is still found in most black-and-white and some color portable TV receivers. A wafer-type switching mechanism rotates the various coils into each RF, mixer, and oscillator circuit (Fig. 8-11). The VHF solid-state tuner might have an FET in the RF stage. Usually the mixer and oscillator transistors are located inside beneath the wafer assemblies.

The snowy picture can be caused by a defective FET or RF transistor. In the earlier solid-state tuners, these transistors were

■ **8-11** *A typical tuner consists of transistors in oscillator and mixer stages.*

mounted on top and were easily replaced. The FET must be handled with extreme care to prevent damage to the transistor. All tuner transistors should be replaced with parts bearing the original part number or a correct universal replacement.

The absence of picture or sound in the TV receiver is the sign of a leaky oscillator. No sound and no picture with a white raster can result from a leaky or open mixer transistor. Replacing transistors inside the tuner can pose a problem, because components are tightly packed. It's difficult to get side cutters or a soldering iron tip down under the various wafer sections of the tuner.

A poorly soldered connection causing intermittent picture and sound can be located with an insulated tool. Simply tune in a station and probe around with the plastic tool until the picture acts up. Often a poor solder connection or broken component lead is located. Suspect a broken oscillator coil lead or dirty switch contact when only one station cannot be tuned in. Inspect the soldered coil leads where they are brought together at the front of the tuner for poorly soldered connections. The tuner assembly should be sent in for service when cleaning and simple repairs fail to fix the tuner (Fig. 8-12).

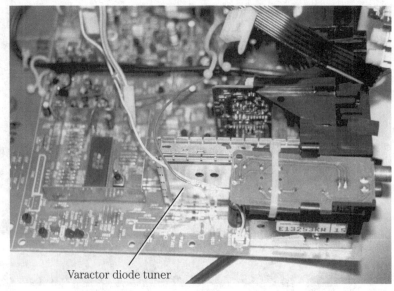

Varactor diode tuner

■ **8-12** *Today, the varactor tuner mounts directly on PC board.*

UHF tuner

In many color TV chassis, the UHF tuner is a separate mechanical tuner, except in those with a combined VHF-UHF varactor assembly. Today, the UHF tuner can be a 70-channel detent, mechanically tuned type. When the VHF channel has been set for UHF operation, B+ voltage is switched from the VHF tuner to the UHF tuner. UHF-VHF switching can be found at the rear or inside of the VHF tuner.

Transmitted signals received by the UHF antenna are fed to a balanced 300-Ω input and coupled to L50 in a typical UHF tuner assembly (Fig. 8-13). The UHF oscillator transistor (Q50) generates an oscillator signal that is 45.75 MHz above the picture carrier frequency of the broadcast station. The incoming signal and the oscillator frequency are mixed by diode D50, which produces a third frequency resulting in the IF signal. This IF signal is fed to the VHF tuner with a 75-Ω shielded cable.

Most problems found with the mechanical UHF tuner are in Q50 or D50. No UHF station will be received if either the transistor or diode become leaky. Q50 should be replaced with the original part number. UHF diode D50 can be replaced with a IN82 UHF diode. Do not replace this diode with an ordinary IN34 type. Check the rubbing of tuning capacitor plates when one station appears erratic while tuning. Low voltage at the UHF tuner terminal can in-

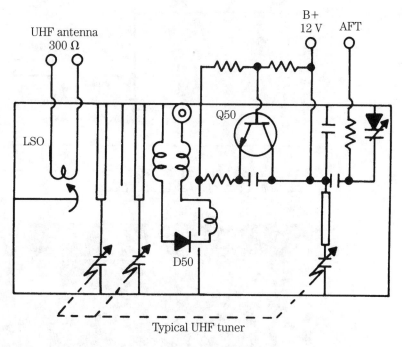

■ 8-13 *A typical UHF mechanical tuner schematic.*

dicate a leaky transistor (Q50) or improper switching voltage. Flex the UHF IF cable for intermittent or drifting UHF signals.

The varactor tuner

Most of the present-day tuners use some form of the varactor tuner within a tuner module. The varactor tuner might contain several varactor diodes that change the frequency of each TV channel. The varactor diode works as a capacitor when a reverse bias is applied to the PN junction. Electronic tuning is performed by a fixed inductance and varying the capacitance with a varactor diode (Fig. 8-14).

When a different dc voltage is applied across the varactor diode, the capacity changes, tuning the inductance circuit. With a different variable resistance on each fixed inductance, each channel can be tuned manually or electronically (Fig. 8-15). You can manually turn each variable resistor with a screwdriver or fingernail and tune in each channel. These small control adjustments are quite critical. Each channel number has a variable resistor. Each variable resistor can tune in more than one station, so make sure you have located the correct broadcast station for a given channel.

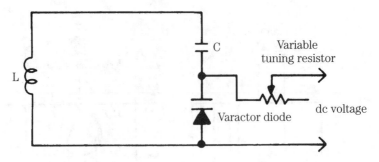

■ **8-14** *Electronic tuning with the varactor tuner is performed by a fixed coil inductor and a varactor diode.*

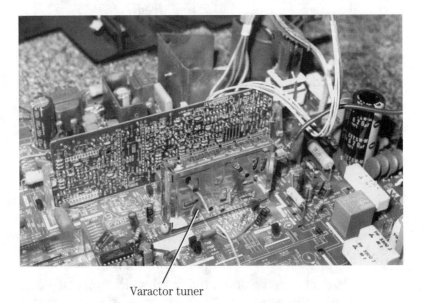

Varactor tuner

■ **8-15** *Stations can be tuned in with the remote or with separate buttons on the RCA CTC146 chassis.*

Because the varactor tuner cannot cover all VHF channels, the entire band is divided into low and high bands. The low band covers channels 2 through 6 and the high band covers channels 7 through 13. A switching diode is used to switch from high to low in an electronic tuner system.

Any varactor tuner can be tested by checking the voltage applied to the tuner (Fig. 8-16). Remember, as the tuner is tuned manually or electronically to a different station, the applied voltage is different on each channel. By checking the applied voltage (variable for each channel), we can determine if the tuner is defective.

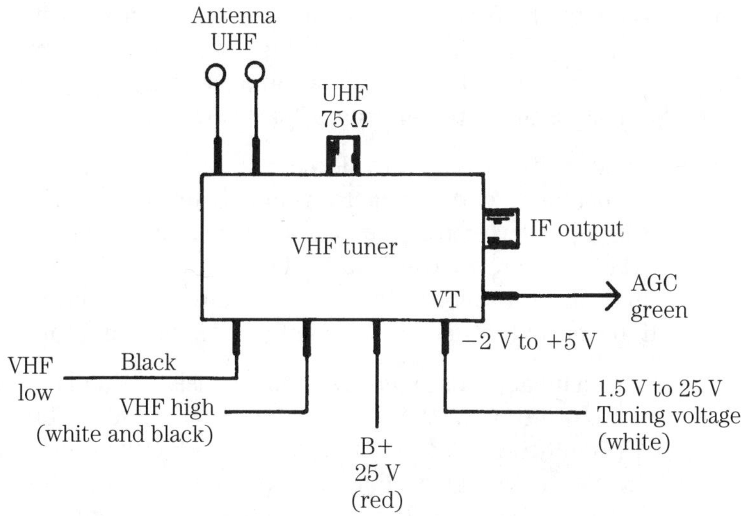

■ **8-16** *Check the varactor tuner with correct supply voltage, AGC voltage, and tuning voltage.*

Another method to check the suspected tuner is with an external dc voltage source. Remove the wire to the tuning voltage post and apply dc voltage to the tuner. Check the schematic for correct voltage polarity. Start at 1.5 V. Slowly bring up the voltage, and notice what stations are tuned in. Suspect a control unit or improper AGC if external voltage tunes in the various local stations. A missing channel can be found in the same manner. Compare the station program with that on another TV.

The UHF tuner can be a separate tuner, or it can be combined with the VHF varactor tuner. Six or more blank channels are provided for tuneable UHF stations with the combined varactor tuner. The UHF stations are tuned in with the same type of variable resistor. Often, with the separate UHF tuner, the voltage is switched from the VHF tuner and applied to the UHF tuner.

Most problems found in the varactor tuner are caused by transistor and varactor diodes. Intermittent problems or tuner drifting can be caused by poorly soldered connections or by a breakdown of components. Because these components are inside and difficult to get to, most tuners are sent to the tuner depot for repair.

J.C. Penney manual varactor tuner

In a J.C. Penney 685-2026 (made by RCA), a manual-type varactor tuner is found with a varactor tuner board and separate tuning assembly. The tuner module can be easily changed if it is suspected of being defective because cables plug into the tuner board assembly. A check of varactor tuning voltage can be made at the tuner assembly to determine if the varactor board is defective.

When stations began to drift or one station cannot be tuned in, suspect a defective varactor tuning board (Fig. 8-17). Sometimes the variable resistor control is so critical that the station drops out when the screwdriver is taken from the control. Sometimes, when manually turning from channel to channel, the station drops out or intermittently tunes in. Suspect a defective varactor tuner board.

The manual tuner indent assembly consists of a plastic shaft and a switch to tune in the various stations. Erratic or intermittent reception can be caused by a dirty or defective switch. It's best to change the variable tuning board assembly. This assembly comes in two different sections and can be ordered directly from your RCA parts distributor or from J.C. Penney Co.

■ **8-17** *Remove the varactor tuner from the PC board by unsoldering the pins from the PC board.*

Varactor tuner drifting

In many of the latest TV chassis, varactor tuners are found both in the VHF and UHF tuners. Station drifting of either the high or low band of the VHF tuner can be caused by a defective UHF tuner. A leaky capacitance diode or component within the UHF tuner can lower the tuning voltage applied to the VHF tuner. Often this condition exists after the TV chassis has operated for several hours.

Tune in the station that drifts or fades out and leave the TV operating. When the station disappears, cut the B+ wire feeding the UHF tuner. If the station pops back in you can assume the UHF tuner is defective and not the VHF tuner. Let the chassis operate for several hours making sure the VHF stations do not fade or drift off channel.

Mark down all color-coded leads or wires to the UHF tuner on a piece of paper (Fig. 8-18). Tape the paper to the TV chassis or cabinet for replacing the tuner after factory repair. Remove the UHF tuner and send it to the tuner repair depot. Besides the UHF tuner, a defective VHF tuner can cause stations to drift off channel.

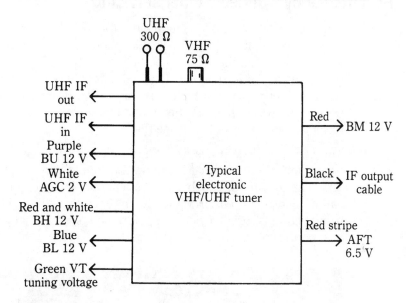

■ **8-18** *Check all voltages on the varactor tuner before removing it and sending it on for repair.*

RCA CTC145E tuner drifting

In a brand new CTC145E chassis, the higher channels would operate for several hours and then drift off. Tuning voltage that was applied to the tuner was fairly normal. All voltage was normal upon the other tuner terminals. A new tuner was ordered, solving the drifting of channels.

RCA CTC177 board-tuner

With intermittent soldered joints in the RCA CTC177 board-tuner, the picture may flop, go out, become snowy, and collapse. This tuner is built on the PC board chassis. Check the shield solder joints and tuner ground connections.

Remove the bottom cover of the tuner. Solder all shields to ground. Renew the old solder joints with new solder. Do not apply too much heat, or you will raise the foil PC wiring from the board. Solder all joints that lead to ground or shield within the tuner circuits. Return the pads and make sure the tabs in both the shields and the cover are clean and can accept solder. Reinstall the tuner shield and tuner cover. Solder all tabs with fresh solder.

RCA frequency synthesis tuner servicing

Many of the latest RCA TV module tuners use a frequency synthesis system. Replacing the MST tuner module or MSC frequency module can determine which module is defective (Fig. 8-19). Most technicians substitute a new module and turn the defective module in for repair. Do not overlook a defective IF cable or module connecting wiring harness. Some manufacturers do not want you to service modules, only replace them.

The CTC131 chassis has an MST multiband tuner and an MSC tuner control module that make up the tuning system, providing 127-channel tuning capability. This channel-lock tuning system can be controlled either manually or remotely. The digital command center is a digitally encoded IR remote system that gives the user control of the TV receiver (Fig. 8-20).

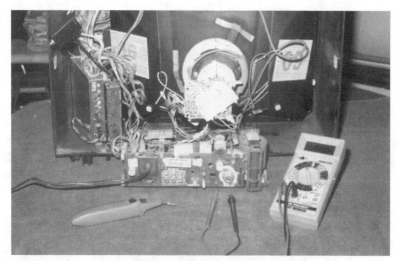

■ 8-19 *The tuner and system control modules in CTC108, 109, and 110 chassis.*

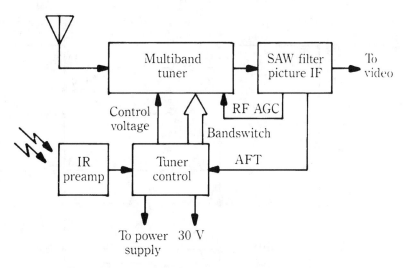

■ **8-20** *A block diagram of RCA CTC131 tuner system.* Thomson Consumer Electronics

RCA system control

Within the RCA CTC156/157/158/159 chassis are system control/ system reset and communications circuits. A serial data bus carries communication between the system control microcomputer IC and AIU processor. Y3301 (4 MHz) controls the clock for timing the system control computer and AIU. The system control microcomputer generates the clock signal at pin 33 for the serial data bus. This clock operates at 125 kHz, and is generated by dividing the 4-MHz clock signal (Fig. 8-21).

Pin 2 of the system control microcomputer IC (U3100) (interrupt request (INT)) is pulled high by the 5-V SW supply. Pin 2 must be high for the system control microcomputer to operate normally. Make sure pin 2 is pulled high before changing the system control IC (U3100).

The M/S (master/slave) at pin 34 is also pulled high by the 5-V supply. This pin must be high before the master mode will function. If the voltage is lost, the system control microcomputer will enter the slave mode.

The XRP DET (x-ray shutdown detect) is at pin 36 and is supplied with a highly developed transistor and resistor network. When the deflection circuit is shut down, the 9-V run voltage disappears, producing a low at pin 36 and shutting down the chassis.

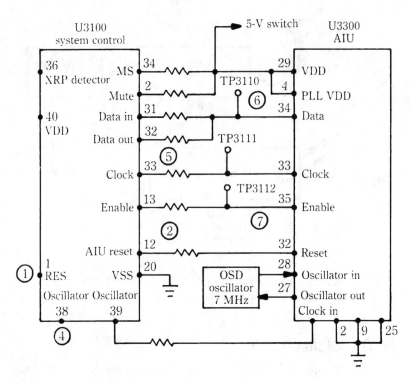

■ 8-21 *The system control microcomputer (U3100) and AIU IC (U3300) control the TV receiver operation.*

The system control ensures that U3100 starts at the same point in the program when the TV is turned on. The 5 V is applied to U3100 and U3300 when set is turned on. Both ICs must be reset each time power is applied to the chassis.

The RCA control system can be checked by taking voltage and waveforms on the pins of U3100 and U3300. Pin 1 should have 5 V. If it does not, suspect the 12-V supply or the 5-V standby regulator. Pin 12 should have 5 V. If not, U3100 is not operating correctly. Suspect U3300 if the voltage does not go high at pin 32.

Check the waveforms on pins 13, 31, 32, and 33 of U3100 and pins 33, 34, and 35 of U3300. When the waveform is missing, remove the pin on U3100 and take another waveform check. Suspect U3100 if the signal is missing. Check and suspect U3300 if the signal is present. Check U3100 and U3300 by the numbers.

Customer control

All customer control functions are generated electronically. The controls from U3300 (AIU) are contrast, brightness, color, tint, sharpness, treble, bass, volume, balance, and on/off. These are controlled by the input of the remote-control or keyboard system. The infrared receiver picks up the infrared signal from the remotes, amplifies the signal, and applies it to pin 38 of U3300 (Fig. 8-22).

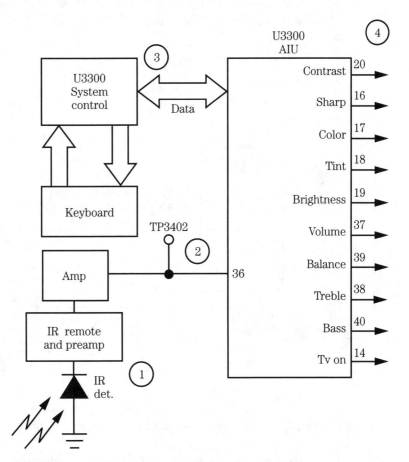

■ **8-22** *The remote and control AIU IC (U3300) controls the customer operations of TV.* Thomson Consumer Electronics

If the keyboard operates correctly and the infrared remote does not, suspect remote or infrared receiver circuits. Suspect the keyboard contacts and cable if the remote works and the keyboard does not. If only one function does not operate, suspect problems with components on the output pin (U3300) of that function. If all functions do not operate, suspect U3100, U3300, or the keyboard or remote-control system.

RCA CTC157 tuner control

The tuning system is a Frequency Synthesis (FS) type with a crystal-controlled phase-locked loop. The tuning system contains tuning control IC (U1001), band decoder and switch (U3600), and varactor diode tuning (Fig. 8-23). The tuner uses a number of varactor diodes in the local oscillator circuits. These diodes act like a variable capacitor when a different voltage is applied from the tuning control IC.

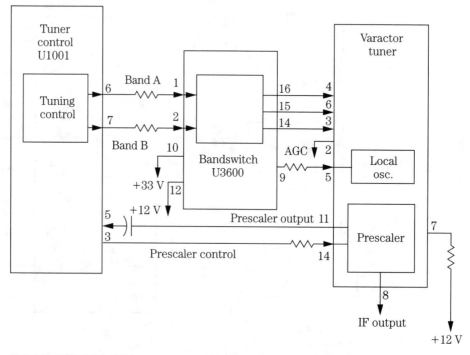

■ 8-23 *RCA CTC157 frequency synthesis tuning control circuits.*

Both the bandswitching and tuning voltage control circuits must operate correctly for proper tuning voltage to be produced. In order for the tuning control circuit to synthesize the correct tuning control voltage, all signals used by the tuning control circuit must be correct.

When the channel is changed, the system control computer IC supplies the tuning control data signal to the AIV. The AIV interprets the data and generates the exact bandswitching signals and feedback control signals. Feedback control signals used within AIV and the PSC signal provided to the prescaler in the tuner.

Check the tuning voltage applied to the varactor tuner (pin 5). If these applied voltages change and no stations, suspect a defective varactor tuner. Take critical voltages upon tuner, U3600 band switch, and tuner control U1001. Check AGC voltage applied to the tuner. Make sure all supply voltages are normal in each IC.

Sylvania TS-15/TS-18 tuning systems

The TS-15 and TS-18 tuning systems are quite similar in operation. The TS-18 tuning system uses an 8-bit microcomputer, a computer interface tuning and control chip (ClTAC), and an on-screen display IC for processing on-screen graphics. The UHF and VHF tuners and CITAC are mounted on the main chassis. The microcomputer, 5-V regulator, on-screen display, clock/RAM, and ROM ICs are mounted on the TS-18 module.

The tuning system receives 13 Vdc from the main chassis. This voltage source comes from the switched mode power supply (SMPS). The 13-V source is applied to a 5-V regulator (lC305) that supplies 5 V for the tuning system and 5 V to the main chassis. The 5-V source keeps the microcomputer alive when the set is turned off (Fig. 8-24).

The microcomputer controls the on-screen display, RF switch, power LED, stereo decoder mode, and audio and video circuits. The microcomputer (IC301) receives information from the remote control and keyboard. The C1TAC IC controls bandswitching, channel tuning, and AFT of the tuner. The volume, treble, bass, and balance are controlled by a digital-to-analog converter (IC312) located on the main chassis. The volume is controlled in 64 steps, and the audio is muted when changing channels or by using the mute control key.

When the set is turned on, it tunes to the last channel viewed. The system should maintain the last channel tuned and return to that channel after power interruption. A lithium battery located on the TS-18 module operates the clock/RAM IC306 for this purpose. The last channel viewed, sound and picture settings, and on-screen clock are maintained within IC306. Crystal Y302 (32,768 kHz) is divided by a 15-stage binary counter within IC306 to arrive at 1 Hz, which is used by the time clock. When the lithium battery goes weak or dead, the on-screen display will show "Battery Failure." The battery works only when the set is unplugged from the wall.

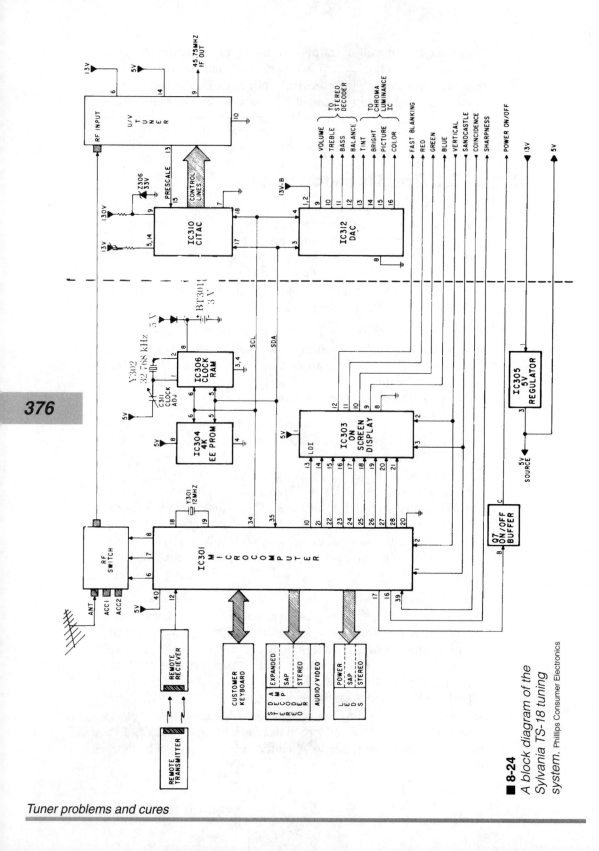

■ 8-24

A block diagram of the Sylvania TS-18 tuning system. Phillips Consumer Electronics

Tuner problems and cures

IC304 stores information such as personal preference, favorite station, caption memory, and other codes. Data is entered into and exits from IC304 and IC306 in serial form on two lines. One line is the clock signal from IC301, and the other is the data signal from IC301, IC304, or IC306. The clock signal is noted as SCL, the data signal as SDA. Both of these lines together are referred to as 1S02TC BUS by some sources in the industry.

The on-screen channel numbers of the TS-18 tuning system can appear in different formats. The normal display area for the channel numbers is in the upper left corner of the picture tube. The lower right corner displays clock information. Both channel and clock numbers are displayed in yellow.

Menu selections

Three major menus are used with the TS-18 tuning system: audio adjustments, color and picture adjustments, and clock and channel adjustments. To view the menus, press the menu button on the remote control or the keyboard. Each time you press the menu button, another menu is displayed (Fig. 8-25).

■ **8-25** *The treble menu is selected to set the treble sound level with the remote control in many large TV screens.*

■ 8-26 *The audio menu to select and set the various operations with the remote control.*

Menu 1 has adjustment selections for base, treble, balance, audio mode, expanded sound, and input selections for the RF (Fig. 8-26). Menu 2 has adjustment selections for brightness, picture, color, tint, sharpness, and tuning. Menu 3 has time set, channel scan, parental control mode code modifications, personal preference programming, options for time/channel/tuning displays, and program memory.

The TS-15 and TS-18 tuning systems have a sleep timer function. The TS-15 does not have parental control and captions. All operational modes for both tuning systems are via the menus.

The TS-18 tuning system can be serviced by replacing the TS-18 module or by checking components located on the main chassis. Check the input and output signals and voltages to and from the main chassis at interconnect plugs P/J312, P/J305, P/J310, P/J304, P/J308, P/J301, and P/J309 on the TS-18 module (Tables 8-1 and 8-2). When taking voltage measurements, set the adjustment menus to mid-range for stereo, expanded sound, antenna, normal tuning, normal time, and all-channel scan. For normal operation, set the volume at minimum.

■ Table 8-1 Sylvania tuning system with voltage tests
at P/J301, P/J302, P/J304, and P/J305 for defective tuner operation.

P/J301		P/J302		P/J304		P/J305	
Pin	Voltage	Pin	Voltage	Pin	Voltage	Pin	Voltage
1	5.0	1	4.9	1	11.5	1	Gnd
2	4.8	2	4.9	2	12.2	2	11.8
3	Gnd	3	4.9	3	11.6	3	2.1
		4	Gnd	4	Key	4	Key
		5	4.6	5	11.8	5	1.6
		6	4.6			6	4.7
		7	4.6			7	4.7
						8	4.8
						9	2.9

Phillips Consumer Electronics

■ Table 8-2 Check the tuning system with voltage checks
at P/J307, P/J308, P/J310, and P/J312 in a Sylvania C9 tuning system.

P/J307		P/J308		P/J309		P/J310		P/J312	
Pin	Voltage	Pin	Voltage	Pin	Voltage	Pin	Voltage	Pin	Voltage
1	11.8	1	8.8	1	Gnd	1	Gnd	1	2.3
2	9.8	2	9.2	2	11.8	2	12.6	2	1.4
3	11.0	3	Key	3	0.0	3	Key	3	Key
4	Key			4	0.0	4	0.0		
5	9.3			5	3.2	5	0.0		
6	9.3					6	0.0		
						7	0.0		

Phillips Consumer Electronics

379

Realistic 16-261 portable tuner troubleshooting

To troubleshoot the tuning system in the Realistic 16-261 portable, follow Tables 8-3 through 8-5. Before replacing any major component, such as a tuner, microprocessor (CPU) IC, prescaler/PLL IC or bandswitch IC, check the tables (Tables 8-3 through 8-5) and block diagrams (Figs. 8-27 through 8-32).

■ **Table 8-3 A troubleshooting flowchart for checking the IC microcomputer (C532).** Radio Shack

---

Troubleshooting on tuning section

Caution: Before the replacement of any major tuning components such as a tuner, microprocessor (CPU) IC, prescaler/P.L.L. IC, or band switch IC, check it according to the following charts to see if the part is defective.

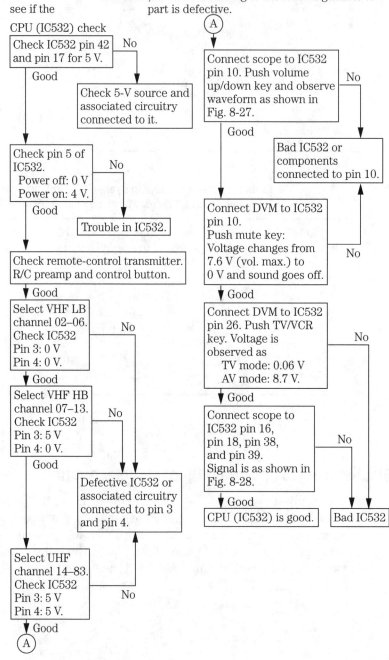

CPU (IC532) check

Check IC532 pin 42 and pin 17 for 5 V. — No → Check 5-V source and associated circuitry connected to it.

Good

Check pin 5 of IC532.
  Power off: 0 V
  Power on: 4 V. — No → Trouble in IC532.

Good

Check remote-control transmitter. R/C preamp and control button.

Good

Select VHF LB channel 02–06. Check IC532 Pin 3: 0 V Pin 4: 0 V. — No

Good

Select VHF HB channel 07–13. Check IC532 Pin 3: 5 V Pin 4: 0 V. — No → Defective IC532 or associated circuitry connected to pin 3 and pin 4.

Good

Select UHF channel 14–83. Check IC532 Pin 3: 5 V Pin 4: 5 V. — No

Good

(A)

(A)

Connect scope to IC532 pin 10. Push volume up/down key and observe waveform as shown in Fig. 8-27. — No → Bad IC532 or components connected to pin 10.

Good

Connect DVM to IC532 pin 10.
Push mute key:
Voltage changes from 7.6 V (vol. max.) to 0 V and sound goes off. — No

Good

Connect DVM to IC532 pin 26. Push TV/VCR key. Voltage is observed as
  TV mode: 0.06 V
  AV mode: 8.7 V. — No

Good

Connect scope to IC532 pin 16, pin 18, pin 38, and pin 39. Signal is as shown in Fig. 8-28. — No → Bad IC532

Good

CPU (IC532) is good.

Tuner problems and cures

Tuner and prescaler IC (IC521) check

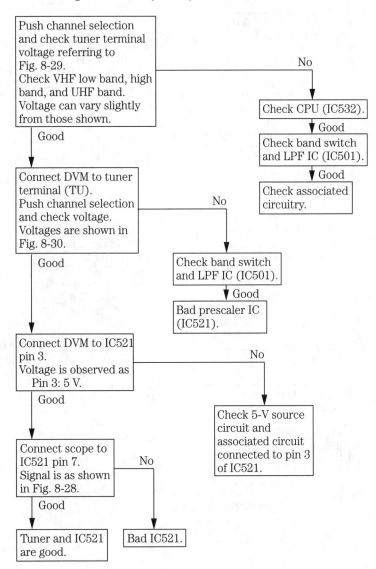

Band switch and LPF IC (IC501) check

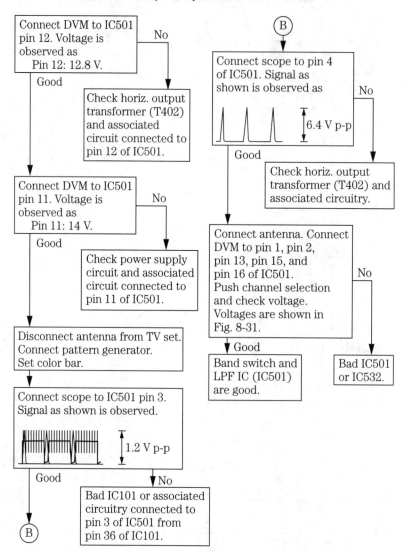

Connect DVM to IC501 pin 12. Voltage is observed as
    Pin 12: 12.8 V.

No → Check horiz. output transformer (T402) and associated circuit connected to pin 12 of IC501.

Good

Connect DVM to IC501 pin 11. Voltage is observed as
    Pin 11: 14 V.

No → Check power supply circuit and associated circuit connected to pin 11 of IC501.

Good

Disconnect antenna from TV set. Connect pattern generator. Set color bar.

Connect scope to IC501 pin 3. Signal as shown is observed.
    1.2 V p-p

Good → B

No → Bad IC101 or associated circuitry connected to pin 3 of IC501 from pin 36 of IC101.

B → Connect scope to pin 4 of IC501. Signal as shown is observed as
    6.4 V p-p

No → Check horiz. output transformer (T402) and associated circuitry.

Good

Connect antenna. Connect DVM to pin 1, pin 2, pin 13, pin 15, and pin 16 of IC501. Push channel selection and check voltage. Voltages are shown in Fig. 8-31.

Good → Band switch and LPF IC (IC501) are good.

No → Bad IC501 or IC532.

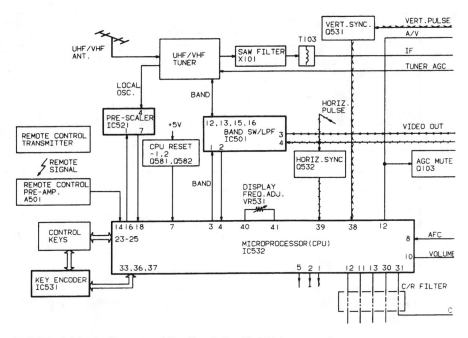

■ **8-27** *A block diagram of the Realistic 16-261 tuner system.* Radio Shack

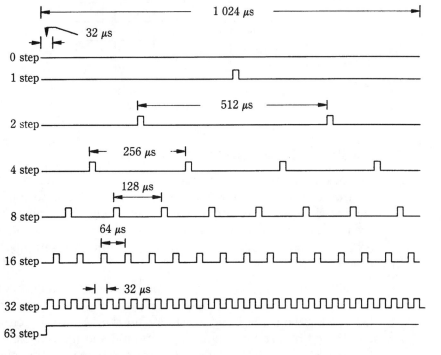

■ **8-28** *The waveforms at pin 10 of IC532 to check the volume up/down key.*
Radio Shack

Realistic 16-261 portable tuner troubleshooting

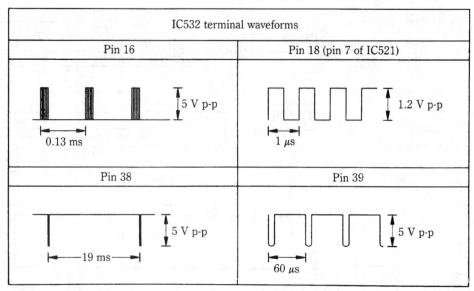

IC532 terminal waveforms	
Pin 16	Pin 18 (pin 7 of IC521)
5 V p-p 0.13 ms	1.2 V p-p 1 μs
Pin 38	Pin 39
5 V p-p 19 ms	5 V p-p 60 μs

■ **8-29** *Scope waveforms at pins 16 and 18 of IC532 to check for the correct signal at IC532.* Radio Shack

384

	TUNER TERMINALS			
	LB	HB	UB	MB
VHF LOW	12 V	0 V	0 V	12 V
VHF HIGH	0 V	12 V	0 V	12 V
UHF	0 V	0 V	12 V	12 V

■ **8-30** *Tuner voltage on the terminals of the low and high bands.* Radio Shack

The electronic tuner keyboard

The electronic tuner functions with a tuner, control unit, remote control receiver, or keyboard assembly. The control unit supplies various voltages to the tuner module or assembly. No moving parts are found in the tuner (Fig. 8-33). The control unit is made up of several ICs and diodes and a frequency-synthesizer with micro-processor control modules.

In the keyboard chassis, you simply press the correct numbers to select the right station. With the RCA chassis, to select any station below 10, you push zero and the station number. Above 10, you select the correct station number. In Admiral and many Japanese keyboards, you push the station numbers and then push the select button, and the selected station pops in.

NORMAL CHANNELS 02-83								CATV CHANNELS 01-47, 95-99					
CH.	VOLT.	CH.	VOLT.	CH.	VOLT.	CH.	VOLT.	CH.	VOLT.	CH.	VOLT.	CH.	VOLT.
02	1.90	23	4.94	44	10.50	65	15.70	01	5.73	22	5.13	43	19.27
03	3.28	24	5.23	45	10.73	66	16.01	02	1.90	23	9.27	44	20.22
04	4.35	25	5.52	46	10.96	67	16.33	03	3.28	24	9.73	45	21.38
05	6.65	26	5.82	47	11.19	68	16.66	04	4.35	25	10.18	46	22.73
06	7.97	27	6.12	48	11.42	69	17.01	05	6.65	26	10.63	47	24.30
07	5.54	28	6.42	49	11.45	70	17.37	06	7.97	27	11.06	95	9.73
08	6.31	29	6.72	50	11.87	71	17.76	07	5.54	28	11.48	96	11.05
09	6.89	30	7.01	51	12.10	72	18.16	08	6.31	29	11.91	97	12.40
10	7.25	31	7.29	52	12.33	73	18.57	09	6.89	30	12.33	98	13.91
11	7.78	32	7.57	53	12.57	74	18.96	10	7.25	31	12.75	99	15.69
12	8.30	33	7.85	54	12.80	75	19.36	11	7.78	32	13.18		
13	8.80	34	8.12	55	13.05	76	19.80	12	8.30	33	13.61		
14	2.44	35	8.38	56	13.29	77	20.27	13	8.80	34	14.04		
15	2.69	36	8.62	57	13.54	78	20.79	14	19.07	35	14.49		
16	2.95	37	8.86	58	13.78	79	21.35	15	22.55	36	15.09		
17	3.22	38	9.10	59	14.04	80	21.97	16	1.57	37	15.44		
18	3.49	39	9.33	60	14.30	81	22.65	17	2.06	38	15.95		
19	3.78	40	9.57	61	14.57	82	23.42	18	2.59	39	16.49		
20	4.06	41	9.80	62	14.84	83	24.37	19	3.15	40	17.08		
21	4.35	42	10.03	63	15.12			20	3.73	41	17.73		
22	4.65	43	10.26	64	15.41			21	4.33	42	18.45		

■ **8-31** *Tuning voltages of normal channels from 2 through 83.* Radio Shack

CHANNEL	IC501 TERMINALS				
	PIN 1	PIN 2	PIN 13	PIN 15	PIN 16
14-83(N)	5 V	5 V	12 V	0 V	0 V
07-13(N) 07-13(C) 16-47(C)	5 V	0 V	0 V	12 V	0 V
02-06(N) 01-06(C) 14,15(C) 95-99(C)	0 V	0 V	0 V	0 V	12 V

(N): NORMAL MODE, (C):CATV MODE

■ **8-32** *Voltages on the terminal pins of IC501 with the various channels.*

The electronic tuner keyboard

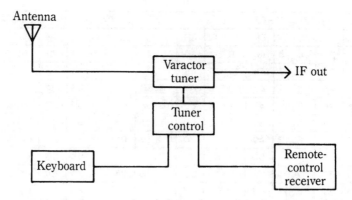

8-33 *The electronic varactor tuner has no moving parts; it is controlled only by a change in the tuning voltage.*

Actually, the keyboard selects the station tied to the control unit and the control unit selects the correct voltage applied to the tuner. When a dc voltage is applied across the varactor diode, the capacitance of the diodes changes the inductance-tuned circuit to the desired station. The control unit operates the same in a remote-control receiver.

The tuner selects the channel with voltage from the control unit. Check the tuning voltage when a wrong station is tuned in, when it refuses to change channels, or when it drifts off channel. Improper or erratic voltage indicates a defective control assembly. Look for various tuning voltages found in the manufacturer's service literature and Howard Sam's Photofacts. Each channel is listed with correct tuning voltage (Fig. 8-34). If the voltage is correct and there is no tuner action, suspect a defective tuner.

Another form of keyboard tuner control is found in Japanese and other foreign TV receivers. The keyboard might be mounted directly on the selector board or there is a separate keyboard (pushbutton) assembly. Each channel has a pushbutton to locate the exact station (Fig. 8-35).

Terminal voltage	VHF tuner	UHF tuner
B + voltage	15 V	15 V
Tuning voltage	1.5 V to 25 V	1.5 V to 30 V
AFT voltage	0.2 V to 15 V	0.2 V to 15 V
AGC voltage	− 2 V to 5 V	1.5 V to 5 V

8-34 *Typical tuner voltages.*

■ **8-35** *Pushbutton selector board for VHF and UHF tuners.*

The keyboard selects each station with the help of the selector board. A channel control and memory IC controls the band-switching and tuning voltage connected to the VHF and UHF tuner. You might find a separate VHF tuner or both tuners can be included in one unit. The IF signal is cabled to the TV chassis.

Some of these selector control boards have plug-in connections for easy removal. Most of the foreign control units are wired directly into the tuner and TV chassis. Determine if the tuner is defective. Check for variable tuning voltage at the tuner. Suspect a defective tuner when all voltages are normal with a different tuning voltage as each channel button is pressed.

Scan tuning

Scan tuning allows the operator to slowly scan up or down the channels for favorite programs. This type of tuning is especially useful when connected to a cable TV system with many stations to choose from. There are many different types of scan and push-button tuning devices, but they all perform the same task.

In an RCA keyboard or scan tuning operation, the up-and-down switch assembly feeds into an MSC tuner control module (Fig. 8-36). The frequency synthesis tuner control module feeds the correct VHF and UHF band voltage to the tuner module. An LED display indicates the channel number.

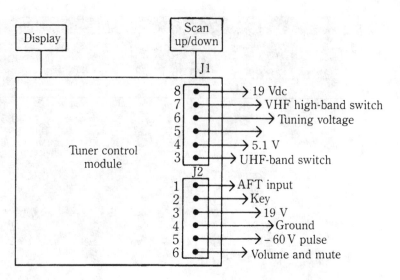

8-36 *An RCA scanning control module in an RCA TV chassis.*

A defective MSC scan module can cause improper (or no) channel up and down action, improper addition or deletion memory of an incorrect or missing LED display, improper (or no) volume and mute control, no tuning voltage to the tuner, or improper (or no) channel change or skipping of channels. Replace the module for the above symptoms.

If the suspected module does not restore a missing LED number, check the cable and plugs going to the LED display assembly. Intermittent display can be caused by a defective cable or wiring. Check for a poor crimped wire at the LED socket. A poor ground connection of the MSC module can cause erratic (or no) scanning action.

A snowy raster with no picture or sound can be caused by a missing 60-V pulse at terminal 5 (Fig. 8-37). Look for broken wiring or a defective plug or connection at L103 on the flyback transformer. Check for a pulse waveform at terminal 6 on the tuner and terminal 7 of flyback transformer.

Tuner control or tuner module

To determine if the tuner is defective, replace the tuner in the modular chassis. Most of these tuners plug into the tuner control or memory module. If the raster is still snowy or has a white screen, replace the control module. In remote TV chassis, replace the power control module when the chassis is completely dead.

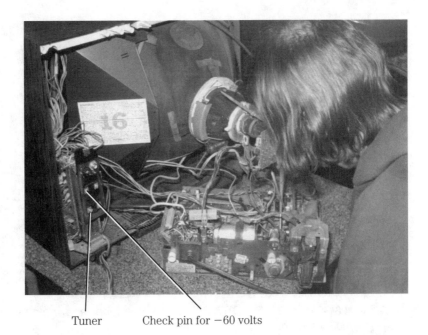

Tuner   Check pin for −60 volts

■ **8-37** *A snowy raster caused by –60 volts at the memory module in an RCA chassis.*

Make sure that all plugs are firmly pressed down on each module. If the tuner and control module plug together, make sure the plug-in assembly inside the module lines up properly. In some modules, you might find that the AGC wire plugs into the tuner separately. This wire can accidently pull off the exchanging tuner modules and can cause a snowy (or no) picture. Press down on the cable wire assembly after the tuner is operating.

Electronic tuner repair

The defective electronic tuner in all TV chassis can be replaced. Some of these tuners are mounted together with a memory or bandswitching module. Simply remove the two metal screws and unplug the tuner module. Another tuner with the same parts number can be substituted to cure the snowy picture. When both units are tightly plugged together, voltage measurements between the control and tuner are difficult to make (Fig. 8-38). Simply substitute each module until the correct one is found. Sometimes when lightning strikes the TV, the tuner, memory, and control units are all damaged.

The quickest method to repair the front end of any TV chassis is with tuner or control module substitution. The suspected module

■ **8-38** *In the RCA CTC108 chassis, the tuner and memory module mount together.*

is unplugged and another one is mounted in its place. Some manufacturers would rather you replaced the module instead of trying to repair it (Fig. 8-39). Unless you have a lot of time, the correct test equipment, and the knowledge to repair these tuners and control modules, they should be returned to the TV distributor or manufacturer. Most tuner repair depots repair many different American and Japanese modular tuners and control units (Fig. 8-40).

As time goes by, the electronics technician might be able to service the electronic tuner, memory, and control modules. Service data supplied by each manufacturer can provide technical data for these repairs. Do not try to service all-electronic tuner and control modules without exact service data. Usually, most service establishments are swamped with many different service problems and repairing these modules is last on the list.

Seven actual tuner problems

Here are seven actual tuner-related problems that happened with various TVs.

Tuner system	Defective component
Laser channels	Replace control module
No picture on 9 and 11	Replace control module Check AFT adjustment
Won't change channels	Replace control module
Searching on all channels	Replace control module
VHF searching	Replace tuner module
Intermittent channel 5	Replace tuner module
Keyboard malfunction	Replace control module Check plug-in connection Check keyboard assembly
Numbers missing on display	Replace control module
Incorrect channel display	Replace control module Check plug-in connection
Intermittent display	Replace control module
No display	Replace control module
Erratic channel 5	Defective pushbutton on keyboard
Weak picture on channel 5	Replace tuner module
Weak picture on channel 13	Replace tuner module
No picture, no sound	Replace tuner module and control module
Flashing picture	Replace tuner module Check IF and input cables on tuner Loose AGC connection
Snowy picture	Replace tuner module Damaged balun coil AGC lead off
No UHF picture	Replace tuner module Check for UHF tuning voltage at UHF tuner
Weak channel 21	Replace tuner module Check UHF antenna cable leads

■ **8-39** *A modular tuner troubleshooting chart.*

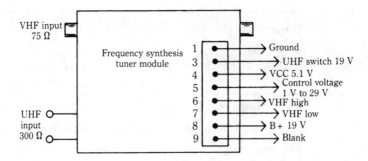

8-40 *Most tuner modules can be returned or exchanged at the manufacturer's distributor, or sent in to a repair center.*

Tuner drifts off channel

Monitor the waveform at output of first IF transistor (Q2301) to determine if the IF or the tuner is at fault. Check to see if the AGC voltage is from 3.1 V to 8 V with a station tuned in. Check the tuner supply voltage on pin 7; it should be 12 V (Fig. 8-41). Monitor the different tuning voltages on pins 4, 5, and 6. If the tuning voltage is applied to the varactor tuner when it drifts off channel, replace the tuner in the RCA CTC145 and CTC146 chassis.

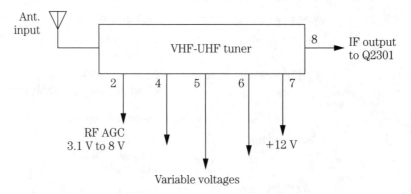

8-41 *Check the various voltages on the VHF-UHF varactor tuner to determine if the tuner is defective.*

Intermittent snowy picture

When the picture became snowy in an RCA CTC140 chassis, you could tap the tuner and the picture might turn good and then fade out. All the tuner connections were soldered around the tuner; all voltages were fairly normal. It was necessary to replace the intermittent varactor tuner (Fig. 8-42).

■ **8-42** *The defective tuner caused a snowy picture in the RCA CTC140 chassis.*

No channel tuning

The tuner would not tune in a station in the Emerson EC10 chassis. The voltage supply source was checked upon the tuner and was normal at 11.3 volts. When the RF pin 5 was checked, no voltage was found on this RF AGC terminal. At first IC101 was suspected, but all voltages were normal. Defective C110 (10 µF) was replaced off of pin 10 and solved the tuning problem (Fig. 8-43).

Flashing picture

Although the outside antenna was old and could have caused the flashing problems in a Sanyo 31C40A, the set was brought into the shop for observation. Often another set in the house or a small TV is used to check the outside antenna reception.

The TV was still flashing when it was connected to the shop antenna. Although the picture was fairly normal with the subber clipped to the IF cable, the picture was still flashing. One thing for sure, the picture was better with the tuner-subber.

The IF and antenna cables were flexed and held in many different positions, with the picture still flashing. A shorted or poorly connected IF cable can cause a picture to flash off and on. All wiring connections were inspected at the various tuner terminals.

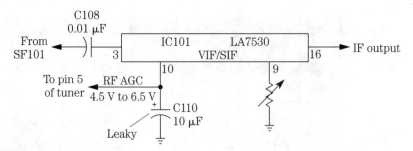

■ 8-43 *Defective capacitor C110 caused a lack of channel tuning in an Emerson EC10 chassis.*

Voltage measurements at the tuner would rapidly change as the picture continued to flash off and on. Even the RF AGC voltage would change rather rapidly. The AGC voltage was clamped with a variable dc voltage power supply at the tuner RF AGC terminal.

The tuner responded with 3 to 8 V applied from the variable dc source. The flashing picture was caused by improper AGC voltage to the VHF tuner (Fig. 8-44). Replacing IC101 (PIF amp-AGC-AFC sound-det) with a direct manufacturer's replacement (M5186AP) solved the flashing picture.

394

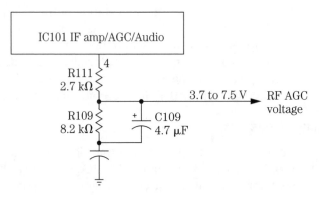

■ 8-44 *A flashing picture was caused by a bad IC101 in a Sanyo 31C40A chassis.*

Unusual tuner problem

A no-raster, no-picture, no-sound, and no-channel-display symptom was found in an RCA CTC118A chassis. Both the memory and tuner modules were replaced, with no results. Normal high voltage was measured at the picture tube. Sometimes when channel 13 was pressed, a flash of light went across the screen. A tuner-subber on the IF cable indicated the rest of the chassis was normal.

Voltage measurements on the tuner (18 V) and memory module were fairly normal. The picture and sound were brought back with an injected B+ tuning voltage of from 1 V to 27.5 V on pin 7 of the tuner. Another memory module (MSC-006A) was installed with the same results.

By rechecking the schematic and again taking voltages on the memory control module, it was found that pin 5 of J2 should have had a –60-V pulse. A quick check with the scope found the pulse waveform was missing from pin 5. This –60-V waveform is taken from a separate winding on the flyback transformer (Fig. 8-45). A continuity check across pin 7 and 14 of T402 was good.

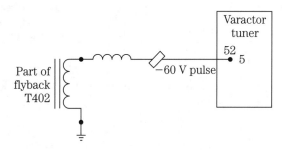

■ **8-45** *No sound, no picture, and no display; the cause is a missing –60 volt pulse in an RCA CTC118A chassis.*

Again the chassis was fired up and pin 7 showed a good waveform. A continuity ohm reading across L104 indicated an open coil. The unusual problem was a poor soldered connection at one end of L104. Resoldering the coil terminals brought back the correct waveform pulse at pin 5. Several of the same chassis might come in with this unusual tuning problem. This trouble can occur with any of the RCA sets using this type of control tuning.

The negative dc voltage (–60 V) on pin 5 cannot be measured with a voltmeter. Check for a waveform pulse at pin 5 with the scope. Another quick method is to use the low-ohm scale of the DMM or VOM and measure continuity from pin 5 to chassis ground. A low measurement will indicate normal continuity. Suspect a bad soldered connection at L104 with an open measurement.

Fuzzy pictures

When tapping around the tuner in an RCA CTC140 chassis the picture would fuzz up and then appear normal. Solder all tuner terminals, IF amp (Q2300) terminals, and pins on U2300. In some

CTC140 chassis, the trouble was the tuner and IF amp, while in others it was the IF amp and U2300 pins that needed to be soldered.

Channel jump from one channel to another

The channel would jump up from one channel to another without touching the remote or the channel selector in an RCA CTC135D chassis (Fig. 8-46). When this occurred, the voltage at pin 42 of U1001 ranged from 1.2 V to around 2 V. Pin 43 was a little low, but pin 42 should be 5.95 V. The 12-V source was normal, but had a poor connection with L312. Although the connection looked good, soldering L312's connections solved the intermittent channel change.

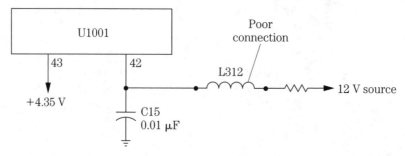

■ **8-46** *A poor connection at the end of L312 changed channels intermittently in an RCA CTC135D chassis.*

RCA on-screen display

Most RCA chassis have similar on-screen display circuits. The OSD signal is output from pin 24 of the AIU (U3300). The black edge signal is provided internally within the CTU processor (U1001). The OSD signal input is found at pin 47 (Fig. 8-47). The on-screen display output signal is found at pins 11, 12, and 13 to the B-Y, G-Y, and B-Y CRT bias drive transistor.

To troubleshoot the on-screen display circuits, check the input signal at pin 27 from pin 24 of U3300. If it is normal, check the output signal at pins 11, 12, and 13. With no output signal on either pin, suspect a defective IC (U1001). For a no-green-OSD problem, check the OSD output from pin 24 of U3300. If it is OK, check Q2903, Q5004, and CR5002. For no black edge, check the black output at pin 21 of U3300, and then check Q2706 and CR2709. For a solid green screen, check the signal at pin 24 of U3300. If the signal is low, suspect U3300. If there is normal input, check Q2903 or Q5004. With good chroma and no luminance, check the output at pin 24 of U3300. If it is low, suspect U3300. If OK, check for a shorted Q2709.

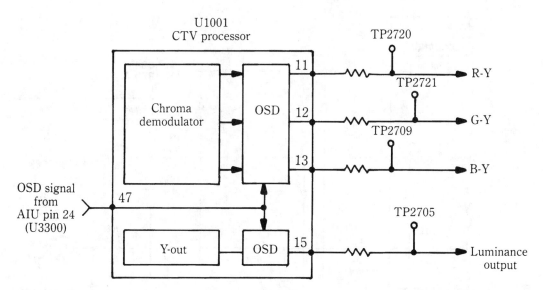

■ **8-47** *The RCA on-screen display (OSD) circuits in the CTV processor (U1001) feeding to each –Y transistor.* Phillips Consumer Electronics

Sylvania C9 on-screen graphic display

The method for input of the on-screen graphics signals are shown in Fig. 8-48. The video signals coming from the chroma/luminance IC640 are positive-going signals. If graphics information is to be placed on the picture tube, the signal from IC640 needs to be inhibited. This is done by character blanking transistor Q640.

Q640 is turned on by a positive-going pulse, fast-blanking signal reviewed from the on-screen display IC. The output signals from IC640 are shorted to ground through the transistor when Q640 is turned on. At the same time, information from the on-screen display IC is input to the RGB buffer transistors (Q643, Q641, and Q642). The control (fast blanking) and input signals are isolated from each other with 1-kΩ resistors (R690, R688, and R689). The RGB outputs from the buffer transistors are sent to the CRT board to drive the picture tube.

To troubleshoot the on-screen circuits, check the normal on-channel video at pins 13, 14, and 15 of IC640 with the scope. If the screen is blank, suspect a defective blanking transistor (Q640). You can see the on-screen graphics without a station tuned in because this information is applied directly to the bases of the buffer transistors.

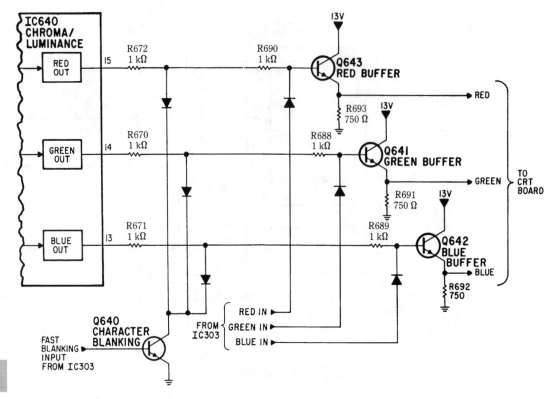

■ **8-48** *Sylvania's C9 on-screen display (OSD) input circuit with character blanking at Q640.*

Make your own tuner-subber from a used or discarded color or black-and-white TV chassis. Select a manual-type tuner. Keep the knob, etchem plate, and numbers so you can identify the channel numbers from the old TV. Clean up the tuner and install it in a separate enclosure or cabinet. How to build the tuner-subber is found in the book *Build Your Own Test Equipment*, published by TAB Books, an imprint of McGraw-Hill.

Conclusion

No matter how complicated it is, the varactor tuner was designed to bring in the desired channel. Simply break down the tuning problem and repair it. A tuner-subber with accurate voltage measurements helps solve most tuner problems. Improper tuning voltage can indicate a defective control unit. Inject dc voltage to the tuning voltage terminal of the varactor tuner to determine if the tuner is defective.

A poor AGC system can produce symptoms like a defective tuner. Clamp the AGC terminal of the solid-state tuner with a variable dc voltage to eliminate the AGC circuits. Remember, the AGC voltage is always positive at the solid-state tuner terminal.

After locating a defective tuner, remove it and send it in for repair. Most tuners or tuner modules can be returned to the manufacturer's distributor, or sent to a tuner repair depot for repair. Leave tuner repair and alignment to the experts unless you have the time and correct test equipment. For easy troubleshooting, check Table 8-6.

■ **Table 8-6 A troubleshooting chart for tuner problems.**

What to check	How to check it
No stations tuned in.	Check for dirty tuner on manual tuners.
Snowy pictures.	Check tuner voltages.
Sub tuner-subber to determine if tuner or IC circuits are defective.	Check IF circuits.
Check AGC voltages on tuner.	Check AGC circuits.
Check tuning voltages on varactor tuner.	Check control IC and voltages.
Tuner normal.	Check IF and AGC circuits.

Tuner repair centers

PTS Corporate Headquarters
5233 Highway 375.
Bloomington, IN 47401
800-844-7871

PTS Corporation
4941 Allison St. #11
Arvada, CO 80002
800-331-3219

PTS Corporation
110 Mopack Rd.
Longview, TX 75602
800-264-5082

PTS Corporation
15042 Parkway Loop.
Suite D
Tustin, CA 92680
800-380-2521

Troubleshooting the color circuits

9

THE TV RECEIVER MUST HAVE A SHARP AND CONTRASTING black-and-white picture before attempting to tune in a color picture (Fig. 9-1). Improper tuner, IF, and video signals can result in a poor color picture. Make all color adjustments including correct fine tuning to improve the color picture. Check for improper color symptoms viewed on the TV screen.

Besides defective color circuits, the color picture can appear a mess, with color impurities on the raster. A defective degaussing section can foul up the color picture. Turn the color controls off and check the black-and-white picture. Degauss the CRT to eliminate patches of color impurities in the corners or at the bottom of the raster. Remember, if the black-and-white picture is not good, the color picture will be the same.

A weak or gassy picture tube can cause a poor color picture. To determine if the CRT is at fault, set the brightness control for average brightness. Now rotate the contrast control and notice if the picture forms deep patches of color. A blotchy colored raster can be caused by a weak picture tube.

Correct test equipment

The color-bar generator and scope are ideal test instruments for locating and servicing the color circuit. Attach the color-bar generator to the antenna terminals of the TV receiver and scope each color stage according to the waveforms found in the manufacturer's literature. A dual-trace scope is ideal for troubleshooting the latest color circuits. Correct manufacturers' service literature and schematics are a must when servicing the color circuits.

Accurate voltage and resistance measurements within the transistor or IC circuits can quickly locate a defective color stage. Im-

■ 9-1 *A technician working on the color section of a TV chassis.*

proper voltage source feeding the color circuits can cause poor (or no) color signal. Check the low-voltage power supply if the voltages are low. Suspect a defective bypass capacitor, transistor, or IC if the voltage is low on a certain terminal.

Color signal injection from the analyst test instrument can help locate a defective color stage. Simply inject the color signal at the various test points throughout the color stage, or on the correct IC terminals. For correct color alignment, a sweep-marker generator and scope must be used (Fig. 9-2).

■ 9-2 *Besides the DMM, the crystal checker is another valuable test instrument for the color section.*

Color-dot-bar generator

The color-dot-bar generator or NTSC color generator are required instruments in setting up and troubleshooting the color circuits. The color signal from the generator is used in Howard Sams Photofacts in scoping the various color circuits. Critical waveforms are taken upon the chroma-luminance IC. Besides color tests, the color dot-bar generator can be used to check vertical and horizontal linearity, correct color bars, and pincushion circuits (Fig. 9-3).

■ **9-3** *The color-dot-bar generator is used for setup, adjustment, and color troubleshooting.*

Color waveform test points

Because IC circuits have been used in color circuits over the last 10 years, critical color test points will help solve most color problems. Some manufacturers include the various block diagrams of color stages within the IC (Fig. 9-4) on the schematic diagrams, while others do not. The main thing is to know where to take the various color waveforms on the IC terminals.

The chroma input signal is applied to pin 3 of IC701 (Fig. 9-5). The 3.58-MHz oscillator is controlled from pins 11, 12, and 13. Color control (R2) adjusts the color with a dc voltage at pin 2. R1 adjusts the dc voltage for tint control at pin 14. The three-color video ma-

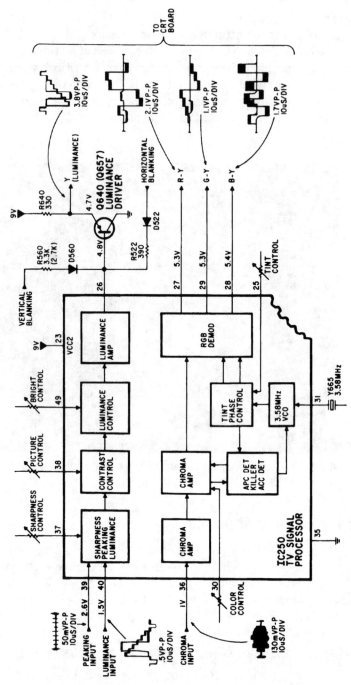

■ **9-4** *The inside view of a TV signal processor IC250 with the different color circuits.*

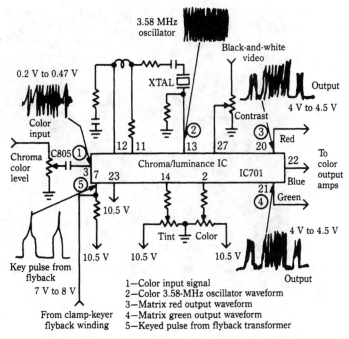

9-5 *The various color waveforms found on the chroma/luminance IC (IC701) in the RCA CTC111 chassis.*

1—Color input signal
2—Color 3.58-MHz oscillator waveform
3—Matrix red output waveform
4—Matrix green output waveform
5—Keyed pulse from flyback transformer

trix signals are fed from pins 20, 21, and 22 to the separate color transistor output stages. A keyed waveform is found at pin 7 from the flyback circuits (Fig. 9-6).

Check for color input signal at pin 3. If a low color signal is noted here, turn R3 to maximum. Improper signal at this terminal can indicate a poor IF response or black-and-white picture. The black-and-white picture waveform can be checked at pin 27. Measure for correct dc voltage at pin 23.

To see if the 3.58-MHz oscillator is oscillating, take the waveform from pin 13. Check pins 20, 21, and 22 for matrix output waveforms. Notice if the keyed waveform is found at pin 7. Although each manufacturer might use a different IC chip to take care of color and luminance circuits, the above waveforms will indicate where the color signal can be found or is missing.

Sylvania C9 chassis chroma/luminance processing

Because luminance and chroma processing are found in the same IC, they are discussed together. The luminance signal from the comb filter circuit is applied to pin 9 of the chroma/luminance IC

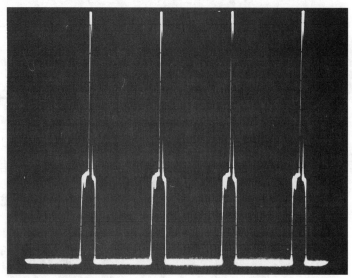

■ 9-6 *The keyed waveform at pin 7 from the flyback circuits.*

(IC640). Peaking is provided at pin 10 with high-frequency components of the luminance signal (Fig. 9-7). Inside IC640, the luminance signal is amplified and receives a peaking and dc contrast control voltage from the picture control at pin 7. The processed luminance (brightness) signal is applied to the RGB matrix to mix with the color difference signals to develop the RGB signals.

The chroma signal is found at pin 3 and is applied to the chroma processor. The chroma processor is controlled by the color control input at pin 6. The color processor exits at pin 24 of IC640, is coupled back with pin 17, and is then applied to the demodulator section. The tint control circuit receives a dc voltage from the tint control at pin 21.

The color oscillator of IC640 operates at a 7.16-MHz frequency instead of the regular 3.58 MHz, to provide better phase control and stabilization of the oscillator. The 7.16-MHz signal is divided by two for use by the demodulator after color synchronization is performed. The color oscillation frequency is adjusted by C625. The color crystal signal is found at pin 22.

The sandcastle pulse input is found at pin 8 of IC640. The chroma/luminance IC has a sandcastle detector to provide gate burst, blanking, and keying of signals for color processing. If the sandcastle pulse is missing at pin 8, no color output (RGB) will be found at pins 13, 14, and 15.

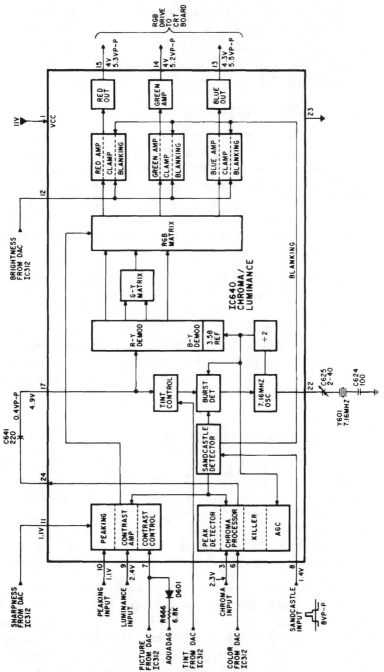

■ 9-7 The chroma/luminance circuitry found in the Sylvania C9 color chassis. Phillips Consumer Electronics

Sylvania C9 chassis chroma/luminance processing

Chroma signals are processed through the demodulator and G-Y matrix, and into the RGB matrix. Here the R-Y, B-Y, and Y signals are mixed to develop the RGB signals for the amplifier. The amplified RGB signals are applied to the red, green, and blue output circuits. The red is outputted at pin 15, green at pin 14, and blue at pin 13. The dc voltage found at pins 13, 14, and 15 varies from approximately 2 V to 4 V (minimum to maximum brightness).

Troubleshooting the chroma/luminance C9 circuits

Check for 11 Vdc at pin 1 of IC64 from the SMPS power supply. Scope for the composite video signal at pin 17 of IC201. Check for an open circuit between IC201 and IC640 if signals are not present. Check for chroma (color) at pin 3 and luminance at pin 9 of IC640. Do not overlook a leaky IC640 shorting the signal to ground. Remove pins 3 and 9 from the circuit and check the resistance between the pins and chassis ground. A resistance under 1-kΩ indicates a leaky IC.

Scope pins 24 and 17 for a chroma signal to see if the input color circuits are operating. If there is no color output at pins 13, 14, and 15, suspect a defective IC640 with normal color input signals. Do not overlook the sandcastle signal. If no sandcastle waveform is found at pin 8, suspect the sandcastle circuits. You will receive no color output at pins 13, 14, and 15 without a sandcastle waveform. Check for color oscillations at pin 22 of IC640. A defective crystal (Y601) or C625 prevents chroma output.

RCA CTC156 luminance processing

The video signal from the video buffer (Q2302) is applied to the contrast preset control (R2716) (Fig. 9-8). The video signal is passed through the phase compensation delay line (DL2701). After the delay line, the signal is divided into two paths (Fig. 9-9). The high-frequency component of the video signal goes to the sharpness circuit within U1001 at pin 52. The low-frequency portion of the signal is applied to pin 53 of the contrast circuit within the IC.

The gain of the sharpness control is controlled by dc voltage at pin 51. The dc voltage is produced by filtering the digital sharpness control signal (BRM) output from U3300 (pin 16). The high-frequency luminance output of the sharpness circuit is input to the contrast circuit. The gain of this circuit is controlled by the dc voltage at pin 8. This contrast voltage comes from the BRM signal at U3300 (pin 20).

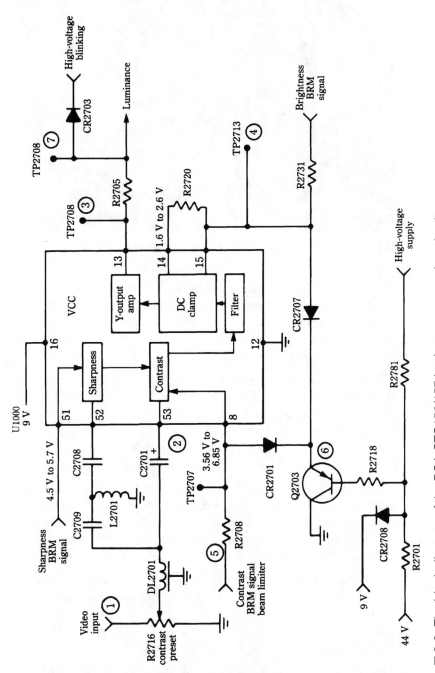

■ 9-8 The block diagram of the RCA CTC156/157 luminance processing circuits. Thomson Consumer Electronics

■ **9-9** *The latest delay line found in a 13-inch portable TV.*

The output of the contrast amplifier is filtered and passed to a dc clamp circuit. The reference level for the video signal is set by the clamp circuit. Pin 19 of the AIU (U3300) develops the brightness control voltage. This dc level will vary as the brightness and contrast control voltages are varied.

The luminance signal from the dc clamp is applied to the luminance output amplifier (Y output) and exits from pin 13 of U1001. CR2703 adds the horizontal and vertical blanking signals. Now the luminance signal is applied to pin 8 of the luminance buffer (Q2901). The luminance information is mixed with the chroma output circuit at the CRT driver/bias transistors.

The horizontal blanking signal is applied at pin 13 of U1001. This blanking signal inside U1001 is used as a timing reference signal. Both chroma and luminance circuits can be affected if the horizontal blanking signal is missing or weak. Both the horizontal and vertical blanking signals are positive pulses, with 9-V peak voltage.

The deflection circuits sync signal is developed in the luminance processing section. Generation of the sync signal is required for proper operation of deflection and chroma circuits within U1001. Check the sync separator at the tint control input of U1001. You should see negative-going composite sync pulses.

The beam limiter circuit reduces the amount of drive to the CRT when a predetermined level of beam current is exceeded. The beam limiter circuit is tied to the brightness and contrast control lines with CR2701 and CR2707. The beam current overrides the contrast and brightness settings generated by U3300 during periods when beam current exceeds the predetermined level.

The high-voltage resupply voltage from pin 6 of T4402 begins to fall as the beam current rises. Now the base voltage of Q2703 goes down. When the base voltage falls, the transistor turns on and pulls down, controlling the brightness and contrast voltages at pins 8 and 15. R2701, R2718, and R2781 determine the point where the beam limiter circuit starts to override the brightness and contrast voltages.

Troubleshooting CTC156 luminance circuits

Check for video signal at the contrast preset control and pin 53 of U1001 (TP2307). Check the CRT driver/bias stages if video signal is present. Scope for missing blanking signals at pin 13 and CR2703 (TP2705). If the video output signal is missing at pin 13 with normal input voltage at pin 53, check the brightness, contrast, and sharpness control voltages while varying the controls. Check for dc clamp voltage at pin 14. The voltage can vary slightly due to adjustments and different test instruments. The dc clamp voltage can vary from 1.6 V to 2.6 V as the brightness control is rotated.

Check pin 16 of U1001 for correct supply voltage (9 V). Do not overlook a poor ground connection of pin 12. Check the control pulse outputs of BRM AIU (U3300). If the BRM signals are normal, with abnormal brightness and contrast voltages, suspect the beam limiter circuit and Q2703. Suspect U1001 when the input signal is present but there is no output signal at pin 13.

RCA CTC130C color processing

The chroma input signal is applied at pin 3 of U701. U701 provides both luminance and color processing. Combining the two signals into one IC allows color and contrast tracking. Many of the latest RCA TV chassis use the same chroma/luminance IC circuits (Fig. 9-10).

The 3.58-MHz local color oscillator is controlled by pins 11, 12, and 13 of the IC. The variable dc input for tint control is at pin 14. Input pins 18 and 19 are used for the phase-shifted 3.58-MHz oscillator that is used for L and Q demodulation of the chroma information.

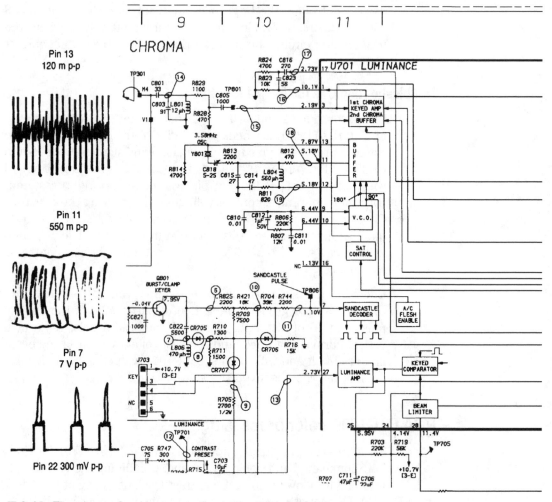

■ 9-10 *The chroma/luminance section in the RCA CTC130 chassis.* Thomson Consumer Electronics

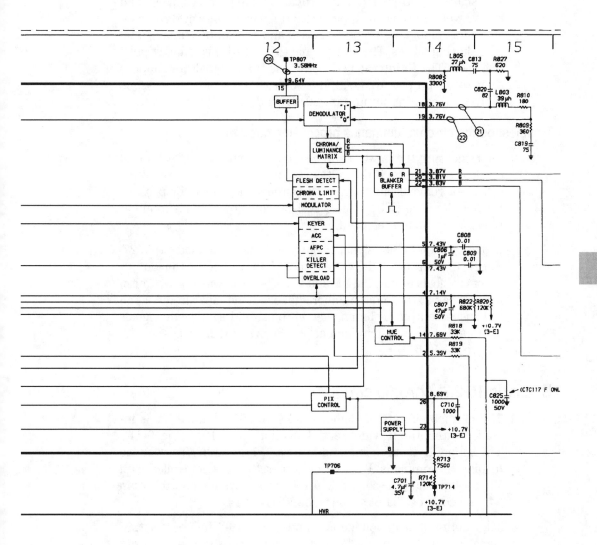

413

The luminance and chroma signals are matrixed together in the matrix amp. The three output color signals (green, red, and blue) are coupled through the blanker/buffer stages. The three color outputs are pins 20, 21, and 22. Horizontal and vertical blanking are provided in the blanker/buffer stages before exiting IC701.

The signals of horizontal and vertical blanking, burst keying, and block level clamping are applied to pin 7. The input waveform consists of a combination mixed vertical and horizontal blanking and burst keying pulses. The input signal (sandcastle) is found at pin 7 of TP806. The IC internally decodes the three signals and applies them to the proper circuits. If the sandcastle waveform is missing, there is no color at output pins 20, 21, and 22.

Troubleshooting chroma/luminance processing circuits

First, measure the power source voltage at pin 23 (10.7 V) for no color or brightness. If the voltage is low, check the power supply source or remove pin 23 from the circuit and check for possible leakage. Replace U701 when a resistance below 1 kΩ is found at pin 23 with the pin removed.

Scope the input color signal at pin 3 and the luminance at pin 27. Check for a sandcastle waveform at pin 7. Scope the output chroma signals at pins 20, 21, and 22. When the waveform is missing or weak at these pin numbers, take voltage measurements on the suspected pin. If luminance, chroma, and sandcastle signals are found at the input and not at pins 20, 21, and 22, suspect a defective IC (U701).

Realistic 16-261 chroma circuits

The chroma circuits of Radio Shack's 16-261 portable TV is included within the IF, video, and deflection IC (IC101) (Fig. 9-11). The supply B+ source is found at pins 9, 11, and 20. The voltage feeding these pins is 9 V. The brightness level is controlled with VR301 at pin 29. The color is controlled from system control at pin 34. The color oscillator signal is found at pin 14.

The subtint control voltage is found at pin 13. The color output signal is found at pins 16, 17, and 18, and it feeds the color output transistors. The luminance (Y) out is found at pin 19, and it feeds to the emitter circuits of the blue, green, and red output transistors driving the picture tube.

Check the three different voltage sources at pins 9, 11, and 20 with no luminance or color output signals. If voltages are very low, sus-

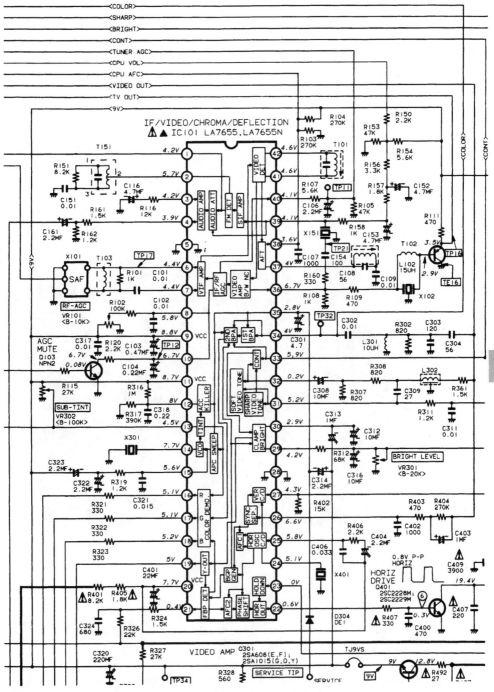

9-11 *The chroma/luminance circuits within IC101 of a Realistic 16-261 19-inch portable.*
Radio Shack

pect the B+ 9-V source or a leaky IC101. Scope the color oscillator at pin 14 and the color output at pins 16, 17, and 18. If there is no color output, but the chroma and luminance signal input is normal, suspect IC101. Accurate voltage measurement on each pin of IC101 and waveforms should identify the defective circuit.

Color IC circuits

The IC component has taken over the color and luminance circuits within the latest TV chassis. You might find one large IC component that contains the color, luminance, video, sync, and AGC circuits (Fig. 9-12) in a single device. Any breakdown in one particular circuit can destroy the IC and cause a loss of color in the chroma section.

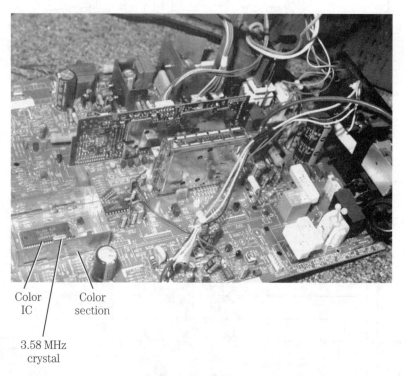

Color
IC

Color
section

3.58 MHz
crystal

■ **9-12** *The chroma stages can be located in one large IC with other circuits in today's TV chassis.*

One of the most important tests points within the color circuits is to check the power voltage source. A low voltage applied to the IC terminal can indicate a defective power supply or leaky IC. Check the schematic for other circuits feeding from the same power source. If the voltage is low at all other circuits, you can assume the power supply is at fault. Lower than normal voltage at the power source of the IC can indicate the IC is leaky (Fig. 9-13).

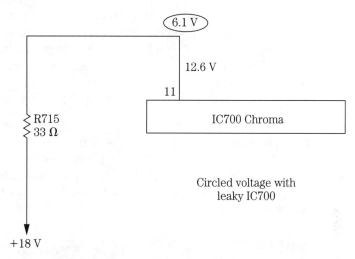

6.1 V

12.6 V

11

R715
33 Ω

IC700 Chroma

Circled voltage with
leaky IC700

+18 V

■ **9-13** *Check for a leaky IC when the voltage is low at the supply pin of the IC.*

To determine if the IC is leaky or the power supply is defective, the PC wiring can be cut at the power source terminal or the IC pin (11) can be removed from the circuit. The quickest method is to remove pin 11 from the PC wiring. Lift the solder from around pin 11 so the terminal is free from the board wiring. A leaky IC will measure under 500 Ω at pin 11.

Now check the dc voltage at the power source or wiring. If the voltage returns to normal, suspect a leaky IC. Check for a burned or open voltage-dropping resistor (R715) and power supply if the voltage source remains low. Often, if the power source is low in voltage, you will find other weak or dead circuits tied to the same power source.

A defective component within the color circuits tied to the IC can cause lower voltage at a certain terminal. Measure the voltages at each color terminal, and compare them with those on the schematic. Leaky bypass and electrolytic capacitors tied to the IC terminals can cause color problems. Open or increased resistances of fixed resistors at IC terminals can produce higher voltages.

A defective IC can cause intermittent, weak, or no color in the picture. The wrong tint or a missing color can be caused by a leaky chroma processor IC. Color bars or colors within the black-and-white picture can result from a defective chroma IC.

Before removing the suspected chroma IC, take accurate voltage measurements on each terminal. Make sure a leaky or open component tied to the IC is not causing the condition (Fig. 9-14). Take accurate voltage measurements at the power source terminal. Check for correct color waveforms at the most critical color test points. If still in doubt, remove and replace the color IC with the correct part number or a universal equivalent replacement. Most universal IC components and transistors work well in color circuits.

Bypass and electrolytic capacitors should be shunted when color ringing is noticed in the picture. Check each capacitor for leakage. Remove one terminal for correct leakage tests after lower than normal measurement is found in the circuit. An intermittent or weak color problem can result from a malfunctioning electrolytic capacitor.

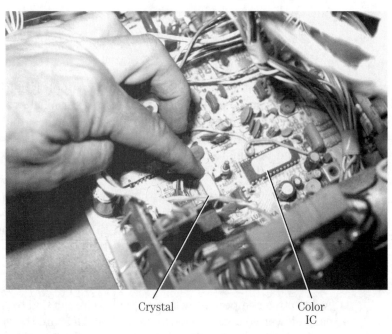

Crystal   Color
IC

■ **9-14** *Look for the color crystal and color IC in the TV chassis.*

Panasonic GXLHM color output circuits

The red, green, and blue color output signals come from the demodulator circuits inside IC301 at pins 25, 26, and 27. The color output signal is coupled by coil and resistance to the CRT board and to each base transistor. The Y or luminance signal is applied at pin 5 to each emitter terminal circuit (Fig. 9-15). Usually the color output transistors operate at a quite high collector voltage (150 V).

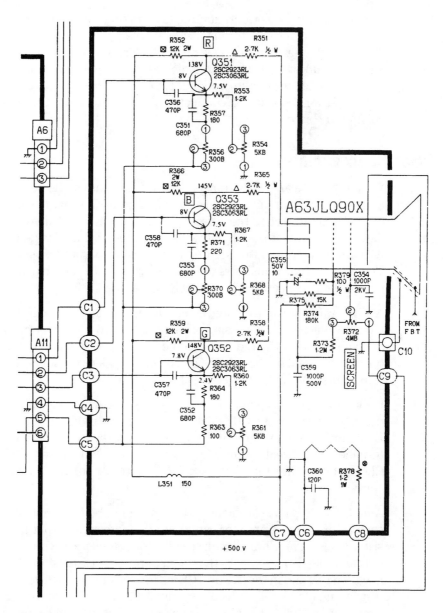

■ **9-15** *The color output circuits in the Panasonic GXLHM color chassis.* Matsushita Electric Corp

The dc supply voltage comes from a half-wave rectifier and capacitor filter circuit conducting a scan-derived voltage from the flyback transformer (T551). The same voltage source is applied to the CRT grid terminal. The color output from each color transistor is directly coupled through an isolation resistance to each cathode of the same color gun assembly.

Troubleshoot the color output circuits by taking voltage measurements of each collector terminal (metal). Very low voltage can indicate a leaky transistor or power source. Very low or no voltage to any color output transistor can indicate a defective poker source. When the screen is of one predominate color, suspect a leaky output transistor. Poor board connections can produce intermittent color on the screen.

Color matrix circuits

Within the tube and early transistorized color chassis, the demodulator, x, and y circuits mixed and amplified each separate color and were connected to the correct picture tube color element. Today, the demodulator and matrix circuits are provided inside the IC, with each color output fed to a single color output transistor. The color output transistors amplify the red, green, and blue signals to their respective gun assembly within the CRT.

A scope waveform at the three color output terminals of the IC will indicate which color is missing or weak (Fig. 9-16). The waveform and amplitude of the red and green color output signal at the IC terminals should be the same. If one color is weak or missing at the IC terminals, suspect a defective chroma IC.

Look for a leaky or open color output transistor when one color is missing or weak at the CRT. Make sure the picture tube is normal. Raise and lower the screen voltage of each screen control to determine if the weak or missing color is indicated on the CRT. Check the picture tube with a good CRT tester for the missing color.

A quick voltage measurement on the collector (body) terminal of each color output transistor can locate the defective output transistor (Fig. 9-17). Normally, these collector voltages are within 5 V of each other. Go directly to the green output transistor if weak or no green is found in the picture. Check for low or very high voltage measurements at the collector terminal. A low collector voltage measurement can indicate a leaky color output transistor or shorted CRT element; a very high collector voltage reading indi-

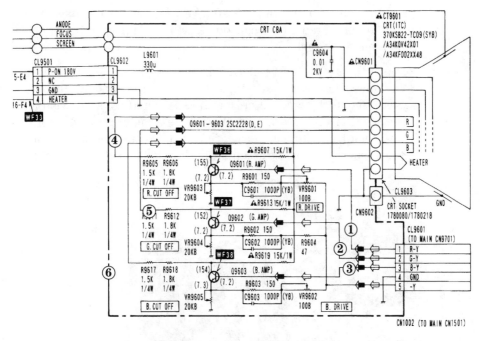

■ 9-16 *Check the color output circuits by the numbers for lost or weak color.*

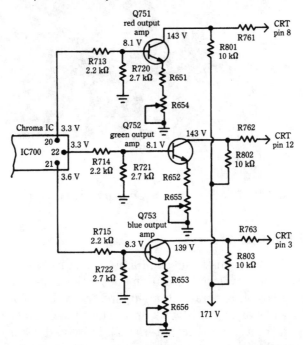

■ 9-17 *Check the voltage at the collector terminals of the color output transistors to locate a defective transistor.*

Color matrix circuits

cates an open color output transistor. Very low collector voltage on all three color transistors can be caused by improper boost or power supply voltage.

Remove the suspected color transistor to be tested for leakage. Sometimes these transistors can be intermittent and when removed, they test normal. Replace the suspected color transistor when the CRT and IC are normal with one weak or missing color. These color transistors are easily replaced, because most of them are located on the chassis assembly of the CRT socket (Fig. 9-18).

Color output
transistors

■ **9-18** *The color output transistors are located on the CRT PC board.*

Color circuits or CRT

To determine if the color amp or picture tube is defective, take a color waveform measurement at the collector terminal of each color output transistor. Go directly to the color output transistor when one color is weak or missing. Take a quick voltage measurement at the collector terminal. Raise and lower each screen control to determine if the picture tube circuits are defective (Fig. 9-19).

Adjustment

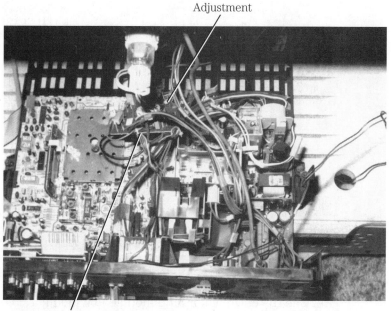

CRT board and socket

■ **9-19** *You may find color bias adjustments on the CRT board.*

Suspect a leaky color output transistor when there is low voltage on the collector terminal. A shorted CRT gun assembly could also be the cause. Remove the transistor and take another voltage measurement. If the voltage is still low, suspect a leaky spark-gap assembly or picture tube element. Remove the CRT socket and notice if the voltage increases. Suspect the CRT if voltage increases.

The screen was all blue with no picture in a Goldstar KMC1344G. The brightness control had no effect and the sound was normal (Fig. 9-20). A voltage measurement at the metal end (collector) of the blue output transistor was low (7.2 V). The normal collector voltage should be about 135 V.

Q509 was removed from the circuit board and it tested leaky. In fact, a 1,152-Ω leakage measurement was found between the emitter and collector terminals. Because high leakage was found between the two terminals with very low collector voltage, R542 was measured for correct resistance. Sometimes these collector load resistors will become hot and change resistance. Replacing Q509 with a universal ECG376 replacement solved the all-blue raster.

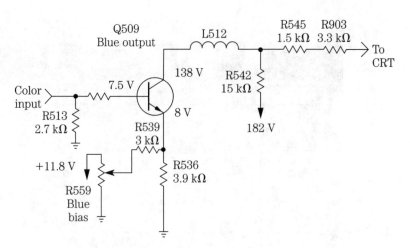

■ 9-20 *The all-blue screen was caused by a leaky blue output transistor (Q509).*

Replace the entire color amp module in a modular color chassis to determine if the color amp circuits are defective (Fig. 9-21). A separate module will be found for each color feeding to the CRT, or they might be combined in one module. A complete color module repair and replacement procedure is found in chapter 10.

No color

Before checking the color circuits for no color, make sure the color control is wide open with the tint control set in the middle of the tint range. Readjust the fine tuning control behind the selector knob to adjust color, picture, and sound. Turn up the color killer and color level controls for greater color in the picture. Make sure the black-and-white picture is normal.

The no-color symptom can be caused by any component in the color circuits. If the 3.58-MHz crystal does not oscillate, no color will be seen in the picture (Fig. 9-22). A leaky transistor or IC can cause the no-color symptom. Bypass and electrolytic capacitors tied to the transistor or IC terminals can cause no color in the picture. Improper or no low voltage from the power supply can produce weak or no color. Check the color waveforms of the 3.58-MHz oscillator. Connect the color-dot-bar generator to the antenna terminals and check for normal color waveforms at the demodulator or matrix stages. Go back to the burst and band-pass stages with no color waveform at the color output transistors. Check the color

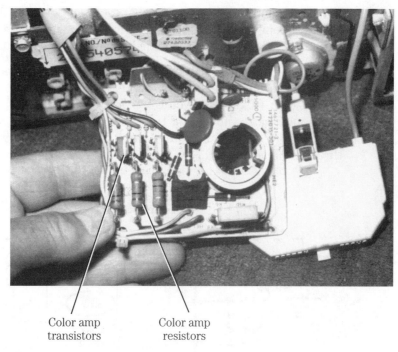

Color amp
transistors

Color amp
resistors

■ **9-21** *Replace the leaky color output transistor and collector load resistor if they are burned or changed in value.*

3.58 MHz
crystal

Color IC

■ **9-22** *Locate the color crystal and large IC to take critical waveforms and voltage measurements.*

at the base and collector terminals of each stage. Check all waveform test points on the chroma IC. Follow the manufacturer's troubleshooting chart, which is included in most service literature.

A quick in-circuit transistor test with the diode transistor test of the DMM located a leaky first band-pass transistor in a Panasonic CT301 (Fig. 9-23). Voltage measurements on all terminals were quite close, indicating transistor leakage. TR601 was removed from the circuit and showed high leakage between all three terminals.

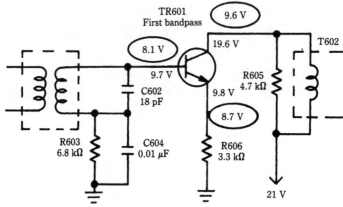

Note: Circled voltages were measured with a leaky TR601.

■ **9-23** *In-circuit transistor tests located a leaky TR601 band-pass transistor in a Panasonic CT301.*

The no-color symptom was found in a Sharp 9B12B with a normal black-and-white picture. Voltage measurements on all IC801 terminals were fairly normal. No color waveforms were found on the IC terminals. IC801 was replaced with the exact replacement part (part number RH-1X0093 CE22). This color IC must be replaced with the exact part, because at this time there is no universal replacement.

Check component C39 and the 3.58-MHz crystal for no-color symptoms in the RCA CTRC97 chassis (Fig. 9-24). C39 has been found leaky in several chassis. All voltages are normal on IC U1 except pins 2 and 3. You might find pin 5 with higher voltage (8.6 V). Check C39 with a resistance measurement at pin 3 to ground. This reading is always under 5 kΩ.

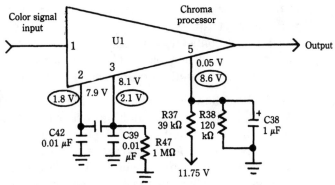

Note: Circled voltages were measured with a leaky C39.

■ 9-24 *Leaky capacitor C39 produced a no-color symptom in this RCA CTC97 chassis.*

Weak color

A weak color symptom can result from a poor black-and-white picture or a defective component in the color stages. Low or improper voltages from the power source can produce weak color. A leaky transistor or IC can cause weak or no color. Check for burned resistors and open bypass capacitors in the band-pass amplifier stages. Suspect leaky or open diodes in the AGC color circuits.

Open coil windings can produce weak color symptoms. Improper color alignment can cause a weak-color symptom, but do not try to adjust these coils or capacitors without proper color test equipment. These adjustments do not jump out of alignment by themselves. Poor board wiring connections can cause weak color conditions.

Intermittent color

The intermittent color problem can take a little longer to locate than the ordinary TV symptom. First check for a dirty tuner. Often, intermittents are caused by heat, poor transistor junctions, and poor board connections. Spray each transistor and IC in the color circuits with coolant. Push up and down and around the color board to make the color come and go. Sometimes heat applied near the component can cause it to act up. When only one color is intermittent, go to the demodulator, the x and y circuits, and the color output transistors. Also, suspect a color gun assembly of the picture tube.

Poor eyelet or griplet connections within the General Electric ABC chassis can produce intermittent color pictures. Solder all double-sided griplet connections around the color circuits. Some technicians place a small bare piece of hookup wire through the eyelet and solder the lead on both sides for a good connection.

Check the following components in an RCA CTC120A chassis when intermittent color conditions occur (Fig. 9-25). Spray C814 and C815 with coolant to see if the color disappears. Often, replacing both capacitors solves the intermittent color problem. Check L804 for open coil connections. Sometimes just moving C818 and Y801 will make the color come and go. You should notice very little voltage change when one of the intermittent components acts up.

Suspect a defective 3.58-MHz crystal when color bars are found in a chroma IC. Poor board or component connections within the crystal circuits can develop into color bars. Automatic frequency phase control (AFPC) color alignment can help prevent the color picture from drifting out of frequency and going into color bars. A defective color processor IC can produce color bars.

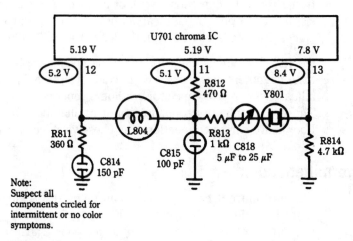

Note:
Suspect all components circled for intermittent or no color symptoms.

■ **9-25** *Check the circled components at the chroma IC if there is intermittent or no color.*

No red

Go directly to the demodulator, x, and red color output transistors when the red color is missing or intermittent. Likewise, when only one color is missing, proceed to the color output circuits and picture tube. Check the picture tube circuits and CRT for a missing color. Raise and lower the screen control to determine if the picture tube or circuits are defective.

The color output circuits can be checked for correct waveforms or voltage measurements. In an RCA CTC87 chassis, the red color was missing from the picture (Fig. 9-26). A normal black-and-white picture was noted with the color control turned down. A quick voltage test at the collector terminal indicated the voltage was high (186 V). Q1 was found to be open in the circuit with an in-circuit transistor test. Replacing Q1 with a GE-27 transistor solved the no-red symptom.

Suspect the low-voltage power supply when no color or weak color is found in the picture of a GE-ECA chassis (Fig. 9-27). With one particular GE-ECA, very little collector voltage was measured at Q403, Q405, and Q401. A voltage check at the boost voltage source indicated no boost voltage. Resistor R979 in the flyback winding of the boost voltage was open. It's wise to also check Y979 for leakiness.

Too much green, just a little blue, and no red was found in a Zenith 14DC15 chassis (Fig. 9-28). The 3.5-MHz waveform was missing at the collector terminal of the 3.5-MHz amp transistor (Q206). A voltage measurement indicated Q206 was leaky. The transistor was shorted on all elements. Replacing Q206 with an SK3122 universal replacement solved the messy color picture.

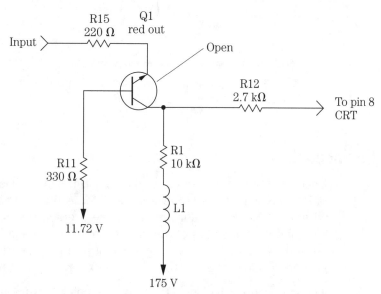

■ **9-26** *Red was missing from the picture due to an open red output transistor Q1.*

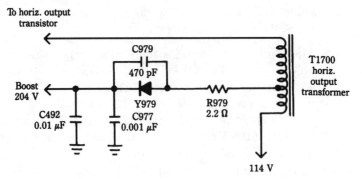

■ 9-27 *Check for weak or absent color at poor boost voltage in a General Electric ECA chassis.*

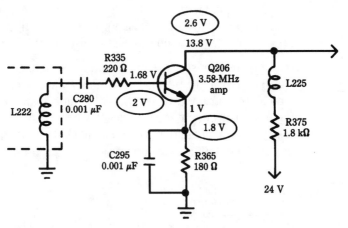

Note: Circled voltages were measured with a leaky Q206.

■ 9-28 *No red was found in the picture of a Zenith 14DC15 chassis, because of a leaky Q206.*

No tint control

Check the tint control circuits when the tint control has no effect on the color picture. Often, one leg of the tint control is open within the tint and color circuits. Take a quick resistance measurement on both sides of the tint control. The voltage on the tint control can run high with an open ground connection to the grounded side of the control. Also, check for an open tint control.

The tint control had no effect on the color picture in an RCA CTC68L chassis (Fig. 9-29). A resistance continuity check on each side of the control found an open between the tint control and

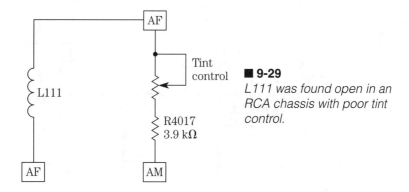

■ 9-29
L111 was found open in an RCA chassis with poor tint control.

chroma module. One side of L111 was found to be open. L111 was removed and repaired. Replacing L111 solved the open connection between AF and the tint control.

All blue raster and shutdown

Suspect a shorted picture tube or spark-gap assembly on the CRT elements when the TV comes on with a bright color and shuts down. The raster gets extremely bright within a few seconds. Often, chassis shutdown can be caused by excessive high voltage at the anode terminal of the picture tube. Test the picture tube for a shorted element. Sometimes the CRT tester might not show a leaky gun assembly. The gun assembly can become shorted only when voltage is applied to the heater and other elements. Pull off the CRT socket and fire up the chassis. If the chassis does not shut down, replace the picture tube. Check for a leaky spark-gap assembly if the chassis does not shut down with the CRT socket removed.

The raster was all blue in an RCA CTC111A chassis before shutdown. Within a few seconds the blue raster became brighter before shutdown. The spark-gap assembly (pin 11) was shorted to chassis ground (Fig. 9-30). Replace with the original part number (146169). Remember in some TV chassis, the spark-gap assemblies can be found inside the CRT tube socket.

Critical color waveforms

There are six color waveforms that are found upon the color IC processor to indicate the IC circuits are functioning. Take a color input terminal to determine if color is at this point (Fig. 9-31). Color waveform (Fig. 9-32) will indicate if the color oscillator is functioning. The color output signals for the three different color

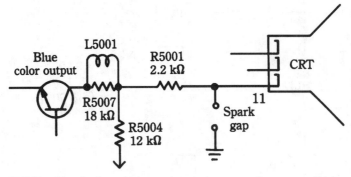

■ 9-30 *Check for a shorted spark gap when an all-blue screen is noted on an RCA CTC111.*

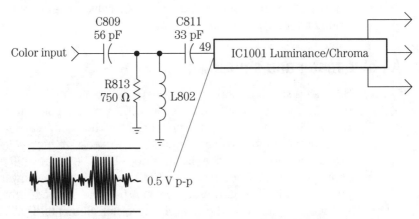

■ 9-31 *Check for the correct color input waveform to determine if color is present here.*

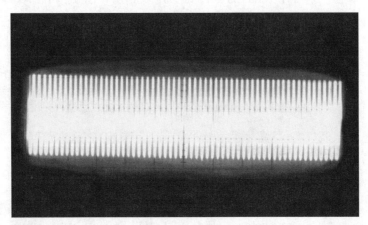

■ 9-32 *The color waveform (3.58 MHz) will indicate that the chroma oscillator is functioning.*

output transistors are taken and all three look somewhat alike (Fig. 9-33). If the color output signals are missing, suspect a defective color IC or improper voltage supply. Last but not least, a color trigger pulse from the flyback is shown in Fig. 9-34. No color will be found in the picture if any one of these waveforms are missing.

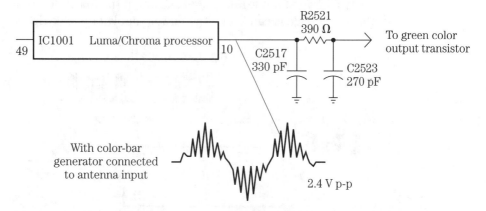

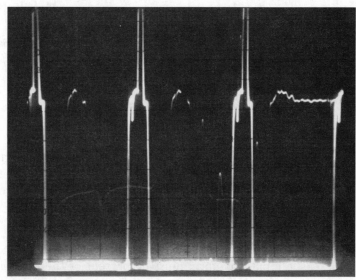

■ **9-33** *Color output waveforms on all three color output circuits.*

■ **9-34** *Trigger pulse from the flyback winding to the color IC.*

Troubleshooting the RCA CTC108 color chassis

Make sure a good black-and-white picture is found before attempting to service the color section. Readjust the color and tint controls. Check the picture for no color, intermittent color, and color bars. In addition to the CTC108 chassis, the following color procedure is the same for CTC107, 109, 111, and 115.

Measure the dc voltage at pin 23 (10.5 V) of U701. If the voltage is between 10.5 V and 12 V, assume the power supply voltage is normal (Fig. 9-35). A lower voltage can indicate a leaky IC U701 or a defective power supply. Remove pin 23 or cut the foil to determine if the low voltage is caused by the power supply or a leaky IC.

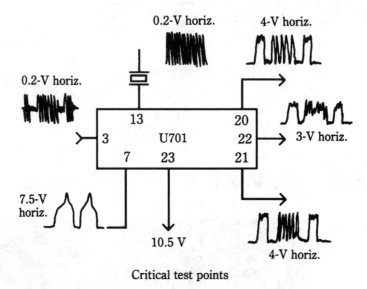

Critical test points

■ 9-35 *Check the power supply pin (23), color input, oscillator, three color outputs, and the trigger pulse if there is loss of color.*

Now check the input color waveform at pin 3. Turn up the color level control (R816) if the amplitude of the color waveform is low. Check the chroma signal at the chroma buffer transistor (Q800) if poor or no color signal is found at pin 3. Check the IF circuits for loss of color signal.

Proceed to pin 7 and check the sandcastle waveform. The sandcastle input waveform must be present or there will be no color output from U701. In most color circuits, the keying waveform comes directly from a separate winding of the horizontal output transformer.

Check the 3.58-MHz oscillator waveform at pin 13. No color will be present in the picture if the oscillator is not working. Measure the voltage at pin 13 (7.65 V). Substitute the crystal (Y801) if the voltage is normal without the oscillator waveform. Replace the crystal with the exact replacement part. Suspect C818 for intermittent or no color (Fig. 9-36).

A normal demodulator waveform at terminals 20, 21, and 22 indicates the color circuits are normal at U701. Check the chroma input signal at pin 1 and 17 if the demodulator waveforms are missing or improper. Measure the voltage on pins 1 and 17. Take a resistance measurement between the terminals and chassis ground. An infinite resistance measurement should be found at terminal 17 (Fig. 9-37).

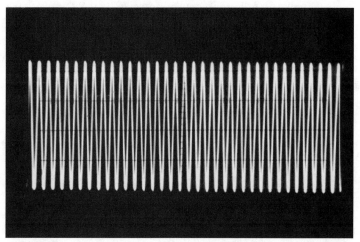

■ **9-36** *The color oscillator (3.58 MHz) waveform taken from IC or crystal pin terminal.*

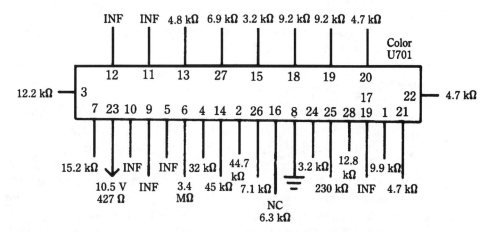

■ **9-37** *Normal resistance taken with a DMM in an RCA CTC107, 108, 110, 111, or 115 chassis.*

Suspect a defective U701 if there is no demodulator output with normal color voltages and signal to the demodulator circuits. With normal demodulator waveforms at U701 and one color missing, suspect the color output amp or picture tube. Take a voltage test on the missing color output transistor collector terminal. Measure the voltage at the metal heat sink at the end of transistor. Compare this voltage measurement with the other two color transistors. If the collector voltage is high, suspect an open output transistor. A low voltage measurement can indicate a leaky transistor or picture tube element.

Check the following components (Fig. 9-38) for color problems related to the chroma processor U701. No color is identified by number 1. For intermittent color, check number 2. For weak color, check number 3. Check number 4 for oscillating and color bars.

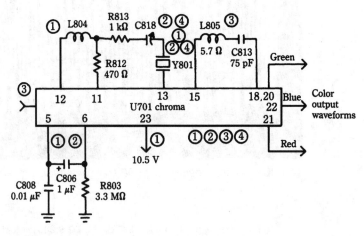

■ 9-38 *Check these components if there is weak, intermittent, or no color at IC (U701).*

Color "tough dog"

An intermittent color problem is perhaps one of the most difficult malfunctions to diagnose in the TV chassis. Monitor the signal waveform at the output terminals of IC1001 at pins 9, 10, or 11 (Fig. 9-39). Suspect problems in the color output circuits if waveforms are normal. If waveforms become weak or disappear, suspect IC1001 and components in the color circuits. Check the color oscillator waveform at pins 4 and 5 when the color disappears. Suspect crystal or components tied to these terminals if the oscillator quits operating. Monitor the color input terminal 49, contrast waveform at pin 53, oscillator frequency at pins 4 and 5, and color

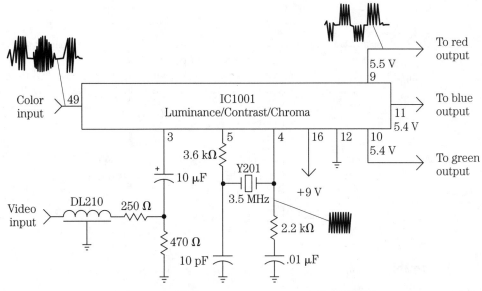

9-39 *Monitor the intermittent colors at pins 9, 10, and 11 of IC1001.*

output at pins 9, 10, and 11. Monitor all voltages on the color circuits, especially the supply voltage terminals 14 (9 V) and 7 (3.5 V). Replace the intermittent color IC (IC1001) when other components and voltages are normal. Take color waveforms with a color-dot-bar generator connected to the antenna.

Unusual color problem

No color was the symptom in an Emerson MS250R model. A quick waveform test at output pins 19, 20, and 21 were not there, indicating trouble within the IC201 chroma IC and circuits. No color waveform was found at pin 16 of the color oscillator (Fig. 9-40). the X201 color crystal was replaced, and still there was no color. The input color waveform was normal at pin 6. All voltages were fairly normal and IC201 was suspected. After replacing TA7644BP with universal replacement ECG1547, the color was still out. When VC201 was accidently prodded, the color came and went out. VC201 (20pF) trim capacitor was replaced and color was restored.

Sylvania C9 chassis digital-to-analog converter

There are no mechanical customer controls to adjust color, volume, tint, picture, and contrast in the Sylvania C9 chassis. A spe-

437

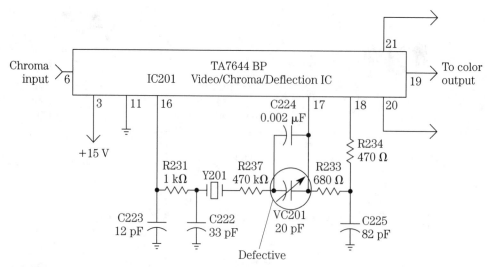

■ 9-40 *The unusual color problem in an Emerson MS250R model was caused by trimmer capacitor VC201 (20 pF).*

cial digital-to-analog converter (IC312) is used to change the picture and sound (Fig. 9-41). The input circuit is from the TS-15/TS-18 tuning system control. The input signal can be made with either keyboard buttons or remote control.

There are eight programmable, six-bit digital-to-analog converters in IC312. The I²C bus slave receiver, with three programmable address lines, receives serial data and clock from the tuning system microcomputer. I²C stands for information interchange, and is a name describing a two-conductor serial data bus. Address pins 5, 6, and 7 of IC312 are grounded, so set the slave receiver address to zero. The DAC (digital-to-analog circuit) will only respond to information that is sent to device zero from the microcomputer.

The DAC power source is found at pin 1 (13 V) and the reference voltage is found at pin 2. The C9 chassis application for the DAC uses the same source voltage for both pins 1 and 2. Each individual DAC can be programmed separately by a six-bit word up through 64 steps. Voltage (V_{max}) determines the maximum output voltage for all DACs. The resolution will be approximately V_{max} 164.

When power is turned on, all DACs are set to their lowest values. As the microcomputer takes control, new or present information is sent to set the DACs. This signal controls the picture and sound levels previously selected by the consumer.

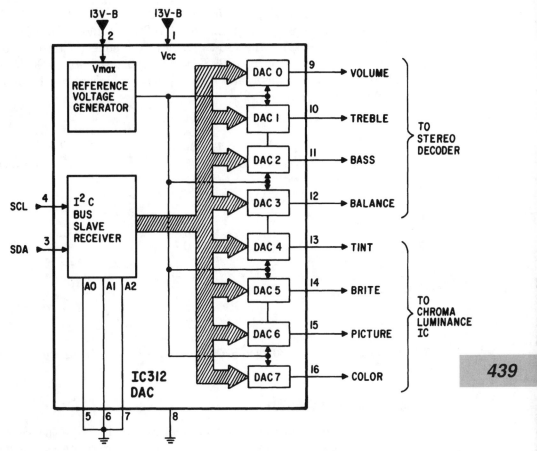

9-41 *A block diagram of the digital-to-analog circuit (DAC) in the Sylvania C9 chassis.* Phillips Consumer Electronics

Four DAC outputs control volume, treble, bass, and balance. These are sent to the stereo decoder module to control the audio circuits. Four other DAC outputs control tint, brightness, picture, and color settings. These are sent to the chroma/luminance circuitry. Sharpness control settings are generated by a DAC located inside the tuning system microcomputer.

Five actual color case histories

What follows are five actual color-related problems that happened to common TV chassis.

No red in picture—Sony SCC-548D

The red output color was found fairly normal on pin 19 of chroma IC301, but there was none in the picture tube. A quick voltage test on the red color output transistor Q703 (located on the CRT "C" board) indicates a leaky transistor. Q703 was removed and found leaky (Fig. 9-42). R709 (15 k) going to the collector terminal of Q703 was burned. Q703 (2SC2276) was replaced with an ECG171 universal replacement, and R709 was also replaced, which cured the no-red symptom.

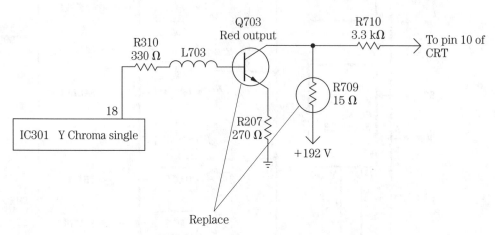

■ **9-42** *Leaky transistors Q703 and R709 (15 K) were replaced to restore the red output in a Sony TV.*

No color—Goldstar CMT-2612

A good, normal black-and-white picture was found in this Goldstar chassis, without any color. A quick color scope waveform check on pins 9, 10, and 11 showed no color. A normal color waveform was found at pin 30 of IC501. Pin 14 of the oscillator had no waveform. The supply voltage on pin 16 was less than 1 volt (Fig. 9-43). Pin 16 was removed from the PC board, and had a 0.15-Ω measurement to common ground. IC501 (LA7629) was replaced with an NTE7008 universal replacement. R550 and D504 were also replaced in the low-voltage (11.2-V) source of the power supply.

Intermittent or no color

Sometimes the color in an RCA CTC156 color chassis (Fig. 9-44) would disappear and return without any remote-control transmitter adjustments. Suspect capacitor C2810 on pin 5 of the color IC (U1001). When the color is tuned in, the voltage should be around 8 V, and should be about 5.2 V on a black-and-white or no-color picture. If the voltage is below 5 V, suspect C2810 is leaky. This ca-

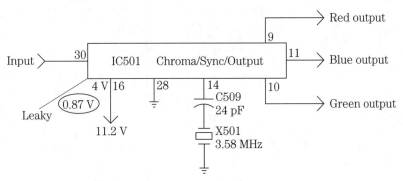

■ 9-43 *A lack of color in a Goldstar TV chassis was caused by a leaky IC501 with low voltage on supply pin 16.*

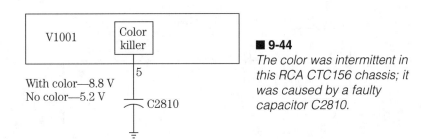

■ 9-44
The color was intermittent in this RCA CTC156 chassis; it was caused by a faulty capacitor C2810.

pacitor has caused a lot of different color problems on several different RCA TV chassis.

No color—RCA CTC107 chassis

For a no-color symptom, suspect the color crystal (Y801) at pin 13 of U701. Scope pin 13 to see if the crystal is oscillating before replacing it. Check L805 for an open coil with a weak color symptom in the RCA CTC109 chassis (Fig. 9-45). Make sure 10.5 V is at the voltage source supply pin 23 of IC U701.

Intermittent color

Suspect C611 in the RCA FJR2020T chassis when the color becomes intermittent, noise appears, and the picture disappears (Fig. 9-46). This capacitor is noted for developing intermittent and open conditions. C611 is located between the video buffer and comb processor (U600).

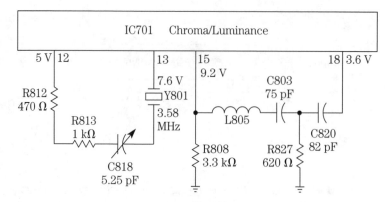

■ **9-45** *Check for an open L805 and Y801 at pins 13 and 15 to troubleshoot weak color in an RCA CTC109 chassis.*

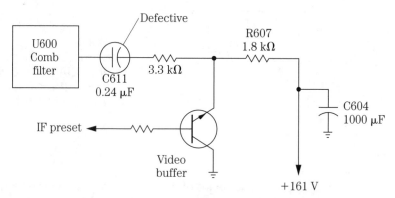

■ **9-46** *A defective C611 (0.24 µF) caused intermittent color in an RCA FJR2020T model.*

Troubleshooting with the color-dot-bar generator

The low-cost portable color pattern generator provides a variety of test signals and patterns for TV and VCR servicing. These testers are usually crystal-controlled, with output on Ch3 or Ch4 of the TV. The video patterns consist of a dot, crosshatch, and color bars, a gated rainbow, three bars, or ten bars. There is also a color raster, full-field color raster, user-adjustable hue, blank raster, purity, white wide bar, and half-screen white bar. The video output is 1 volt peak-to-peak (Fig. 9-47).

The color-dot-bar generator is ideal when setting up and adjusting the TV set, or for making adjustments after a new picture tube has been installed. This tuner is also connected to the TV antenna for color troubleshooting. Most manufacturers and Howard Sams Photo-

■ **9-47** *The color-dot-bar generator can help troubleshoot the color circuits with the oscilloscope as an indicator.*

facts use the color bar or NTSC color generator for color waveforms within the color circuits.

Conclusion

Checking the color section waveforms can solve the most difficult color problems. After locating the missing color signal, take voltage and resistance measurements. Testing the suspected transistor within the circuit can help. Besides the chroma IC, check for leaky capacitors or a change in resistance between the ground and terminal pin.

Intermittent color symptoms can be caused by poor board connections. Chassis with double-sided board wiring can have intermittent feed-through connections. Do not overlook the 3.58-MHz color crystal when no or intermittent color symptoms occur. Always replace the crystal with the exact replacement part. Use coolant and heat on suspected transistors or ICs to see if the symptoms change.

Color alignment should not be attempted unless the required test equipment is available. Follow the manufacturer's color alignment procedures for each TV chassis. Do not tamper with color adjustments. The trouble might be elsewhere, and the adjustments do not move by themselves. If you feel that color alignment is needed, take the chassis to a qualified TV technician or to a factory service center that does color alignment. For more information on troubleshooting the color circuits, see Table 9-1.

■ Table 9-1 Troubleshooting the color circuits.

What to check	How to check it
No color in picture.	Adjust color killer.
Check color oscillator waveform.	Check supply voltage to color IC-sub crystal.
Check for color input waveform.	Check luma and input circuits.
Still no output color.	Check waveforms on color output IC. Check waveform from flyback.
No Red.	Check output of red IC. Check red output transistor. Check voltage on red output.
Weak color pictures.	Test picture tube—red, green, and blue guns.

Practical convergence circuit troubleshooting

10

MOST CONVERGENCE PROBLEMS ARE NOTICED AT ONCE after TV repairs. Make touch-up adjustments just before the TV is ready for delivery or pickup. Poor purity can show up as discolored areas in the corners of the picture. One or more colored lines can appear around the subject, indicating improper receiver convergence with a very poor picture. A poor color picture results from improper convergence and poor purity.

Before replacing the back cover, the TV chassis should be checked for purity and convergence. Sometimes, only a touch-up is needed. Poor purity and convergence problems can occur after setting the TV chassis on the side for bottom chassis repair or from accidentally brushing against the picture tube assembly. A loose yoke can cause purity and convergence problems when the chassis or cabinet is moved (Fig. 10-1). The yoke can come unglued from the CRT and cause poor purity. Often, the TV screen can become magnetized when the vacuum cleaner is shut off directly in front of the TV set. Also, the picture tube can be magnetized by a large stereo speaker placed next to the TV cabinet. Purity and convergence adjustments must be made after installing a new picture tube.

Although the TV chassis is degaussed each time it is turned on (by the automatic degaussing network), the CRT should be degaussed with an outside degaussing coil or ring after servicing. Just pass the coil around the picture tube area four or five times and pull it away from the front tube area. Do not shut the degaussing coil off in front of the tube, or you might magnetize the metal shadow mask and have to do it all over again.

Complete color setup should include preliminary checks of the receiver, degaussing, color purity, black-and-white or gray scale, static, and dynamic convergence. The color-dot-bar generator is a

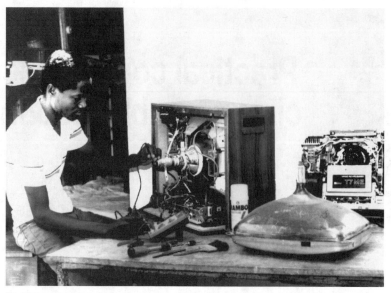

■ **10-1** *Checking for an open filament heater in the CRT. Be sure to fasten the yoke-tightening screw after replacement and purity adjustments.*

must for color convergence adjustments. Use the color-dot-bar generator for a quick convergence touch-up or for a complete convergence job after picture tube replacement. The dot-bar generator can also be used to set the level of the picture, correct the vertical height, and make linearity adjustments.

Manual degaussing coil

The manual degaussing coil is made up of several hundred turns of 24-, 22-, or 18-gauge enameled wire in a large circle (Fig. 10-2). This degaussing coil plugs into the 120-V power line and removes any purity or color areas from the front of the picture tube. A carpet sweeper shut off or started up in front of the TV screen can cause accidental magnetization and leave part of the raster with different colors. The manual degaussing coil is much stronger than the one inside the TV set.

Suspect magnetization of the screen when odd-looking colors appear on the screen. Try to degauss the screen by shutting off the TV for several minutes and then turning it back on. If this does not clean up the raster after several times, use the manual degaussing coil.

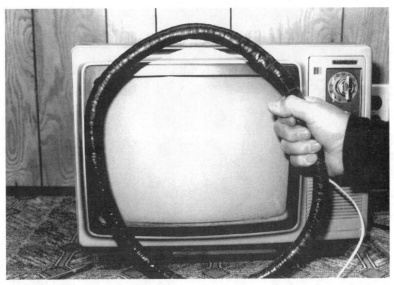

■ 10-2 *Rotate the manual degaussing coil several times in front of the screen to demagnetize the screen.*

Plug in the coil and place it against the face of the picture tube. Go around and around the frame of the tube and through the center area. After a few trips around the tube, slowly rotate the degaussing coil and back up away from the TV screen. Keep rotating the coil as you back away. When you are 6 or 8 feet away from the TV, turn the coil perpendicular to the TV screen and unplug the ac cord.

Check for a patch of color or corner areas that still have some color. Often the corner areas are where the unwanted color begins to appear. Sometimes the yoke assembly will pop loose and cause poor purity or improper convergence. After a new picture tube is installed, automatically degauss the front screen (Fig. 10-3).

Poor purity

Different-colored areas found in the TV raster when the shadow mask is magnetized are collectively referred to as "poor purity." Proper operation of the color tube depends on each color gun assembly striking its own color dot at the front of the picture tube. Improper yoke setting, convergence, or a magnetized screen can produce poor purity. It's possible you will never eliminate a small patch of impurities or have a complete convergence in some pic-

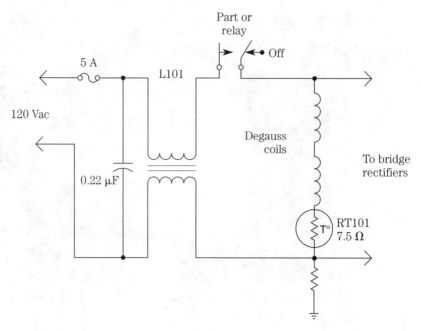

■ **10-3** *The degaussing-coil circuit is in the ac power-line circuits of the low-voltage power supply.*

ture tubes. This can be due to a misaligned gun assembly, improper convergence assembly, or loose shadow mask inside the picture tube.

Usually poor purity and convergence exists at the corners or bottom of the picture tube. Correct degaussing procedures and yoke settings might take care of the most difficult purity problems. First, degauss the entire front area and sides of the picture tube. Now check for impurities on the raster. Turn the service switch to the service position to eliminate the incoming signal, and turn down all screen or bias controls. Now turn up the red screen control for a complete red raster.

Move the yoke assembly back and forth until the whole screen is red. A touch-up adjustment of the purity ring might be needed with those rings found on the picture tube. Check for each color on the raster, noticing any patches of unwanted color. Always keep the yoke assembly level. On some picture tubes, the yoke assembly is glued to the glass area of the tube. When these yokes pop loose, complete purity and convergence adjustments must be made.

Color-dot-bar crosshatch lines

Rotate the color-dot-bar generator control to generate an all-blue-and-green screen for purity adjustments. Some manufacturers want a green field for purity adjustments. Switch the dot-bar generator to crosshatch for conveying the various colored lines. The crosshatch lines can be used to adjust the yoke for level conditions and vertical and horizontal linearity (Fig. 10-4).

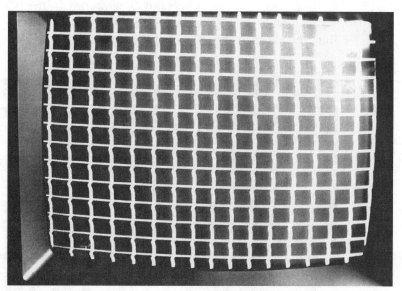

■ **10-4** *The color-dot-bar crosshatch pattern will indicate if there is poor convergence at the ends of the CRT.*

Convergence problems

When convergence is completed, the three color beams must pass through the proper hole or slot in the shadow mask, striking the exact color dot. Because the picture tube glass or dot area is not perfectly flat, the color beams must be corrected with static and dynamic convergence methods. Poor convergence results when colored lines around the subject can be seen at all times. Improper convergence is noted in the corners, top, and bottom of the raster.

Always use the color-dot-bar generator for convergence adjustments. Some TV technicians are experts at making these convergence adjustments without the test instrument, but don't try it. Connect the dot-bar generator to the antenna terminals, turn the tuner to channel 3 (or the channel recommended by the manufacturer), and switch to color lines. Set the generator to the cross-hatch symbol (Fig. 10-5). Convergence adjustments should only be made after purity adjustments.

Notice which color is out of line. You might find one or two colors in line, and one out. Most manufacturers have correct convergence adjustment information in the service literature. Often, the R-G control at the top will correct the red and green lines at the top of the raster. A blue line can be shifted with the blue lateral magnet, while the R-G control at the bottom will converge the red and green lines at the bottom of the screen.

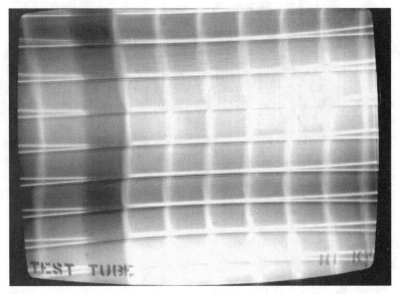

■ **10-5** *Converge the color lines together with the color-dot-bar generator connected to the antenna terminals.*

Black-and-white adjustments

Color temperature, or a normal black-and-white raster, must be achieved for a good color picture. All brightness and contrast controls should be set with a local tuned channel. Pull out the service switch (if found on the chassis), turn the color bias controls counterclockwise, and turn the drive controls fully clockwise. Rotate all screen controls counterclockwise.

Adjust the red screen control to produce a thin horizontal line. Adjust the screen and color bias controls to produce a white horizontal line. Usually one color bias control is found completely counterclockwise. Return the service switch to normal position. Readjust the color drive controls to obtain a warm black-and-white picture. Remember, color bias controls adjust low light or dark areas, and color drive controls adjust high light or light areas of the raster. Follow the manufacturer's literature for complete setup.

Panasonic GXLHM purity and convergence

Adjustment of the picture tube is only needed after replacing the CRT, installing a new yoke, or if the TV has been dropped or someone has moved the original settings. Start with initial center static convergence, then purity adjustment, and then final static convergence.

Place the new yoke on the neck of the CRT. Do not tighten the yoke at this time. Position the purity/convergence assembly (Fig. 10-6). Release the lock-in assembly and place like tabs of the purity device together at 12 o'clock to reduce its magnetic field effect. Now manually use the degaussing coil.

NOTE: On an assembly with no locking nut, cut the wax seal with a razor blade to free the adjustment magnets.

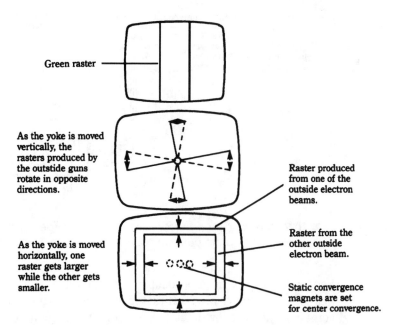

■ **10-6** *Purity and yoke adjustment of the Panasonic GXLHM chassis.* Matsushita Electric Corp of America

Turn the set on and slide the yoke assembly back and forth on the neck of the picture tube. Manipulate the yoke until a nearly white raster is seen over the entire screen. Now connect a dot/cross-hatch signal generator to the antenna terminals and observe the convergence at the center of the screen.

Adjust the four poles on the two magnets (center rings). Separate the tabs and rotate the device to converge blue and red. Now adjust the six poles on the three magnets (rear rings). Separate the tabs and rotate the device to converge blue and red (magnets) with green.

Purity adjustment

For purity adjustment, operate the TV for 15 or 20 minutes to stabilize the CRT with full brightness. Degauss the CRT with an external degaussing coil. Short TP14 to ground for a blank raster. Set the brightness control for normal brightness level.

Turn red and blue low-light controls fully counterclockwise to obtain a green field (Fig. 10-7). Adjust the drive controls if a green field is not obtained. Loosen the deflection yoke clamp screw and move the yoke back as close to the purity magnet as possible.

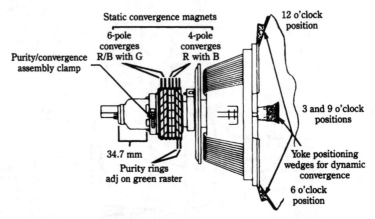

■ **10-7** *The six-pole and four-pole static convergence magnets are found behind the purity rings.* Matsushita Electric Corp of America

Loosen the purity magnet locking nut and adjust the purity magnet to set the vertical green raster precisely at the center of the screen. Adjust the low-light controls for a red screen. Slowly move the deflection yoke forward and adjust it for the best overall red screen. Now tighten the deflection yoke screw clamp.

Produce a blue and then a green raster using the low-light controls. Make sure that good purity is obtained on each respective field, and that a uniform, white raster is obtained by adjusting the red, green, and blue low-light controls. If the screen is not uniformly white, repeat the procedure.

Final convergence adjustment

Before attempting convergence adjustment, vertical size and focus adjustments must be completed. Connect a dot-pattern generator to the set. The brightness level should be no higher than necessary to obtain a clear pattern.

Converge the red and blue dots at the center of the screen by rotating the R-B static convergence magnet. Align the converged red and blue dots with the green dots at the center of the screen by rotating the (R-B)-G static convergence magnet. Tighten the convergence magnet lock nut. If there is no lock nut, melt wax with a soldering iron (not at maximum temperature) to reseal the assembly.

Slightly tilt (do not rotate) the deflection yoke vertically and horizontally to obtain good overall convergence. After the vertical adjustment of the yoke is complete, insert wedges at the 12 and 6 o'clock positions, then make the horizontal tilt adjustment.

Secure the deflection yoke by inserting wedges at the 3 and 9 o'clock positions. Apply adhesive between the tab (thin portion) of the wedge and the picture tube (Fig. 10-8).

453

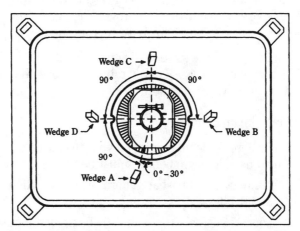

■ **10-8** *Secure the wedges to hold the deflection yoke with the final convergence adjustments in a Panasonic GXLHM chassis.* Matsushita Electric Corp of America

RCA CTC130C center static convergence

The center convergence is accomplished with two pairs of concentric magnets (Fig. 10-9). The center green color is stationary. The other adjustable magnets move the red and blue lines in order to converge them on the green line. The rear pair of magnets (four poles) near the beam binder clamp move the red and blue both vertically and horizontally. Spreading the tabs apart causes the blue and red to move in the opposite direction of the green line. Likewise, moving the tabs closer together causes the blue and red to move closer to the green. Rotating the pair of tabs together moves the red vertically.

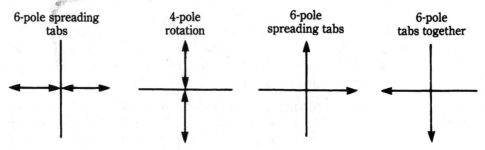

| 6-pole spreading tabs | 4-pole rotation | 6-pole spreading tabs | 6-pole tabs together |

■ **10-9** *Center convergence is adjusted using two concentric magnets in the RCA CTC130C TV chassis.* Thomson Consumer Electronics

The center six poles (tab magnets) move red and blue as a pair both vertically and horizontally. Spreading the tabs apart causes the red and blue to move to the right and up away from the green. Likewise, moving the tabs together causes the red and blue to move to the left and down away from the green.

Do not rotate the six-pole magnets (center) together in any direction. Rotating these center tabs will move the green center line away from vertical center.

Magnetic tape beam bender

The magnetic beam bender on your television set is not adjustable. An adjustable beam bender must be used for these adjustments.

1. Connect the color-bar generator to the VHF antenna terminals and display a crosshatch pattern.
2. Loosen the locking ring on the beam bender assembly. Loosen just enough so it can be moved for adjustments.
3. Start with all tabs at the 12 o'clock or upward position. Rotate the rear four-pole magnets with spread of the center six-pole

magnets as required to converge the line in the center of the CRT.

4. Rotate the movable locking ring counterclockwise to secure the adjustments. Do not overtighten the beam bender locking ring.

Static convergence (magnets)

Static convergence is accomplished with the three center color magnets (Fig. 10-10). Static convergence affects the center of the TV screen. The three color beams must hit the same spot on the face of the tube at the same time. The three magnets are adjusted with a color-dot-bar generator so the three colors converge as one dot in the center of the screen. When all three colors are located inside the dot area, the circle should be a clear white dot. Interaction can occur between purity and static convergence. You might have to touch up the purity after making static adjustments.

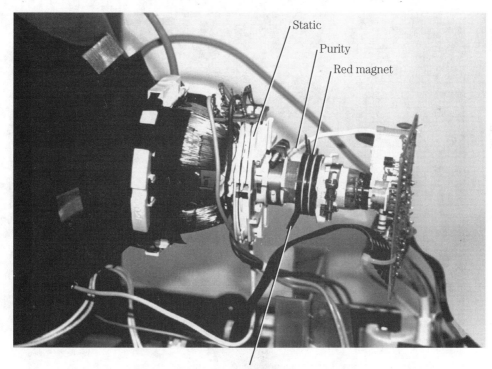

Static

Purity

Red magnet

Blue magnet

■ **10-10** *The static convergence magnets are found on the neck of the picture tube.*

Slotted shadow mask convergence

In present-day picture tubes with a slotted shadow mask, static and dynamic convergences and purity are done with beam magnets mounted at the rear of the picture tube (Fig. 10-11). The yoke assembly is cemented to the base of the tube or bolted to the neck of the CRT. The adjustable beam magnets are mounted on one assembly and bolted to the neck of the picture tube.

The purity magnets are found next to the yoke assembly or at the front of the magnet assembly (Fig. 10-12). The blue pole is in the center with the red pole at the rear of the picture tube. A large lock ring is found at the rear and holds the magnet in position. Slightly loosen the lock ring to adjust the various magnets. Make sure the beam magnet assembly does not turn when making these adjustments.

Yoke assembly   Rubber wedge

Purity magnet

■ **10-11** *The beam magnet is mounted at the rear of the neck of the CRT in the latest TV chassis.*

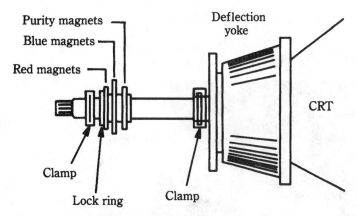

■ 10-12 *The purity magnets are found next to the yoke assembly with a lock ring to hold them in position.*

Shadow mask purity adjustment

Face the TV directly north or south before starting these adjustments. Let the TV receiver operate for 10 or 15 minutes before purity, static, and dynamic adjustments are made. The two front magnets are designed to uniformly move the three beams in both directions and amplitude. Spreading the tabs of the purity magnets lets the three beams pass through the slots in the shadow mask at different angles to provide normal purity adjustments.

Start with a pure white screen. Instead of centering red as in the older TV chassis, most slotted shadow mask adjustments begin with a centered green raster. Turn up the green bias or screen control for a green raster. Loosen the beam magnet lock ring just enough to turn the magnet rings. Spread the purity magnet tabs an equal distance apart until the green vertical raster is centered on the screen (Fig. 10-13). The magnets should never be rotated together in the same direction. These magnet adjustments are very critical, so proceed slowly. Tighten the lock ring.

The yoke should be moved for dynamic purity adjustment. Dynamic purity adjustments affect the outside edge of the TV screen. Move the yoke backwards for an overall green raster. Keep the yoke level at all times. In some chassis, rubber wedges can be used to hold the yoke in place for overall dynamic purity adjustment. Now check for a solid color raster. Touch up the purity rings and yoke assembly for complete purity adjustment.

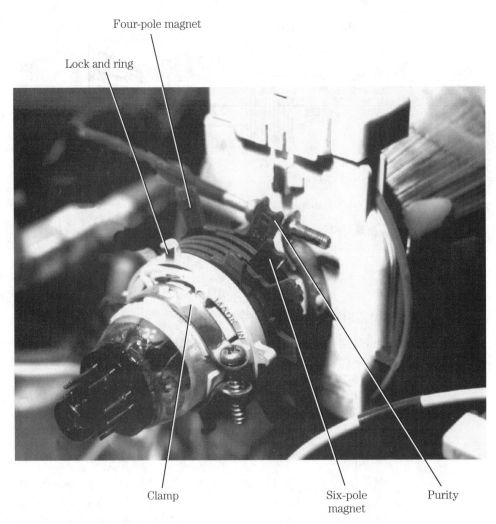

Four-pole magnet

Lock and ring

Clamp

Six-pole magnet

Purity

■ **10-13** *Adjust the purity magnet's tabs apart so a vertical center raster is found on the screen. Now move the yoke backwards and forwards until you obtain an overall green raster.*

RCA CTC130 dynamic edge convergence

Edge convergence is done by tilting the front of the deflection yoke up and down (y axis) and left and right (x axis). Center crosshatch convergence is necessary before attempting edge convergence. Center crosshatch convergence procedures require an adjustable-type beam bender (Fig. 10-14).

Remove the three wedges from under the front of the deflection yoke assembly. Apply a vertical rocking motion (up and down) to

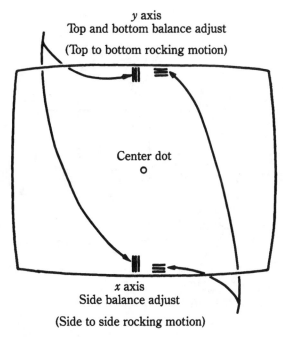

y axis
Top and bottom balance adjust

(Top to bottom rocking motion)

Center dot
o

x axis
Side balance adjust

(Side to side rocking motion)

■ **10-14** *The dynamic edge convergence is done by moving the deflection yoke in the RCA CTC130 chassis.*

the deflection yoke to balance the convergence error of the center red and blue vertical lines. Do this for both top and bottom so that they are separated equally. Do not let them cross over each other on opposite sides (top and bottom). Replace the wedge at the extreme top of the CRT (Fig. 10-15).

Apply a horizontal rocking motion (left to right) to the deflection yoke to balance the convergence error of the center red and blue horizontal lines. Do this so the top and bottom are separated equally, with either the red or blue lines high or low, but not either one high at the top or low at the bottom. Replace the wedges at the 8 and 4 o'clock positions.

Recheck the overall convergence and purity. For best results it might be necessary to repeat the procedures. This only takes a few seconds. Now place anchor tape over the three wedges.

Shadow mask dynamic convergence

Often only a touch-up is needed after repairs. Turn on the cross-hatch pattern of the dot-bar generator. Determine which color is

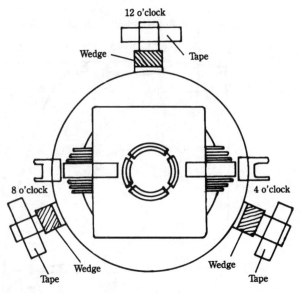

12 o'clock

Wedge

Tape

8 o'clock

4 o'clock

Wedge

Wedge

Tape

Tape

■ **10-15** *Replace the wedges in the yoke assembly at 12, 4, and 8 o'clock.*

out of alignment. Adjust only the magnet ring of either the red or blue to converge the color back in line. Tighten the lock ring at the rear of the magnets. Dynamic convergence is not needed unless the yoke assembly is loose or a new tube has been installed.

Dynamic convergence is made with the crosshatch pattern of the dot-bar generator. Edge convergence is done by tilting the yoke up and down, and left and right. Simply remove the wedges from beneath the yoke assembly (Fig. 10-16). Convergence of the vertical red and blue lines is done by moving the yoke from top to bottom. Move the yoke horizontally to balance the red and blue horizontal lines. Make sure the yoke wedges are held in position with strong adhesive tape. Follow the manufacturer's complete convergence procedure for the slotted shadow mask when proper convergence cannot be obtained.

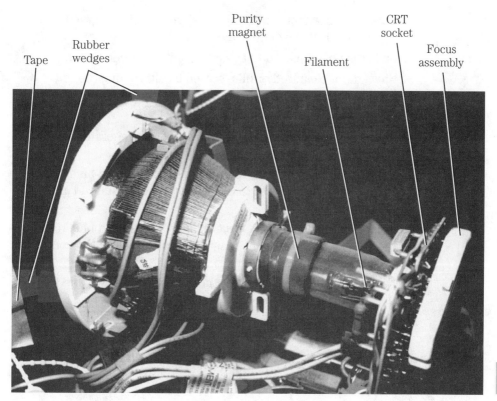

Tape   Rubber wedges   Purity magnet   Filament   CRT socket   Focus assembly

■ **10-16** *Wedge the rubber yoke spacers after each adjustment and hold them in place with strong adhesive tape.*

A typical slotted shadow mask convergence

The following procedure is used to adjust slotted shadow mask convergence.

1. Degauss the front of the picture tube.

2. Obtain a green raster. Turn up the green bias control with the red and blue bias controls turned down.

3. Loosen the yoke and pull it back towards the purity magnet assembly.

4. Rotate the purity magnet rings to obtain a uniform green vertical bank at the center of the screen.

5. Slide the yoke forward to achieve a uniform green raster.

6. Check for a clean red or blue screen by alternating the red and blue bias controls.

7. Perform black-and-white setup.

8. Apply a crosshatch pattern from the dot-bar generator for static and dynamic convergence.

9. Rotate the magnet to converge the vertical red and blue lines. Rotate the magnet to converge the horizontal red and blue lines.

10. Rotate the magnet to converge the vertical red and blue to the green lines. Rotate the magnet to converge the horizontal red and blue to the green lines.

11. For dynamic convergence, tilt the yoke up and down and left to right to converge the outer lines of the screen.

12. Alternately tune spacing with the yoke wedges for proper dynamic convergence.

13. Secure the rubber wedges with strong adhesive tape.

14. Touch up the purity, if it's needed.

Shadow mask static and dynamic convergence

When making the center converge, the red and green tab magnets are moved while the green beam is stationary. The middle pair of magnets move the blue beam while the rear magnets adjust the red beam on the green at the center of the screen. Each set of magnets can be rotated by spreading the tabs of the magnets apart from each other.

Connect the color-dot-bar generator to the antenna terminals. Loosen the lock ring at the rear of the magnet assembly. Adjust the spacing of the center magnet tabs (blue) to converge the blue and green center dots. Separate and adjust the rear magnets to converge the red center dot to the blue and green dots. Touch up both the blue and red magnets for correct center convergence.

Difficult convergence

Some picture tubes will not converge completely, no matter how much time is spent on total convergence. Try to determine if the picture tube is at fault or if there is a defective convergence circuit. Either the CRT or the beam bender magnets will be defective. First try a new beam magnet; next, replace the picture tube, if it's needed.

In older TV chassis, suspect a defective convergence circuit or picture tube. Degauss the front of the tube several times. If the corner or bottom of the raster has a blotchy color that cannot be removed, suspect a loose metal shadow mask. Replace the picture tube. When only one or two colors will not converge, suspect a defective component on the convergence board.

Degaussing circuit problems

The automatic degaussing (ADG) circuit is simple and common to all color TV chassis (Fig. 10-17). A positive coefficient thermistor (R701) has a low resistance and will let current flow through the degaussing coil when the TV set is first turned on. After a few minutes, R701 will increase in resistance and shut off any flow of current through the ADG coil. The coil placed around the front edges of the CRT screen demagnetizes the metal shadow mask to remove any colored or magnetized areas on the raster. Each time the receiver is turned on, degaussing action starts. It will not start again until R701 cools down (at least 5 minutes or so).

463

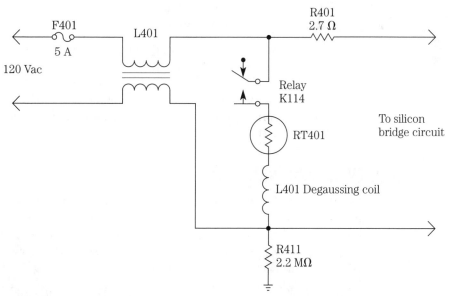

■ **10-17** *The automatic degaussing coil comes on and degausses the picture tube each time the set is turned on.*

A shorted degaussing coil to the metal picture tube mounting bracket can blow F701 (4 A) each time the set is turned on. Simply remove the degaussing coil plug to eliminate the possible short. Suspect an open or increased-resistance R702 in older TV chassis when you have a hot smell from thermistor R701, and bars with color move up the raster. The total ac line (120-Vac) is applied across the degaussing coil and R701. The resistance of R701 can be 4 Ω, 7.5 Ω, 8.5 Ω, 14 Ω, 20 Ω, or 25 Ω.

If the complaint is that a colored or blotchy area remains on the screen at all times, suspect an open thermistor (R701), the degaussing coil, or the degaussing coil connections. Check the ADG circuits by using an outside degaussing coil to clean off all impurities. Take a small permanent magnet and magnetize a small area on the screen. The same method can be used with the external degaussing ring. Let the TV rest for 5 minutes, turn on the receiver, and notice if the magnetized area is removed. If not, take a resistance measurement on R701 and L702.

Convergence hints

Always check for total convergence before replacing the back cover after each TV repair. Only a touch-up may be needed. With the crosshatch generator connected, check the picture level, vertical linearity, and height adjustments. After installing a new picture tube, go back in 30 days to touch up the purity and convergence (a very good customer-relations idea).

Leave the TV on for several hours or go back and tackle the difficult convergence adjustments the next morning. Remember, some older TV chassis will never converge more than 85% of the whole raster. Suspect a defective picture tube when there is poor purity and improper convergence in one extreme corner. Make sure the customer wants the CRT replaced.

Small TV screens with VCR

Small-screen TV sets can have a VCR unit installed in the bottom of the TV chassis. Most screen sizes are from 13 to 20 inches in color-TV/VCR combinations. The same tuner and audio circuits can be used for both the TV and VCR, while separate power supplies are found. The VCR has front-loading features (Fig. 10-18).

Often, the VCR tape mechanism is a separate unit and can be removed without disturbing the TV chassis. Service the tape deck

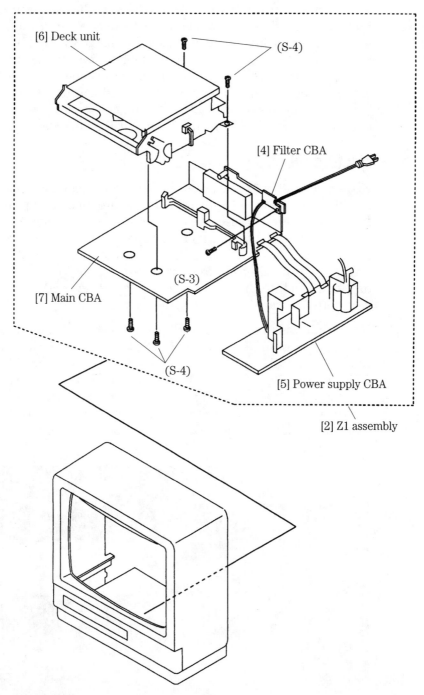

[6] Deck unit

(S-4)

[4] Filter CBA

[7] Main CBA

(S-3)

(S-4)

[5] Power supply CBA

[2] Z1 assembly

■ **10-18** *Remove the VCR mechanism from the TV cabinet after removing the TV chassis.*

with extension service fixtures, obtainable from the manufacturer. Both units (the TV and the VCR) should be removed from the cabinet before replacing the picture tube. Discharge the CRT anode before attempting to remove any assemblies. Remove the anode cap, focus wire, screen wire, and various sockets before attempting to remove the TV chassis. You may find several PC boards (such as the main, power supply, CRT, and filter boards) with several sockets and cables to tie the TV and VCR units together.

Broken picture tubes

Extreme care should be exercised when working around or hauling the TV set, so as not to damage the picture tube. Most picture tubes are broken because of carelessness. Simply letting the TV back drop down on the neck of the tube when removing screws from the back cover can cause the tube to break (Fig. 10-19). Always leave two loosened screws at the top, so the cover will not drop down on the tube. Remove the last two screws with your fingers, with the other hand holding the TV's back cover.

Use both hands to remove a small picture tube from the cabinet. Lay the front of the TV on a pad or blanket so not to scratch the screen. It may take two or three people to remove the large 31- to 35-inch glass tubes from the cabinet. Do not carry the picture tube with the screen area against one's stomach. Be careful while working around picture tubes. Just breaking one picture tube may take away the profit from three or four repair jobs.

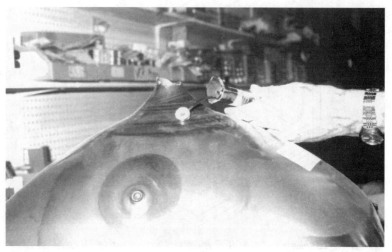

■ **10-19** *Use extreme care while working around the picture tube so as not to break off the neck of the tube and cause extensive damage.*

11

Servicing the remote-control circuits

SERVICING THE REMOTE-CONTROL UNITS IS JUST AS EASY as repairing the horizontal stages of a TV receiver. Most TV receivers sold today are of the remote-control variety. With many additional cable stations, the remote receiver has moved rapidly into most homes. The remote-control circuits offer easy operation of the TV from a distance. Like any other TV component, the remote circuits break down and need repair.

The remote circuit consists of a transmitter and receiver to operate the various functions of the color receiver. In the very early remote TVs, the mechanical transmitter operated the on/off, volume, and channel selector. The remote receiver picked up the transmitted signal, operating the various functions with several small ac motors. Later the mechanical transmitter was replaced with a supersonic unit (Fig. 11-1). Today the infrared transmitter covers many different operations, including scanning stations up and down, selecting individual channels, on/off, selecting volume, and muting the sound. Besides controlling all TV functions, you can find some remote transmitters that control the various operations of a VCR.

467

Basic remote transmitters

One of the first methods used to generate a supersonic wave for remote control was a mechanical or physical method of tapping a metal rod with a hammer or trip-type plunger (Fig. 11-2). To generate three different signals, each rod was of a different length. Each rod represented a different frequency to control the on/off, channel up, and channel down. The supersonic signal was picked up by the remote transducer of the remote receiver to control the various operational functions (Fig. 11-3).

■ 11-1 *Many different sizes of batteries are found in the different remote transmitters.*

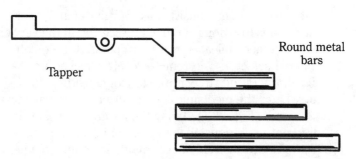

Tapper

Round metal bars

■ 11-2 *The mechanical plunger used in many early remote transmitters.*

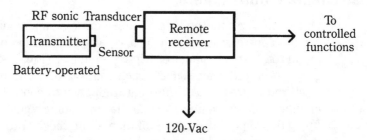

RF sonic Transducer

Transmitter

Sensor

Battery-operated

Remote receiver

To controlled functions

120-Vac

■ 11-3 *A block diagram of the sonic transmitter, with the transducer in the receiver.*

Another transmitting method was to generate an electronic signal to control certain frequencies to be radiated through a speaker or transmitter (Fig. 11-4). The frequency zone chosen was between 44.75 kHz and 47 kHz. This frequency avoids most erroneous signals that might trigger the remote receiver. Of course, garage door openers, door and telephone bells, and other sources of supersonic signals can trigger the TV's supersonic receiver.

In a Sharp transmitter circuit you will find the on/off and volume are controlled at 41.5 kHz, channel-up at 40 kHz, and channel-down at 38.5 kHz. Likewise, in an RCA CRK19A remote transmitter, the on/off and volume operate at 44.75 kHz, with channel-up and -down at 41.75 kHz. Other manufacturers use different frequencies for volume and channel selections. You might find that one frequency controls the on/off and volume in one model, while in another model the same frequency is used for channel-up or -down. This helps prevent the interaction of remote transmitters when two or more remote-control TVs are in the same home.

Although the supersonic remote transmitter was a big improvement over the mechanical transmitter, there were a few drawbacks in operation. The biggest disadvantage was when several remote TV sets were found in a large apartment house. Your TV set might be turned on by a neighbor with another remote TV. Erroneous signals from various electrical and RF-generating devices can trigger the remote-control receiver. The design of the infrared remote transmitter has solved most of these signal problems.

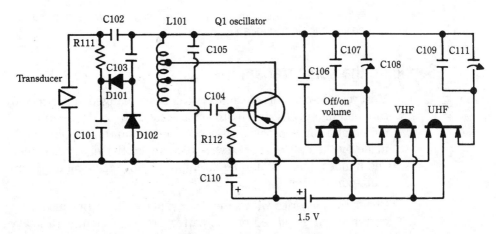

■ **11-4** *A typical sonic transmitter, controlling on/off/volume and VHF/UHF tuning.*

Testing supersonic remote transmitters

The small sonic remote transmitter can be checked with another TV receiver using the same type of remote. Always bring the remote to the shop when on a house call and the remote control does not function. Most remote problems involve the remote transmitter. The remote receiver in the TV set can be checked with a new remote transmitter to determine if the receiver is defective. If battery replacement does not repair the remote transmitter, the unit should be sent in for exchange or repair.

The supersonic remote transmitter can be checked with the help of the remote receiver amplifier alignment procedure (Fig. 11-5). The oscilloscope is connected to the test point of the receiver with the remote transmitter close to the receiver transducer or microphone. Each coil is peaked for maximum indication on the scope with the remote transmitter operating. Repeat the adjustments with the transmitter several feet from the TV receiver. Follow the manufacturer's alignment procedure for remote amplifier adjustments.

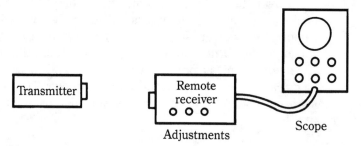

■ **11-5** *Check the remote transmitter signal with remote receiver and scope as an indicator.*

Servicing the sonic transmitter

Most remote-control transmitters can be exchanged or serviced by the manufacturer's distributor or the factory, or they can be repaired at a tuner repair center. Simple transmitter repairs can be made by the operator or service technician. Battery replacement and minor repairs should be attempted before packing up the remote-control unit.

Check the batteries with a meter or replace them. Because heavy current drain is placed on the battery, replace them with heavy-duty types. Today, the remote transmitter is often used constantly. Inspect the battery terminals for corrosion (Fig. 11-6) and for broken wires.

■ **11-6** *Check for corroded or loose battery terminals in the remote transmitter.*

The small transmitter has only a few working components and is easy to repair. In an RCA CRK26E, replacement of a few components can solve the defective remote (Fig. 11-7). The remote might have been dropped, tearing the small RF coil loose from the PC board. T1001 can be replaced easily, because it comes factory-tuned and locked. Check transistor Q1001 for open and/or leaky conditions (Fig. 11-8). Sometimes small children poke sharp objects into the small sonic transducer elements, damaging the transducer unit. This component can be replaced by removing only two connecting wires. Small channel buttons can stick. A plastic button can bind against a plastic front piece and not engage the switch area. Spray the sticky buttons with a silicone cleaner/lubricant. Replace all remote transmitter parts with the original replacement part.

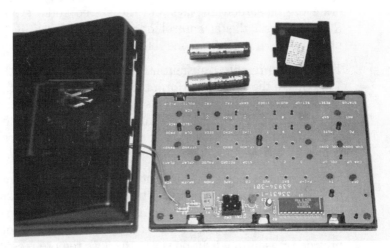

■ **11-7** *Check the infrared diode, IC, transistors, and voltage measurements in the infrared remote.*

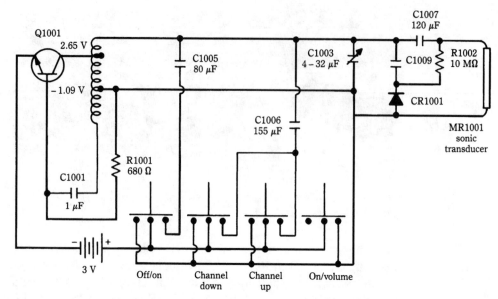

■ **11-8** *Check the transistor and take voltage measurements with a tuning button pressed down.*

Infrared remote transmitter

The infrared remote transmitter comes in many sizes and shapes (Fig. 11-9). Some of today's remotes can change the remote-controlled VCR, as well as the color TV. At first the digital control transmitter might appear quite complicated, but it is easy to operate. Besides operating the regular TV controls, the remote can operate the on-screen displays and the stereo audio system. The RCA Dimensia digital control operates the TV, VCR, AM-FM, phono, tape, and CD from one remote (Fig. 11-10).

The infrared remote transmitter must be pointed directly at the TV to make the remote perform the different operations. The remote receiver will not respond if someone stands between the operator and the TV receiver. An infrared remote will not interfere with any other TV receivers in the house. The infrared remote is generally very accurate and dependable.

Most infrared remotes are constructed like the regular sonic transmitter, except that LEDs serve as transducers. A regular keyboard assembly feeds into an encoder IC processor with crystal-controlled driver and amp transistors (Fig. 11-11). One or more LEDs provide transmitting light power to the TV's remote receiver. The various operations of separate channel selection, channel-up and

■ **11-9** *The infrared remote transmitter may operate a TV, VCR, receiver, and CD player.*

■ **11-10** *The digital remote can operate the TV set and VCR unit.*

-down, on/off, volume, mute, and recall can be found within the infrared remote transmitter. Most infrared transmitters are operated by 4.5 to 9 V.

RCA CRK39K remote transmitter

The RCA CRK39K infrared transmitter operates on two AAA batteries with a total of 3 V. IC U1001 is crystal controlled by Y1001

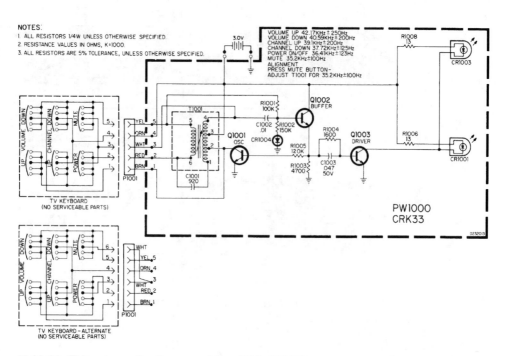

NOTES:
1. ALL RESISTORS 1/4W UNLESS OTHERWISE SPECIFIED.
2. RESISTANCE VALUES IN OHMS, K=1000.
3. ALL RESISTORS ARE 5% TOLERANCE, UNLESS OTHERWISE SPECIFIED.

VOLUME UP 42.17KHz ± 250Hz
VOLUME DOWN 40.59KHz ± 200Hz
CHANNEL UP 39.1KHz ± 200Hz
CHANNEL DOWN 37.72KHz ± 125Hz
POWER ON/OFF 36.41KHz ± 123Hz
MUTE 35.2KHz ± 100Hz
ALIGNMENT
PRESS MUTE BUTTON-
ADJUST T1001 FOR 35.2KHz ± 100Hz

■ **11-11** *The schematic diagram of the RCA CRK33 remote transmitter.* Thomson Consumer Electronics

474

at input pin 11 and output pin 12 (Fig. 11-12). The drive and sense inputs are controlled with a momentary-contact switch. For instance, pressing the power-on button connects the sense 1 (pin 18) to the drive 5 input (pin 5). The infrared output signal (pin 8) is applied to Q1001, and the transmitting LED is found in the collector circuit of Q1001. This infrared light is pointed towards the TV receiver, which in turn operates the infrared remote receiver.

Testing the infrared transmitter

Check the output of the infrared transmitter to determine if the transmitter or the receiver in the TV is defective. The infrared transmitter can be checked on another TV using the same remote. If the remote changes all functions on the TV, suspect a defective remote receiver in the TV. Another method is to use an infrared indicator card in front of the remote transmitter. The infrared indicator card will change to a different color if the transmitter is working. The infrared indicator card is obtained through TV dis-

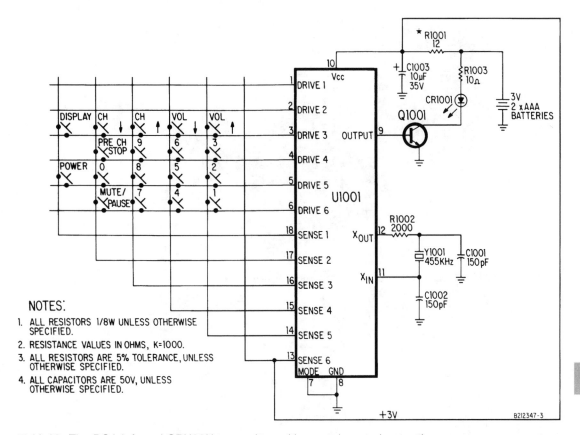

NOTES:

1. ALL RESISTORS 1/8W UNLESS OTHERWISE SPECIFIED.
2. RESISTANCE VALUES IN OHMS, K=1000.
3. ALL RESISTORS ARE 5% TOLERANCE, UNLESS OTHERWISE SPECIFIED.
4. ALL CAPACITORS ARE 50V, UNLESS OTHERWISE SPECIFIED.

■ **11-12** *The RCA infrared CRK39K transmitter with crystal-control operation.* Thomson Consumer Electronics

tributors. The RCA infrared indicator for checking the output of remote-control TV transmitters is stock number 153093. The infrared remote can also be checked with a portable radio. Each time a button is pressed, you will hear a gurgling sound in the portable radio (Fig. 11-13).

The infrared remote transmitter can be checked as any remote unit. Check and replace the 9-V battery. Determine if the remote transmitter is defective before removing the back cover. Take voltage measurements on the various transistors (Fig. 11-14). Check each transistor with the diode-transistor tests of a DMM. The LEDs can be checked with the diode test function. The keyboard assembly should be replaced instead of attempting to repair it. Send the remote transmitter in for repair or exchange when simple repairs will not cure the malfunction.

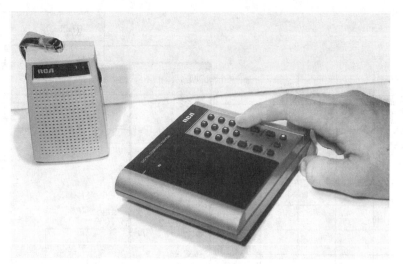

■ **11-13** *Check the remote out beside a radio.*

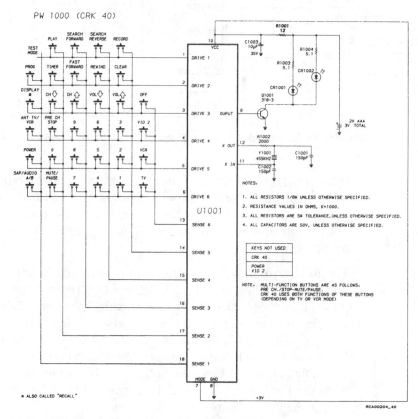

■ **11-14** *Schematic diagram of the RCA CRK40 remote transmitter.* Thomson
Consumer Electronics

Infrared remote-control tester

You can check hand remote transmitters in minutes with the infrared remote-control tester (Fig. 11-15). Place the remote a few inches from the tester and press any button on the remote. You should hear a chirping or audio tone when any button on the remote is pressed. Some infrared remotes produce a chirping noise, while others produce a loud tone from the piezo buzzer. This remote tester will check all infrared remotes for TVs, VCRs, and CD players. Move the remote transmitter back and forth in front of the tester until the loudest sound is heard. Check the remote for a weak signal by moving the remote away from the tester. With this

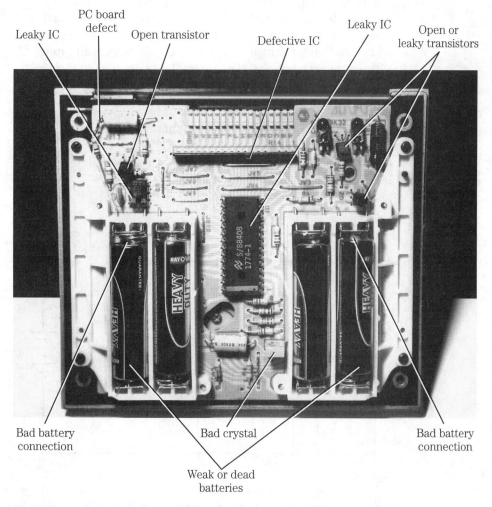

Leaky IC   PC board defect   Open transistor   Defective IC   Leaky IC   Open or leaky transistors

Bad battery connection   Bad crystal   Bad battery connection

Weak or dead batteries

■ 11-15 *Check the following parts in the remote transmitter for defects.*

477

Infrared remote-control tester

checker, some remote transmitters can sound off at 2 feet. Keep the remote pointed towards the tester at all times.

Check each remote button for correct operation. If the button appears erratic or intermittent, suspect a dirty button. Suspect a broken or poor battery connection when the remote operates intermittently. When you have to press hard on the button or it takes a little time for the remote to act, suspect weak batteries. Check for poor connections or loose batteries when the remote control must be tapped or shocked into operation. No sound indicates a defective control unit or dead batteries.

The circuit

The small remote transmitter tester circuit is built around an npn infrared photo transistor. These photo transistors can be picked up at almost any electronics supply store. Sometimes the transmitting LED and photo transistor come as an operating pair (Fig. 11-16). The infrared beam is picked up by Q1 with the input signal applied to IC1. The infrared npn silicon phototransistor used here provides high-speed photosensitivity.

IC1 is a general-purpose amp IC with low power consumption. The op-amp (741) can be found at almost any electronics store. The eight-pin IC amplifies the audio signal to a piezo buzzer. R1 adjusts the input signal. Greater distance is possible with IC1 in the circuit. A complete parts list is given in Table 11-1.

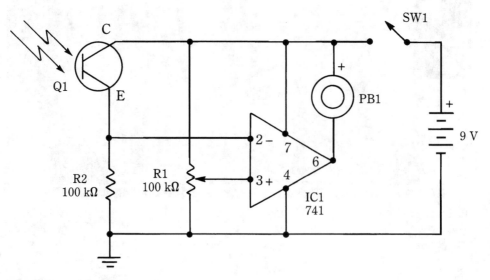

■ **11-16** *The schematic of the remote transmitter tester you can build in one evening.*

478

■ Table 11-1 A parts list for an infrared remote control tester.

Q1	Infrared phototransistor npn type—276-142 or equivalent
IC1	741 general-purpose op amp IC 8-pin type—276-007 or equivalent
PB1	PC mount piezo buzzer—273-065 or equivalent
R1	100-kΩ, PC board PAT, thumbscrew or screwdriver adjust—271-220 or equivalent
R2	100-kΩ, 0.5-W resistor
SW1	Submini, SPST toggle switch—275-645 or equivalent
Battery	9 V
Case	3.25 × 2.125 × 1.125 project box—270-230 or equivalent
Miscellaneous	Piece of perfboard, 9-V battery, lead and socket, hookup wire, 8-pin DIP socket, solder

RF remote-control receivers

There are many types of remote receivers. The receiver circuit consists of a transducer or sensor fed to a high-gain remote amp. Some remote receivers have a separate gain control, while others do not. The high-gain signal is fed to the various stages to control the many functions of an older-type system (Fig. 11-17). One frequency controls the tuning motor; another signal, at a different frequency, controls the on/off and volume.

In other remote RF receivers, channel-up is controlled with one frequency, channel-down by another frequency, and on/off and volume are controlled by yet a third RF signal. The on/off signal is applied through a multivibrator circuit developing a negative pulse. This is applied at two separate transistors, so the set can be switched on and off. The on/off signal is applied to a transistor amp with a power relay. The relay applies on/off ac voltage to the TV chassis. Although each manufacturer has a different-looking re-

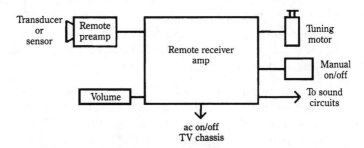

■ 11-17 *The older remote receiver consisted of a remote amp, preamp, transducer, and motor.*

mote receiver system, they all do the same thing: They make TV operation easier for the operator.

Standby power circuits

The standby power supply circuits must operate all the time so the remote transmitter can turn on the TV chassis. A standby voltage must be applied to the infrared (IR) receiver so the remote can turn on the TV set. The standby voltage circuit in an RCA CTC157 chassis supplies 12 volts to the IR receiver and 5 volts to the system control microprocessor (U3100) (Fig. 11-18).

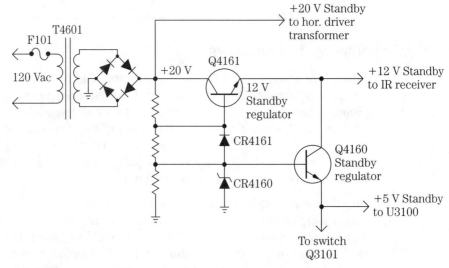

■ **11-18** *A standby power supply for the remote receiver in the TV chassis.*

The power line voltage is connected directly to the primary winding of T4601. The secondary voltage is applied to a bridge rectifier circuit providing a +20-V standby voltage to the horizontal driver transformer. The +20 V is regulated by transistor Q4161 down to the infrared receiver (IR). Q4160 is a 5-V standby regulator that is furnished to the system control microprocessor.

Low or improper 12- and 5-V supplies can be caused by a leaky or open voltage regulator (Q4161). Notice the two zener diode regulators in the base circuits. The absence of a 12-V source can result from an open voltage regulator. If the 5.6-V zener diode (CR4160) opens, both the 12- and 5-V standby supplies will go higher. The 12- and 5-V sources are very critical for proper standby operation.

Infrared remote receiver

The infrared rays transmitted by the infrared transmitter are picked up by the light sensor of the remote amplifier feeding the remote receiver circuits. The photodiode is called a light sensor, pin diode, or remote sensor (Fig. 11-19). The preamp is called a photo amp board, remote amplifier, or preamp board. The pulse-modulated infrared rays provide remote control for changing channels, volume-up and -down, power on/off, and many other functions. Most manufacturers request that the remote-control preamplifier unit be completely replaced instead of repaired. Some manufacturers do not supply parts for them.

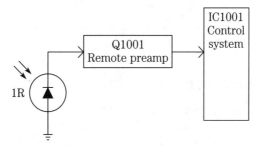

■ **11-19** *A block diagram of the infrared preamp board with the emitting diode.*

From the high-gain preamplifier board, the FM signal is connected by cable to the remote-control receiver circuit. There are many different kinds of remote receivers found in each chassis (Fig. 11-20). Some of these remote receivers are of the modular type. This kind of complete remote receiver unit can be interchanged by unplugging a few cables, removing the original unit, and substituting a new one.

The new infrared remote receiver can have many functions. Besides channel selection, a scan-up and -down control might be found. Some units have a separate on/off switch. Besides controlling the volume, a remote might also mute the sound. A channel-recall button is provided in some remote-control systems.

The remote-control receiver can be exchanged at the manufacturer's distributor or sent in for repair at various tuner repair depots. The remote receiver can be serviced by the TV technician if correct test instruments are available. Do not attempt to repair these circuits without the manufacturer's service literature and a schematic that provides voltages.

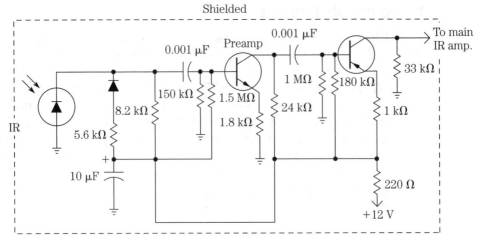

■ **11-20** *A typical transistorized remote preamp found in the receiver of the TV chassis.*

Simple remote-control receiver

You might find the remote receiver as a part of the memory or control module, or as a separate chassis in the color TV (Fig. 11-21). The remote receiver can be quite simple or rather complicated, depending on its functions. The simple remote receiver contains an IC amp, an infrared LED, and very few components (Fig. 11-22).

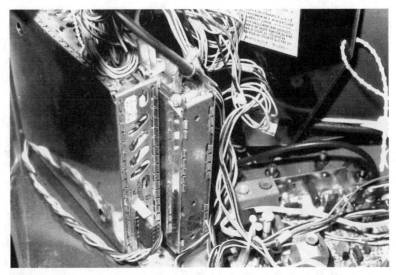

■ **11-21** *The remote receiver is located in the control module of the RCA chassis.*

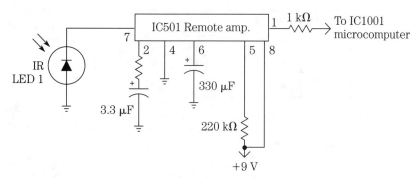

■ **11-22** *A typical remote receiver used ahead of the microcomputer in the latest TV chassis.*

The infrared transmitter signal is picked up by LED1 and coupled directly to pin 7 of IC501. The amplified output signal appears at pin 1 and is fed through plug PG241 to the microcomputer within the control module. The 9-V power source is fed from the TV circuits.

RCA MCR015A/B remote receiver

In the larger remote-control receivers, the remote module consists of an infrared LED (CR1) and several preamp transistors (Fig. 11-23). Volume-up and -down, channel-up and -down, on/off, and mute will have different function frequencies. The remote decoder IC (U1) receives these frequencies at pin 2. Volume-up and volume-down signals enter pins 5 and 6 of U1. Pin 7 controls the on/off, while pin 11 controls the channel-up and pin 9 controls the channel-down.

The volume-up and -down control is found at pins 12 and 13 of U1 with Q6. With minimum volume, the base voltage of Q6 is around 0.74 V. With maximum volume, the base voltage is 0.06 V. The collector voltage at minimum volume is 14 V and the maximum is 6.5 V. The base and collector voltages of Q6 vary with the volume level.

U1 is crystal-controlled by Y1101. The oscillator-in signal (pin 4) and -out signal (pin 3) can be scoped. Check the supply voltage source at pin 16. The remote receiver circuits can be serviced by checking to see if the receiver works on either remote or manual operation. If the receiver operates on manual and not remote, check the remote transmitter and receiver input circuits. Check transistors with the DMM or transistor tester. Take scope signals on remote decoder IC U1. Measure critical voltages and resistance on operating components.

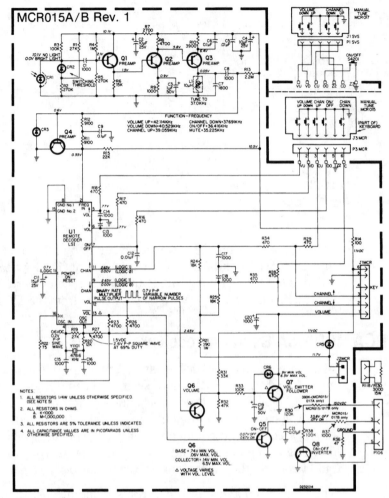

■ **11-23** *The schematic of the remote-control receiver in the RCA MCR015A/B unit.* Thomson Consumer Electronics

RCA CTC157 infrared receiver circuits

Two different inputs of customer control consist of a remote control and keyboard. The keyboard operates the system control (U3100) and in turn controls a microcomputer (U3300). The keyboard information is input at pins 21–24 and 25–28 of the system control IC (U31ll). All customer controls are generated electronically (Fig. 11-24).

The second input to the control interface (U3300) is an infrared receiver (IT). The IR detector picks up the remote control's signal and feeds into a preamplifier circuit. Q3401 amplifies this signal

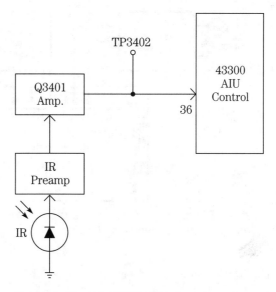

■ 11-24 *The RCA CTC157 infrared remote receiver schematic.* Thomson Consumer Electronics

and applies it to pin 36 of U3300. In turn U3300 relays or controls each function selected by the remote control transmitter.

When the remote transmitter will not control the TV chassis, check the batteries in the remote. Does TV operate with keyboard controls? Try another remote. Notice if all functions are inoperative. Check for remote input at TP3402 for scan pulses with oscilloscope. If all customer controls do not operate, suspect system control (U3300). When the keyboard controls the TV chassis and neither remote functions nor the set can be turned on, check the standby voltage circuit.

TV control modules

You will find separate remote receiver control modules to the remote amp, manual tuning assembly, on/off switch, ac power, and control plug (Fig. 11-25) that can be easily replaced by removing plug-in cables. The entire remote receiver module can be substituted with another module to determine if the receiver is defective. Also, the complete module can be sent in for repair.

First, check the most common problems before sending the module in for repair. Intermittent operation of the on/off button can be caused by dirty contacts. Place a clip wire across J1104 to determine if the TV chassis turns on and off. Make sure ac power is ap-

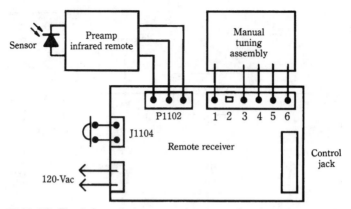

■ **11-25** *The infrared sensor and preamp feeds to the remote receiver in a frequency synthesis tuner.*

plied to the remote primary windings of the power transformer inside the module.

If the volume can be turned up and not down, suspect a defective manual tuning assembly, socket, or remote receiver. Replace the remote module to isolate the manual tuning assembly (Fig. 11-26). Suspect contacts within the volume-down area of the manual assembly if switching the receiver module does not help. Unplug the manual tuning assembly and short pins 1 and 4. The sound should go down. Likewise, if the channels will not scan up, but will go down, short pins 4 and 6. Each function can be checked in the same manner by using pin 4 as common ground.

■ **11-26** *Small pushbutton microswitches feed into the microcomputer IC to operate each selected station.*

In the latest TV chassis, you will find that the remote preamp assembly plugs into the frequency synthesis tuner circuit board assembly. No separate remote receiver is found in these models. The tuner circuit board module must be exchanged when the remote-control transmitter is normal, but the set has no remote-control functions.

On/off circuit tests

One of the major functions of the remote is to turn the TV off and on. With a no-sound, no-picture symptom, the TV chassis or remote receiver might be defective. When the color receiver is entirely dead, check for 120 Vac at the input and output terminal of the remote receiver relay switch connections. An ac voltage at both pins (21 and 22) of a Samsung power circuit indicates that the remote-control relay is normal (Fig. 11-27).

If no voltage is found at the anode terminal of D801, the remote receiver unit might be defective. Listen for a click of the relay each time the set is turned on. In case the relay clicks with no ac at D801, suspect burned or corroded relay contacts, or poor wiring connections at the remote circuit board. Check for ac voltage at pin 22 and then at pin 21. A normal 120 Vac at pin 22 and no voltage at pin 21 indicates open remote wiring. If normal ac voltage is found at pin 21, suspect problems within the receiver power circuits. Check components R801 and L801 for open conditions.

Temporarily clip a jumper wire across remote pins 21 and 22. If the receiver begins to operate, the defect is in the remote wiring, relay, or remote receiver circuits. Take voltage measurements and test each transistor in the remote receiver assembly.

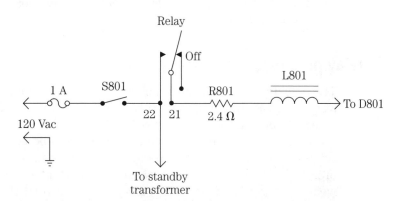

■ **11-27** *Shunt the relay contacts 21 and 22 to isolate the remote standby circuits.*

100-W light bulb test

A 100-W light bulb can be used in series with the power line to reduce the power line voltage to determine if the TV chassis has high-voltage shutdown. Of course, the 100-W bulb cannot be used with remote control sets. The bulb will only flash for a second and then remain off. With a large bulb in series (clipped across the ac fuse holder), not enough voltage is applied to the remote receiver to make it function properly.

To control the ac line voltage, always use a variable or variac power-line transformer with a TV chassis that employs remote control or power circuits (Fig. 11-28). This eliminates the remote-control circuits entirely by applying 120 Vac directly to the low-voltage power circuits. Now the ac line voltage can be controlled with the adjustable line transformer to determine if the TV chassis is shutting down with high-voltage or chassis shutdown.

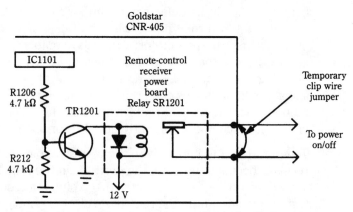

■ **11-28** *Poor relay contacts in a Goldstar CNR-405 remote produced a dead chassis.*

TV relay problems

The defective relay can turn the TV chassis on without a remote or control function, might not turn the TV off or on, might cause a dead or intermittent TV chassis, and might cause a loud chattering noise. An open relay solenoid or shorted diode across the solenoid can cause a dead operation. Intermittent connections or dirty switch contacts can cause intermittent conditions. A weak spring or hung-up switch arrangement can let the chassis operate all the time. The relay is found in the ac and dc power supply circuits.

When the relay will not click in, suspect a defective filter capacitor. Relay chatter may also be caused by a dried-up filter capacitor (Fig. 11-29). The relay may come in by itself and let the receiver become hot unless someone turns it off. Relays that operate in transistor circuits may turn on by themselves, resulting in overheated components.

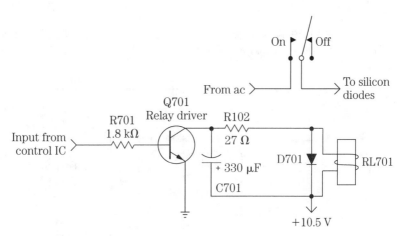

■ **11-29** *The power relay may have pitted or dirty contacts, an open solenoid, or no voltage applied to the solenoid.*

Five actual remote case histories

The following are five real cases of remote-control-related problems in TV chassis.

No standby voltage—remote

Very little standby voltage was found at the remote receiver IC1101 in a Panasonic CTL1032R. Manual operations were normal, but there was no remote action. Upon checking the remote voltage source in the power supply circuits, the standby supply voltage was missing. Regulator IC841 (7BL56) was found leaky (Fig. 11-30). This voltage supplied to the regulator is on all the time, and it comes directly from the power line's 3-A fuse.

No remote—RCA CTC166

The supply voltage was found to be normal at the receiver remote IR preamp IC (U3401), with no remote channel selection. The IR indicator and parts tied to the preamp IC seemed good. Replacing the defective IR preamp solved the remote problem (Fig. 11-31).

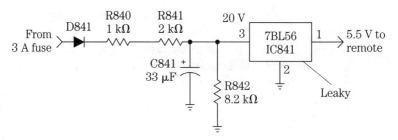

■ **11-30** *A lack of standby voltage was caused by a defective IC regulator IC841 in a Panasonic TV chassis.*

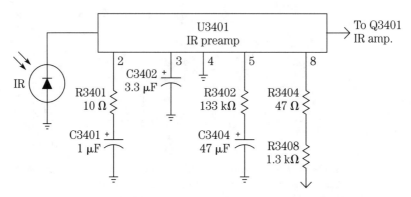

■ **11-31** *The dead remote in an RCA CTC166 chassis was caused by a defective IR preamp IC (U3401).*

Remote dead—Toshiba CF317

The supply voltage was zero volts at the remote signal amp ICR01 on pin 8. Upon checking the standby voltage (+5 V), it was found to be missing at CR06, but high at DE02 rectifier. RR05 (47 Ω) was found open, with a leaky CR06 (47 µF) electrolytic capacitor (Fig. 11-32). Replacing CR06 with a 160-V electrolytic capacitor and replacing RR05 restored remote operation.

No channel-down control

All functions worked except for the channel-down operation in a Sylvania E24-7 chassis. When the channel-down button was pushed, nothing happened. Also, the manual channel-down button on the chassis (SW1030) had no effect on lowering the TV channels (Fig. 11-33).

After locating the manual low button and channel-down circuits on the remote receiver board, several voltage tests were made.

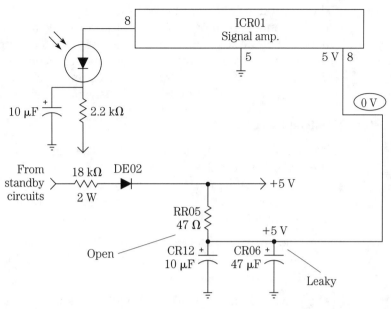

■ 11-32 *RR05 (47 ohms) was found open in the standby power supply of a Toshiba CF317 TV.*

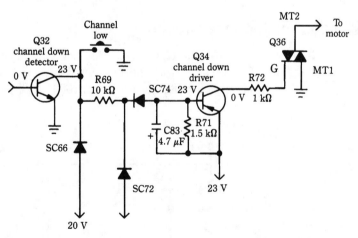

■ 11-33 *A defective triac in a Sylvania E24-7 chassis prevented the remote from scanning the tuner downward.*

The dc supply voltage was 23 V, with the same voltage at the base of Q34. Because the channel-down driver transistor tested off-value in the circuit, it was removed. Q34 tested good, but was replaced. The results were the same.

The power-line voltage (120 Vac) was found on the MT2 terminal of the channel-down triac (Q36), with no voltage at the gate terminal. The triac was defective or there was insufficient gate voltage. Q36 was removed from the circuit, with no continuity across any terminals. After replacing Q36 with a universal ECG 5604 replacement, the channel-down circuits began to operate.

RCA CTC140 remote problem

The remote control would not operate when the set warmed up. The remote transmitter checked normal. After the TV cooled down for several hours, the remote would turn the set off and on. After checking several circuits, the trouble was found to be in the system control circuits.

The voltage would increase on the IR-in (pin 36) of AIU IC (U3300). Upon checking the IR circuit, U3300 was suspected, but everything else seemed to work (Fig. 11-34).

After several hours of working, U3300 was sprayed with coolant when the remote would not shut off the TV. Still nothing occurred. When diodes CR3302 and CR3301 were checked in the IR circuit, CR3302 seemed to change value when warm. CR3302 was found to be leaky, and was replaced with a 4.9-V zener diode.

RCA control-interface circuits

In the RCA CTC156 and CTC157 chassis, the system control microcomputer (U3100) controls the customer keyboard and remote-control input circuits. The system control IC (U3100) and AIU

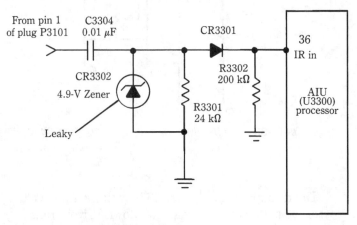

■ **11-34** *A leaky CR3302 in an RCA CTC140 chassis caused the remote to not operate after 3 or 4 hours.*

(U3300) provide interface between the operator and the TV chassis. All of the interface functions are electronically controlled. U3300 controls the contrast, sharpness, color, tint, brightness, volume, balance, treble, bass, and on/off functions (Fig. 11-35).

The remote-control signal is picked up by the infrared detector LED and preamped to pin 36 of U3300. The manual keyboard signals are sent to U3100 and data is transferred to the AIU (U3300). The remote-control command signals are processed internally by U3300, and are then sent to the system control IC (U3100), which decodes the information and outputs any necessary instructions back to U3300.

The AIU (U3300) interprets this control data and outputs the correct control signals. All control operation signals are output from U3300 in the form of a binary rate multiplier (BRM) signal. These BRM signals are different than pulse-width-modulated signals (PWM). The BRM duty cycle remains constant, while the repeti-

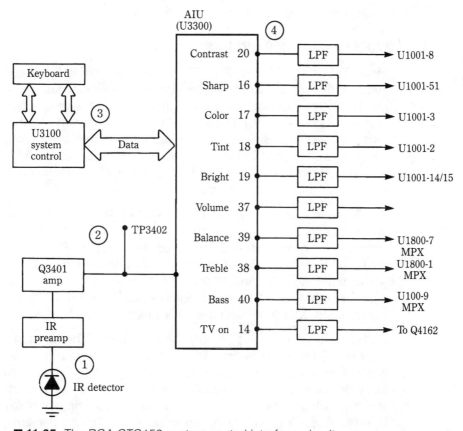

■ **11-35** *The RCA CTC156 custom control interface circuit.* Thomson Consumer Electronics

tion rate is changed to increase or decrease the desired dc control voltage to each low-pass filter.

Troubleshooting interface circuits

Determine if either the remote or keyboard controls operate. If the remote transmitter does not work, check the remote-control unit (1). When the remote is working, check the IR input at pin 36 of U3300 or TP3402. If only one function is operating, check the low-pass filter or the signal processed by that circuit (4).

If the keyboard is not functioning, check the keyboard and system control IC (U3100) (3). If the digital communication signal between U3100 and the AIU does not change when a function control is pressed, suspect U3100. If the BRM outputs are not present, suspect U3300. Do not forget to check the voltage source at U3100 and U3300. When all signals are not working, suspect the system control IC (U3100), AIU (U3300), keyboard, or remote receivers. If only one or two controls do not work, suspect a low-pass filter or the corresponding signal-processing circuits.

RCA keyboard interface circuits

In the RCA CTC156 and CTC157 chassis, the keyboard buttons and system control microcomputer (U3100) provide signals to the AIU (U3300) processor. The remote models do not have the digit keys (0–9) on the front panel of the TV. However, the front panel keyboard does have access to the color, tint, brightness, contrast, and volume. The digit key (0–9) operation can be accessed in remote models with the hand unit. The on-screen display is found in the CTC156 and CTC157 chassis.

The keyboard code consists of four bits (B0-B3). Each bit can be low or high. If the pull-up resistor is in the circuit the bit is high, and when the resistor is not in the circuit the bit is low (Fig. 11-36). Bit 1 (B1) is the manual/remote bit. If R3437 is in the circuit, about 5 V is found at KD1 (pin 23) from the 12-V standby supply. The high at KD1 indicates the set is a remote model. When R3437 is not in the circuit, KD1 is low and the set is a manual model.

Bit 2 (B2) and bit 3 (B3) identify the functions of the individual keys at the keyboard matrix. When changing the state of these bits, determine the function of each key. Switch S1 (B2) is used to allow the same keyboard to be used in a horizontal or vertical installation. SW51 is not found in all keyboards. Determine if SW51 is installed and if it is closed or open before servicing a defective keyboard (Fig. 11-37).

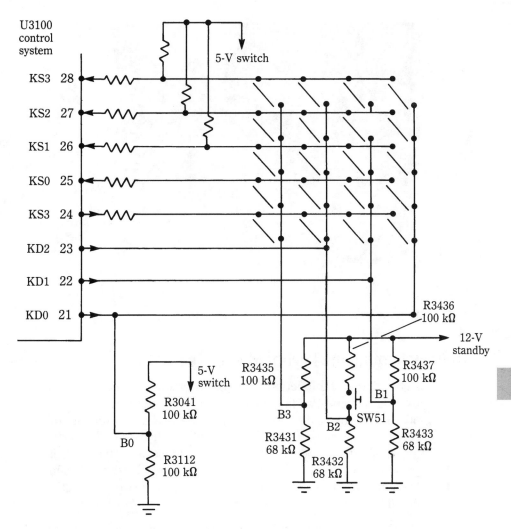

■ 11-36 *The RCA CTC156/157 keyboard interface circuits.* Thomson Consumer Electronics

Bit 0 (B0) has a different function determined by the state of bit 1. If bit 1 is high, then bit 0 identifies the tuner as broadcast or cable. If bit 1 is low (R3142 installed/R3441 removed), then the chassis is equipped with a broadcast-only tuner. If bit 0 is high (R3141 installed/R3442 removed), the chassis is equipped with a cable tuner. The microcomputer outputs the correct control signals to the AIU during tuning to provide band switching and tuning voltages. When bit 1 is low (remote chassis), bit 0 is used to identify the chassis as a mono or stereo chassis. Remember, all remote chassis have cable tuners, while all manual chassis have mono audio systems.

RCA keyboard interface circuits

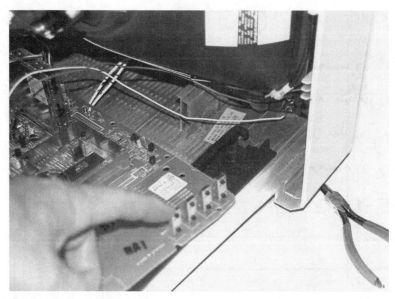

■ **11-37** *Pushbutton switching in the keyboard of today's TV chassis.*

Troubleshooting the keyboard interface

Determine which functions are not operating, check the continuity of all switches, and check the scan pulses of the defective keyboard with the scope. To isolate a single keyboard problem, take continuity measurements. To make sure each keyboard is operating, scope the scanning system while holding the defective front-panel keyboard button and monitor the correct drive and sense lines. If the front-panel keys do not control the correct functions or fail to tune cable channels, suspect a defective keyboard circuit.

Check the continuity of each keyboard with the ohmmeter from the keyboard switch to the power source and system control pin. Of course, the chassis should not be on with ohmmeter measurements. Check all of the connections of the keyboard switch to the 12-V standby power supply. Voltage measurements of the 12-V power supply when each switch is pressed can indicate the circuit is normal.

Universal remote control

Just about every manufacturer of TVs, as well as other electronics manufacturers, make a universal remote control that will operate almost all remote TV, VCR, audio and video components, and CD players. The General Electric RRC500 does the work of three remotes, while the RRC600 model does the work of four remotes. The RRC600 controls up to four infrared audio/video products, with over 200 key combinations, program sequencing, LCD display, and a low-battery indicator included (Fig. 11-38).

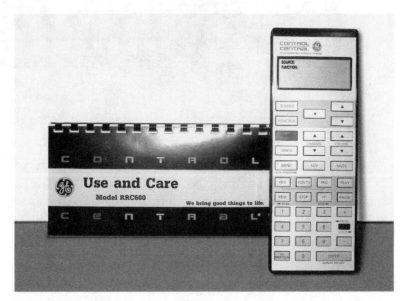

■ **11-38** *The General Electric RRC600 remote controls up to four different machines with over 200 key combinations.*

The Memorex CP8™ Turbo universal remote control can operate home entertainment systems without the user being home. The 16-event programming function allows the user to operate any audio or video component without being present at the scheduled operating time. For example, users can program the television to come on at 6:30 a.m. every day to serve as a wake-up alarm. The CD player can come on when they arrive home at 6:00 p.m. and go off at 7:00 p.m. when the TV is programmed to start. You do not have to touch the audio/video system.

The CP8 Turbo unit has an advanced high-speed microchip that learns and memorizes the infrared codes of each of eight audio or video components. It operates more quickly and accurately than conventional universal remote-control units. The LCD display has a real-time clock and dual display light with large, easy-to-read characters, and it displays the commands in sequential steps. The remote even tells you when to change the batteries.

The universal remote operates on batteries like any other remote (Fig. 11-39). The remote can be checked like other remotes with

the remote tester (discussed earlier in this chapter). Most manufacturers request that the universal remote be sent back to the factory for major repairs.

Conclusion

Although there are many different remote chassis, servicing the remote-control system can be made easy. First, make sure the remote transmitter is working with comparison tests. Isolate the service problem to the TV chassis or the remote receiver by placing a shunt wire across the on/off relay switch assembly. Make accurate voltage and transistor tests on the defective remote receiver chassis.

Replace a faulty remote receiver with a new remote receiver, or substitute the control module to determine which component is defective. Remote-tuning components must be replaced in the event of extensive lightning damage. The remote transmitter, receiver, and modular components can be exchanged or repaired through manufacturer's outlets. Tuner repair centers service most remote-control transmitters and receivers.

When checking the keyboard circuits, measure the resistance across the button switch to the correct pin on the system control IC. Take voltage measurements on pins with the button pressed. Scope the signal and sense waveforms at the system control or IC pins with each keyboard button pressed to determine if the IC and keyboard is functioning. See the troubleshooting chart (Table 11-2) for remote controls.

■ Table 11-2 A troubleshooting chart for remote control circuits.

What to check	How to check it
No remote action.	Check batteries in remote.
Erratic operation.	Check battery terminals and dirty buttons.
Suspected remote.	Check remote with infrared tester.
Sub another remote.	Check remote receiver circuits.
	Check standby voltage.
	Repair remote receiver circuits.
Remote defective.	Send in for tuner repair or exchange remote.

Troubleshooting the sound circuits

12

REPAIRING THE SOUND CIRCUITS IS RELATIVELY EASY. BEsides applying test instruments, you can hear the sound symptoms from the speaker. The sound can be weak, garbled, distorted, intermittent, or just missing. The function of most sound stages is easy to understand. The solid-state sound circuit consist of all transistors, ICs and transistors, or one complete IC sound component (Fig. 12-1).

Sound problems can occur in or out of the sound circuits. Often, a combination of poor picture with garbled sound can occur because of a problem in the video circuits. An unstable picture with poor sound can develop in the AGC circuits. Running sync with a buzz in the sound can occur in the sync or AGC stages. A normal picture and raster with no sound generally occurs in the sound circuits. Check out all other possible sources for combination picture-and-sound symptoms before tearing into the sound circuits.

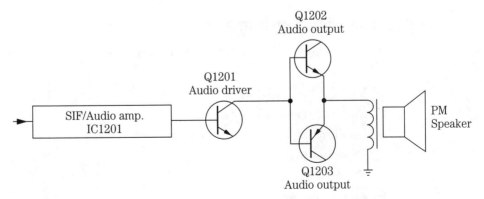

■ **12-1** *The sound circuits may consist of a SIF/driver IC with a transistor amp and output transistors.*

In the tube chassis, the FM sound is taken from the plate of the video amplifier tube. In the solid-state chassis, the sound is tapped at the second or third IF video stage. A typical transistorized sound circuit consists of an IF sound detector, detector amp with audio driver, and push-pull audio output (Fig. 12-2). An IC can serve as an IF sound amp, detector, and audio amp to drive transistors in the audio output stages. The latest TV chassis can have only two IC circuits in the whole chassis, with the complete audio stage combined with other receiver circuit functions (Fig. 12-3).

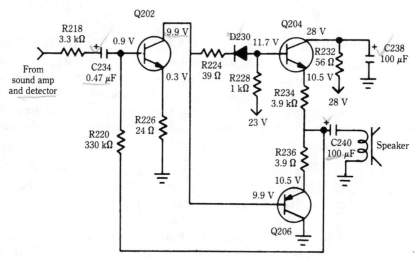

■ **12-2** *A typical solid-state sound circuit audio driver and push-pull output transistors.*

Required servicing instruments

The following test instruments can be used in servicing the sound circuits:

- ☐ VTVM, VOM, or DMM (Fig. 12-4)
- ☐ Sweep and signal generators
- ☐ Oscilloscope
- ☐ Capacitor tester
- ☐ External amp
- ☐ Audio analyzer
- ☐ Miscellaneous PM speakers, test leads, etc.

The VOM or DMM will solve most sound problems with accurate voltage, resistance, and transistor tests. The external amp is

Sound
discriminator coil   SIF coil   Sound power IC
with heat sink on top

■ **12-3** *You may find the SIF/audio driver in one IC and a power output controller in another IC in the latest TV chassis.*

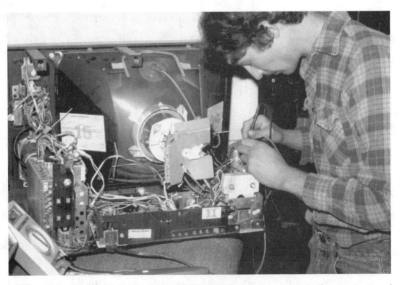

■ **12-4** *The digital multimeter and soldering iron can repair most sound problems.*

Required servicing instruments

handy for locating dead, weak, intermittent, and distorted stages. For critical sound alignment, the scope and generators are a "must" item. Although sound alignment is only needed when critical parts are replaced, a touch-up of the sound coil might be all that is needed to cure the distorted weak sound symptom.

Transistorized sound stages

The sound stages were one of the first circuits to be transistorized in the hybrid TV chassis. Today there are many different sound circuit transistor arrangements, but basically they all accomplish the same thing. The IC can eliminate most of the transistors found in the early sound stages. A typical IC and transistor output circuit is shown in Fig. 12-5.

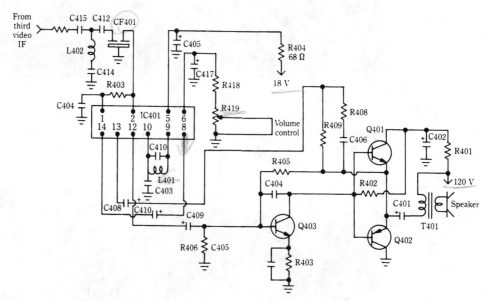

■ **12-5** *A typical IC and transistor sound output circuit.*

The audio signal is taken from the third IF video transistor and coupled by C412 and C415 through a sound take-off ceramic filter (CF401). A 4.5-MHz output signal from CF401 is fed to pins 1 and 2 of IC401. Internally, the limiter circuit is connected to the FM detector circuits. The sound coil (L401) is connected to the sound detector circuit on pins 9 and 10.

The audio amp circuit connects internally to the sound detector circuit. R419 varies the volume of the sound circuits at pin 6 of

IC401. The audio output signal at pin 8 is connected to a driver transistor (Q403).

The npn and pnp junction transistors operate in a push-pull fashion (Q401 and Q402). The audio signal is transformer coupled to the speaker through T401. Notice that two separate voltage sources from the power supply are fed to the IC401 circuits (18 V) and the audio output transistors (120 V).

Transistor sound circuits are back

Often, sound circuits consist of either an IC IF and audio preamplifier, with IC output or one large IC in the sound circuits. Recently, in the lower-priced portables and small screen sizes, transistors are back. The IF/SIF sound systems, combined in one large IC with chroma/AFT/vertical and horizontal circuits, drive three audio transistors (Fig. 12-6).

The first audio transistor is capacity-coupled to the IC with a small electrolytic capacitor. This audio amp provides audio amplification and drives two transistors in push-pull operation. The audio output transistors consist of one pnp and one npn output transistors. R204 and R205, emitter-biased resistors, couple the audio signal to a small FM speaker through a 100 µF electrolytic capacitor.

Signal-trace the sound circuits with an external audio amplifier or scope. Test each transistor within the circuit and remove the tran-

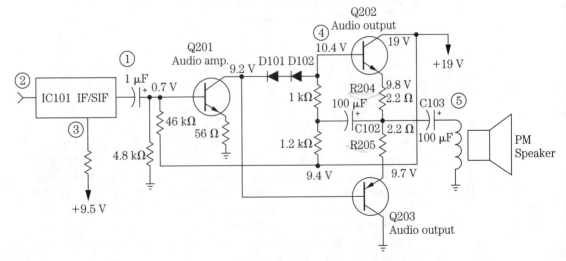

■ **12-6** *Check the audio output circuit by the numbers with external amp and voltage measurements.*

sistor that tests open or leaky. Critical collector and bias voltages can help locate a defective transistor. Forward bias on the silicon transistor between emitter and base terminals should be 0.6 volts, while the pnp transistor has 0.3 volts. Improper bias voltage can identify a defective transistor.

IC sound output circuits

IC401 combines all of the different sound stages into one IC component. The sound signal is taken from the IF video IC and fed to the IF sound coil (L402) (Fig. 12-7). L402 connects the sound signal directly to pins 14 and 15 of IC401. Internally, the IF sound signal is detected and tuned with coil L401 at pins 10 and 11. The volume of the sound circuits is controlled at pins 12 and 16 of IC401.

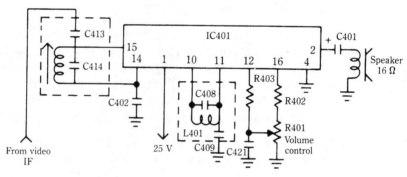

■ **12-7** *You may find only one IC as SIF/driver amp/power output.*

The AF amp and audio output amps are contained inside the sound processor (IC401). C401 couples the amplifier and audio signal to the 16-Ω speaker. A 25-V source supplies power to the sound circuits of IC401.

Inject an audio signal at pin 9, or place the tip of a screwdriver blade (with your finger on the metal blade) on pin 9 and listen for a hum or tone in the speaker. If there is no sound, clip a speaker across the suspected speaker. Keep the volume control wide open. Because the volume control is nothing more than a voltage divider network, hum or tone tests cannot be made at the volume control as in other sound circuits. Check the voltage at pins 1, 2, 4, 7, and 9 of IC401. Replace the sound IC when improper voltages are found with a no-sound symptom.

506

A quick touch-up of coil L401 can eliminate sound buzz or distorted sound. This sound adjustment should be made with a TV station tuned in; it will adjust L401 for maximum or clear sound. Often, only a slight turn is all that's required to touch up the garbled sound. A defective IC401 can cause distorted, weak, intermittent, and motorboating sound problems. Complete sound alignment should be made when replacing an input IF sound transformer or sound coil. Always follow the manufacturer's sound alignment procedures.

Latest IC sound circuits

The sound output circuits may consist of an audio and signal IC driving a larger-power-output IC, or stereo demodulation and MPX stereo circuits. You will find the stereo circuits in the more expensive TV chassis. The sound output IC is capacity-coupled to the SIF/audio signal IC with a small electrolytic capacitor and mono reception. A PM speaker from 16 to 35 Ω is coupled with an electrolytic capacitor to the audio IC. Usually, higher dc voltage is applied to the sound output IC from the secondary voltages of the flyback (Fig. 12-8).

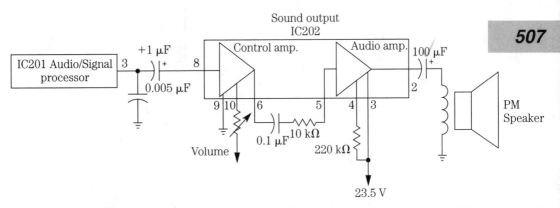

■ **12-8** *The latest IC circuits contain an audio signal processor and power output IC for greater volume.*

Troubleshoot the audio stages by checking the input audio signal at pin 8 of IC202. Trace the audio signal with an external amp or scope at pins 8, 6, 5, and 2. Take a critical voltage measurement on pin 3 (connected to the supply voltage source). Suspect an open coupling capacitor or speaker voice coil when audio signal is present at output pin 2. Do not overlook a defective IC202 when audio is heard at the input 8 terminal and there is no output at pin 2, with normal supply voltage.

Signal tracing sound circuits

The sound circuits can be signal-traced with the audio signal generator and scope, audio signal generator and speaker, TV signal and external amp, or while doing sound alignment. Inject the signal at the input of the audio stages, and use the scope to determine at any point in the audio stages to check for weak or dead circuits.

An external audio amp can be used to pick up the TV signal at various points in the sound circuits to locate distorted, weak, or dead stages. The audio signal can be checked up to the speaker terminals. Start at the speaker terminals and work back towards the input circuits of the audio stages. The audio signal can be checked on each side of a coupling capacitor, or from the base to the collector of each transistor stage. Check the input and output terminals of a suspected IC sound circuit with the external audio amp (Fig. 12-9).

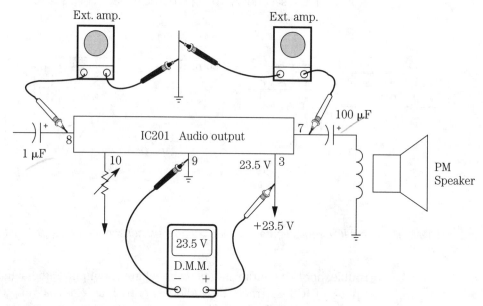

■ **12-9** *Check the input and output terminals of the suspected sound IC with external amp and voltage tests.*

Realistic 16-261 IC audio circuits

The mono audio circuits of the Realistic 16-261 consists of one audio output IC (IC871). The audio input enters pin 2 to the dc volume control circuits, and is applied to the audio output amplifier circuit (Fig. 12-10). The audio output is found at pin 6, and is coupled to the speaker and headphone circuits with C878 (470 µF). The short-circuited earphone jack places an isolation resistor (R877) in the headphone ground circuit.

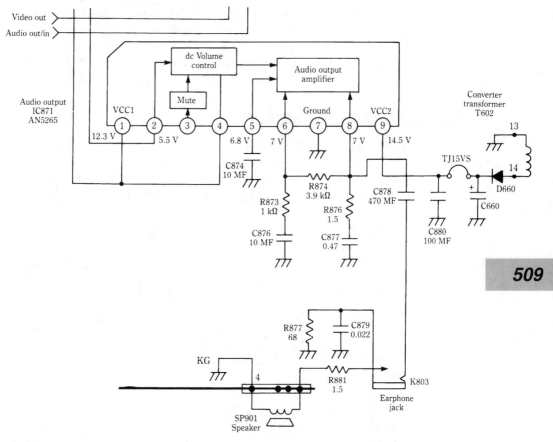

■ 12-10 *The Realistic 16-261 portable audio output circuits in one IC.* Radio Shack

There are two different voltage sources feeding the sound power IC (IC871). Pin 1 is 12.3 V, and pin 9 is 14.5 V from a secondary winding on the converter transformer. D660 provides half-wave rectification with filtering capacitor C660 (470 µF).

Troubleshoot the audio IC using the audio signal tracer with audio signal in at pin 2 and out at pin 6. When audio signal is found at pin 2 and no or low audio is found at pin 6, take voltage checks at pins l and 9. If the normal 14.5 V is missing at pin 9, suspect a defective D660 or C660. Low dc voltage at pins 1 and 9 can indicate a leaky

IC871. Remove VCC pins and take a resistance measurement to ground to check for a leaky sound IC.

Stereo sound

Today you might find that over half of the new TV receivers have stereo sound. You may find a stereo/SAP decoder, expander, switching IC, IC sound processor, audio control, and audio output power ICs. In other stereo channels you will find stereo broadcast, stereo demodulator, and video/audio in/out circuits. Often, the TV receiver will indicate when it is receiving stereo sound by lighting a red LED.

Sylvania C9 sound processing

The IF signal is bandpassed through a SAW filter (Y201) and applied to pins 1 and 18 of the IF sound detector (IC202). The IF signal is amplified and supplied to the synchronous demodulator (Fig. 12-11). The reference signal is adjusted by L213, a 45.75-

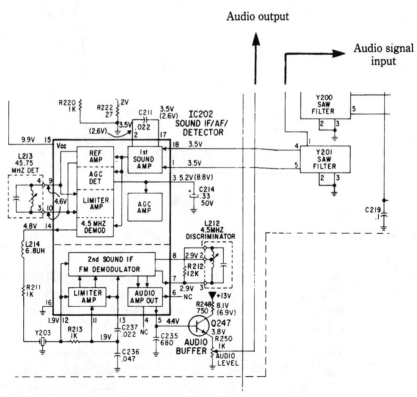

■ **12-11** *The sound/IF/AF/detector circuits of IC202 in the Sylvania C9 chassis.* Phillips Consumer Electronics

MHz detector adjust, and the picture carrier. After detection, the composite signal exits at pin 14. Here it goes through a 4.5-MHz band-pass filter (Y203). It is applied at pins 11 and 12 of the limited amplifier (IC202). The 4.5-MHz signal is demodulated, and the baseband audio signal is amplified by the audio amplifier at pin 5.

The baseband audio signal exits the C9 chassis at P/Y1 and is applied to the stereo decoder module. The audio output stages are located within the stereo decoder module and not in the large PC chassis. The stereo decoder module is repaired by replacement. The major components of the stereo decoder module are stereo/SAP decoder (IC201), the DBX expander/L + R amp (IC202), the sound processor (IC400), audio control (IC501), and left/right audio outputs (IC502 and IC503).

The baseband audio is coupled from P/J101-1 to the stereo decoder. The stereo portion of the signal is passed through the low-pass filter (L201), and the SAP signal is fed through the band-pass filter (L203). Frequencies below 50 kHz (stereo) are fed to the stereo decoder (IC201) at pin 20 (Fig. 12-12).

The L + R signal is buffered and output at pin 16, while the L – R signal is detected and applied to the mode switch. The mode switch selects either stereo or SAP signal with output at pin 8. Both the L + R and L – R SAP output go through the 15-kHz low-pass filter before being applied to IC202.

The L + R signal is amplified at IC202 and output at pin 15. Because the L – R SAP signals are compressed before being transmitted, they must go through the DBX expander to restore them to their original wideband frequencies and spectral expansion before being applied to the audio circuits.

The L + R and L – R SAP signals are fed to the matrix circuit (IC201) with right and left buffer amps output at pins 13 and 14. The right and left audio signals are then applied to the jack panel switching. IC1 switching selects external audio from the stereo decoder circuits. The audio output of the switching IC is applied to the sound processor (IC400).

IC400 provides stereo, pseudo, and spatial mode selection for the audio. The audio input is at pins 2 and 17 of IC400. The sound processor receives high and low logic at pins 11 and 12 to place the IC into the expanded, pseudo, or stereo mode.

The two audio channels are coupled to the audio control IC (IC501). Here the left and right channels receive volume, bass, and treble control. These inputs are controlled from the TS-15/TS-18 tuning

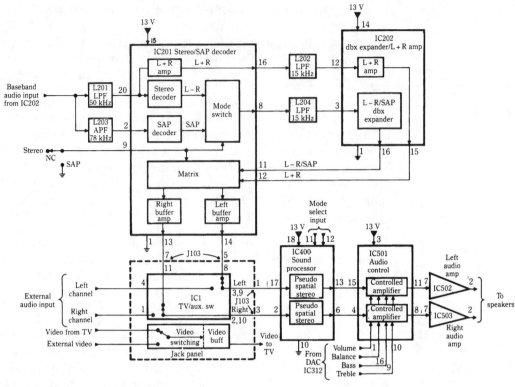

■ 12-12 *The stereo decoder/AV10 jack panel circuitry in the Sylvania C9 chassis.* Phillips Consumer Electronics

system with a special IC (digital-to-analog converter). Treble control is at pin 10 and bass is at pin 9. Volume input is at pin 1 and balance is at pin 16. The output controlled signal (pins 11 and 8) is coupled to the right and left audio output ICs (IC502 and IC503). The amplified audio is applied to the PM speakers.

RCA CTC130C stereo circuits

The CTC130B/C chassis contains a TV broadcast FM stereo/audio B assembly. This chassis provides the capability of selecting a secondary audio program (audio B) besides off-the-air TV broadcast stereo. The secondary audio B program is a switch-selectable audio channel transmitted by the broadcast station containing alternate audio information.

The stereo/audio B transmission is comprised of a wideband composite audio signal. This audio signal contains subcarriers for the regular monophonic L + R channel, the stereo difference (L − R) channels, and the second audio program channel (audio B). The stereo subcarrier is twice the horizontal scan rate frequency, and is an AM signal with suppressed carrier (Fig. 12-13).

The second audio program (audio B) channel is an FM signal centered at five times the horizontal scan rate frequency. Both the L – R stereo difference channel and the second audio program channel (audio B) signals are compressed. This compounded signal at the transmitter is in accordance with the DBX television noise-reduction system. The pilot CW tone, which is crystal-controlled, is transmitted at the horizontal scan rate frequency, indicating stereo audio.

The audio signals cannot be processed by the IF video circuits because a wide baseband audio signal is required for those circuits. The IF signal output from the tuner assembly appears in two separate IF processing circuits. The IF picture processing circuits are located on the main chassis, and the sound processing circuits are on a new sound system assembly.

The 45.7-MHz IF output signal from the tuner assembly is applied to the sound IF/detector stage, where a 4.5-MHz IF sound signal is developed. The IF sound signal is then applied to the sound detector stage, where the 4.5 MHz is detected. The detected 4.5-MHz signal containing the monophonic (L + R) and the stereo (L – R)/audio B signals are applied to the demodulator IC (U3) and U6. The stereo and audio B demodulator circuits recover the stereo

513

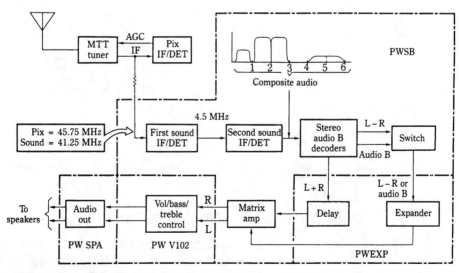

■ **12-13** *The RCA CTC130C chassis stereo broadcast block diagram.* Thomson Consumer Electronics

(L – R), audio B, and monophonic (L + R) signals (Fig. 12-14). The expander stage further processes the stereo (L – R) or audio B signals. The monophonic (L + R) signal is applied to a delay stage, also located with the expander circuit.

During the transmission process, only the stereo (L – R)/audio B signals require DBX compounding. The expander circuits restore the applied signal, either stereo (L – R) or audio B, to its original state prior to compounding for transmission. The monophonic signal (L + R) is applied to the audio delay within the expander circuits to maintain a proper phase relationship with the signal requiring expansion.

The output of the expander and delay stages are then applied to the matrix amp circuits. The L and R signals of the matrix stage are routed to the stereo power amp (PWSPA) circuit board.

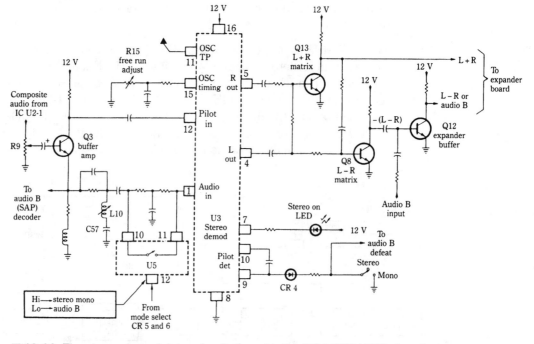

■ **12-14** *The stereo demodulator circuits found in the RCA CTC130C chassis.* Thomson Consumer Electronics

RCA CTC130C video/audio in/out circuits

The control signals for switching both video and audio signals are supplied to the video/audio module (PWV12) from the tuner control assembly on logic control lines A1 and A2. These logic signals are supplied to switching transistors Q7 and Q8 (Fig. 12-15). The

514

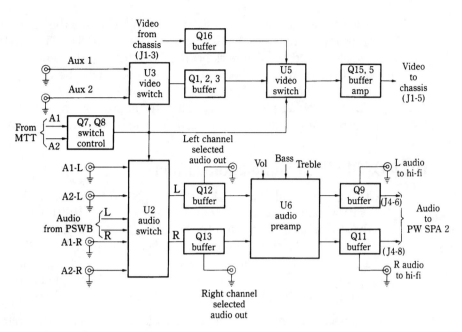

■ 12-15 *The RCA CTC130 video/audio in-out block diagram.* Thomson Consumer Electronics

transistors control the switching mode of video switching ICs U3 and U5 and audio switching IC U2.

External video input signals from the AUX1 and AUX2 jacks are fed to video switch U3. The video input signal is routed through U3 when either input is selected. This signal is applied to the buffer/preemphasis stage (which consists of Q1, Q2, and Q3). The video signal is fed through the video switch IC (U5) and applied to the buffer amp. The video signal is then fed to the main chassis circuit board for normal video processing.

The audio signals are inputed at the AUX1 or AUX2 audio jacks, or from the stereo decoder (PWSB). These audio inputs are switched by IC U2. The selected output signal passes through the left and right channel buffer transistors (Q12 and Q13). The output of Q12 and Q13 go to the rear panel, to selected audio output jacks. This audio level signal is not affected by the volume, mute, or tone controls.

The audio output signal is also fed to the audio preamplifier IC (U6). Here the audio signal is controlled by the mute, tone, balance, and volume controls. The right and left channel audio signal of U6 is applied to the buffer transistors (Q9 and Q11). The audio output of Q9 and Q11 are fed to the audio power amp circuits. The L and R hi-fi audio signals are also found at the output of Q9 and Q11.

The AUX input logic control signals (A1 and A2) from the tuner control module are fed to the base of transistors Q7 and Q8. These logic signals are fed through resistors R127 and R128 to the audio switching IC (U2). When either line A1 or A2 goes high, the appropriate transistor turns on. This supplies a low to the corresponding pin of IC U2. When either pin 9 or 10 goes low, the appropriate switch section closes, routing the AUX1, AUX2, or TV audio through the switch.

RCA CTC157/156 mono audio circuits

The video signal from U1001 (pin 47) is fed into a 4.5-MHz BPF (CF1201) and into the IF sound limiter stage at U1001. Output of the IF sound limiter is fed to the SIF detector circuit. The audio signal is tuned with an LC circuit connected to pins 32 and 33. Just a touch-up of the sound coil can cure slight distortion or drifting audio. The audio signal goes out of pin 34, and is coupled to pin 31 through the attenuator stage (Fig. 12-16).

The volume control is connected to pin 30, and varies the voltage from 1.5 V to 4.8 V at maximum volume. The audio-out stage amplifies the audio signal at pin 28 (TP1202). Here the sound is capacity-coupled to the audio output circuits.

Instead of a regular audio IC, three separate power output transistors are used for amplification. The audio signal from pin 28 of U1001 is coupled through R1919 and C1903 to the base of the audio preamp (Q1901). Q1901 amplifies the audio and is directly coupled to the push-pull audio power amps (Q1902 and Q1903). Q1902 is an npn-type transistor, and Q1903 is a pnp-type. The audio is coupled at the junction of R1909 and R1910. C1904 capacity couples the audio to the 32 Ω PM speaker.

Start at pin 28 of U1001 and TP1202 when troubleshooting the audio output circuits. If there is no audio, check the video output at the band-pass filter network or pins 36 and 35 with the scope. If a video signal is present, check for audio at pins 31 and 34. Go to TP1201 for an audio-controlled signal. Check the 9-V source at pin 37 if there is no audio output.

Signal-trace the audio signal at the base of Q1901, Q1902, and Q1903. Check for an audio signal at the speaker, or at TP1808. For distorted audio output, suspect transistors or bias diodes CR1901 and CR1902. Before replacing leaky output transistors, check for burned bias resistors. Double-check R1910 and R1909 for accuracy (3.3 Ω). Do not overlook C1904 when there is weak, intermittent, or dead sound.

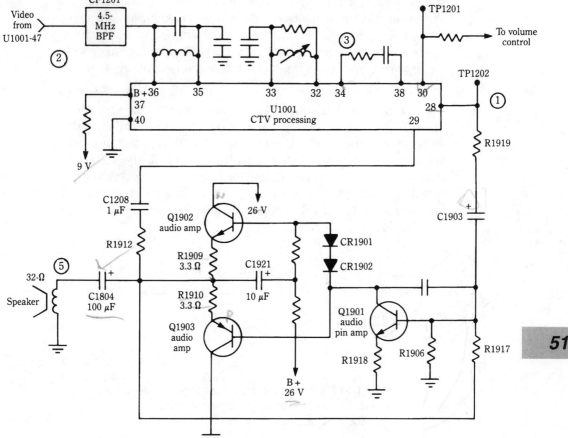

■ 12-16 *Checking the RCA CTC157 mono audio circuits by the numbers.*

Servicing the sound circuits

Although the sound circuits are one of the easiest circuits to service, there are many different problems that can occur in the sound circuits of the solid-state TV chassis. Here are the various sound symptoms, with quick practical methods to locate and repair the different sound stages.

No sound

A no-sound symptom is a quick and easy repair, if the problem is in the sound circuits. The defective component can be quickly located with a few voltage and resistance measurements on the sound transistor and IC components. Another speaker clipped across the old one will locate a defective speaker. Signal-tracing the audio signal with an external audio amp can locate where the sound stops. A small electrolytic capacitor and speaker with a clip wire can help signal-trace the audio output stages (Fig. 12-17).

Clip another speaker across the old one to see if the speaker is open. Take a voltage measurement at the collector terminal of the audio output transistor (Q401) (Fig. 12-18). This voltage measurement will indicate if the transistor is leaky (if the voltage is low), or if there is an open transistor (when the collector voltage is high according to the schematic). Also, a voltage measurement here will indicate if T401 is open or if improper voltage is coming from the low-voltage power supply.

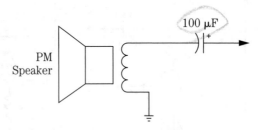

■ 12-17 *Connect a small electrolytic capacitor and PM speaker together to signal-trace output sound circuits.*

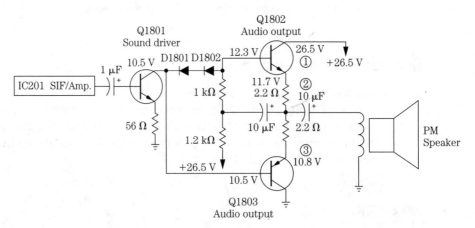

■ 12-18 *Critical voltage measurements at the collector and emitter terminals of output transistors can indicate an open or leaky transistor.*

Take a voltage test at the emitter terminal of Q401. Zero voltage indicates the output transistor or R406 is open. A voltage measurement between the base and collector terminals indicates the normal beta bias voltage. An npn silicon transistor has a bias voltage of 0.6 V. To prove the transistor is defective, take an in-circuit transistor test with the DMM.

The third voltage test at pin 14 of IC301 indicates proper supply voltage and normal IC operation. If the voltage is low, suspect a leaky IC component. With the three quick voltage tests you can quickly isolate most sound problems.

Another method to check the dead or weak stages is with the external audio amp. Check for audio sound at points A, B, and C. Suspect a defective earphone jack or speaker when good sound is noted at the output transformer at point A. Do not overlook a bad shorting pin inside the earphone jack. These earphone jacks act up after frequent earphone usage. A check at pin 10 (C) of the sound IC (IC301) will indicate if sound is normal at the driver or preamp stage of the sound IC.

Signal tracing can be accomplished in the output stage with a small electrolytic capacitor and speaker. Clip one side of the speaker to common ground. Use one end of the capacitor lead as a probe and check for signal at the output transformer and transistor. Normally this signal will be loud. Now go to the output terminal (10) of IC301. Although this audio signal is weak compared to the output terminal, you can still determine if IC301 and the input signal on the base terminal of Q401 are normal. Check the audio signal through the earphone jack to the speaker in the very same manner.

No sound in IC circuits

Suspect a defective IC, speaker, output coupling capacitor and no power source for a dead IC sound circuit. First check the voltage at the supply pin terminal of IC. No voltage indicates a defective low voltage source. This voltage source may develop in the scan-secondary circuits of a flyback winding. Check for an open isolation resistor or regulator transistor. Very low supply voltage can be caused by a leaky sound ouput IC.

Sub another PM speaker when voltage measurements are fairly normal. Shunt a 100 μF, 50-volt electrolytic capacitor across the speaker coupling capacitor. Check the input sound terminal with an external amp or scope. Suspect a defective sound output IC if there is no sound at the speaker and normal supply voltage (VCC).

Do not forget to check a possible defective earphone jack or broken terminal wires (Fig. 12-19).

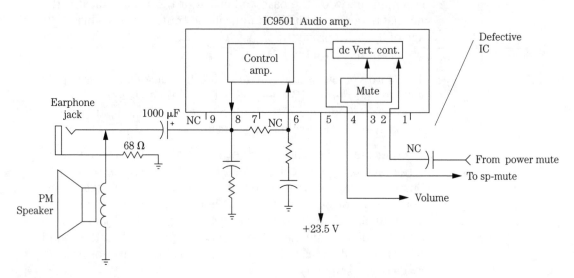

■ 12-19 *Check these sound components that can cause a dead symptom.*

Weak sound

Weak sound can be caused by open or leaky transistors or electrolytic capacitors, a change in resistance, or an improper voltage source. A speaker with a frozen cone will produce weak sound. Weak sound with distortion can be eliminated by touching up the sound coil.

Besides voltage and resistance measurements, audio signal tracing with the scope or external audio amp can locate the very weak audio sound. The input and output audio signal of an IC output component can be signal-traced with the scope or external amp (Fig. 12-20). Check pins 1 and 11 of IC501 with the volume control halfway up. If a signal is present at pin 1 and a very weak signal is at pin 11, suspect a defective IC or parts tied to the IC terminals.

Measure all voltages on each IC terminal. Suspect a leaky IC or an improper voltage source with lower voltage at pin 14. Determine if any other stages are fed from the same source. If these stages are normal, the voltage supply source might be OK. Remove pin 14 from the circuit if the voltage is very low. Notice if the supply voltage returns to normal. Now measure the resistance from pin 14 to common ground. A low resistance (under 500 Ω) indicates a leaky IC.

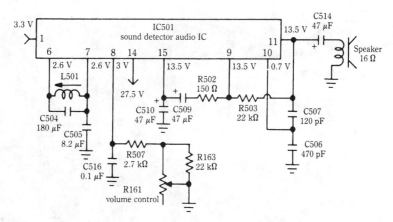

■ **12-20** *Use the scope or external amp to signal-trace the sound output circuits.*

When the voltage at pin 14 is fairly normal, suspect either the IC or a component tied to one of the IC terminals if there is weak sound at pin 11. Take a resistance reading from each pin to common ground to determine if a capacitor or resistor is defective. Remove one end of the suspected component for accurate resistance measurements.

Check each transistor with the DMM for open or leaky conditions in transistorized audio amplifier circuits. Take accurate voltage and bias resistance tests of each transistor. A change in resistance of the collector and emitter resistors can produce weak audio. Check each side of the input and speaker coupling capacitors for weak sound. Do not overlook the possibility of a dried-up electrolytic capacitor. The audio signal can be signal-traced from the volume control to the speaker with the external audio amp.

Distorted sound

Transistors, ICs, resistors, speakers, and sound alignment produce most distortion problems in the sound circuits. Often, sound distortion is located in the audio output stages. Sound distortion can be a signal traced with the scope and external audio amplifier.

A frozen voice coil or dropped cone of a speaker can cause many different sound problems. Lower the volume control until you can barely hear the sound. Notice if the sound is tinny or mushy. The speaker cone might be frozen to the center pole. Now raise the volume quite high for blatting or loose vibrations, which indicate that the voice coil or speaker cone is loose (Fig. 12-21). Check the

521

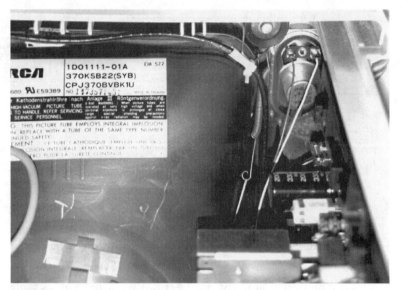

■ **12-21** *Inspect the speaker for a torn or loose coil, dragging voice coil, and/or an open winding.*

condition of the speaker by removing one voice-coil wire and clipping a good speaker across it. Replace the speaker if the distortion disappears. If not, troubleshoot the audio circuits.

Leaky or shorted audio output transistors cause more distortion problems than any other components in the sound stages. Check each transistor in the circuit with either a transistor or diode tester. After locating a leaky transistor remove it from the circuit for accurate leakage tests. You might find one push-pull transistor to be open and the other to be leaky, producing extreme distortion (Fig. 12-22). Suspect the audio output transistor when there is distortion in single-ended audio stages.

Always check the bias and base resistors with the transistors out of the circuit. In many cases, the emitter bias resistors are burned because of shorted audio output transistors. Remove one end of a bias diode for leakage tests. These bias diodes can cause a low level of distortion after the audio output transistors have been replaced.

In a K-Mart KMC1984A, extreme distortion was found in the speaker (Fig. 12-23). Q301 was found to be leaky via in-circuit transistor tests of the DMM. Voltage measurements on the base and emitter terminals indicated a leaky transistor. After Q301 was replaced with a universal ECG198 replacement, some distortion was still present. The base resistor (R302) (10 kΩ) was found to be open in the base circuit of the audio output transistor.

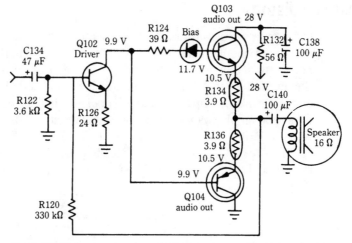

■ 12-22 *Take in-circuit transistor tests to locate a leaky or open transistor. Check circled components for distortion.*

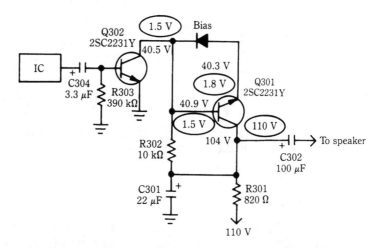

■ 12-23 *Transistor Q301 was found leaky in a transistorized audio output circuit.*

Suspect a defective IC or speaker if there is still audio distortion after other possibilities have been exhausted. Substitute a new speaker. Take accurate voltage and resistance measurements on the IC terminals to determine if the IC component is leaky. Check the input and output signal of the IC for signs of distortion. If the input sound signal is normal and extreme distortion is noted at the output terminal, replace the defective IC.

Extreme distortion

Often extreme distortion can be caused by a leaky output transistor or IC component. Most extreme distortion is found in the audio output circuits. Signal-trace the audio from input of output IC and on the output terminal. If the input audio is normal and distorted at the output terminal, suspect a leaky IC, open electrolytic capacitors, and improper supply voltage. Leaky output transistor Q1902 and emitter resistor R1909 were burned in an RCA CTC 157H chassis (Fig. 12-24). Check Q1903 for open or leaky conditions. Do not overlook the possibility of a defective PM speaker as a cause of distortion.

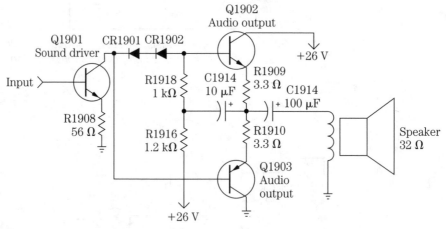

■ **12-24** *Check for leaky audio output transistors or output IC and leaky bias resistors if the problem is extreme distortion.*

Intermittent sound

Practically any component within the sound stages can produce intermittent sound conditions. The most obvious parts are transistors, power ICs, speakers, wiring, and component board connections (Fig. 12-25). Determine if both picture and sound are intermittent. If so, the defective component is likely in the video or AGC circuits instead of the sound stages.

Try to isolate the sound stages by checking for intermittent sound at the volume control and output stages. Signal-trace the audio stages with the external audio amp. Notice if the sound is normal at the top side of the volume control in transistor circuits. Now go from the base to the collector terminal of each driver, AF amp, and output transistor until you notice the audio signal acting up.

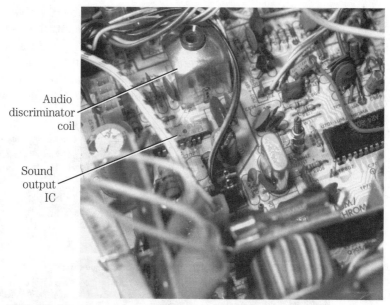

Audio
discriminator
coil

Sound
output
IC

■ **12-25** *Check these components for intermittent sound.*

Intermittent sound can be caused by poor board wiring or bad component connections. Push up and down at the various sections of the sound board. Notice if the points of pressure cause the sound to act up. Take an insulated plastic tool and push each component to determine if a poor soldered connection exists. Pull or flex the speaker and volume control connecting wires for loose connections. Poor socket and lead terminals on remote-control sound modules can produce intermittent sound.

Motorboating noises

A low or loud putt-putt sound is called motorboating. Often, motorboating noises occur in the input and output audio stages. A motorboating sound in the speaker can be caused by broken board wiring in the horizontal circuits. Suspect other circuits creating motorboating sounds when the picture blinks off and on with a motorboating sound in the speaker.

Check for defective AF or audio output transistors creating motorboating noise in the speaker. Short the emitter and base terminals of the AF transistor to eliminate the transistor thought to be causing the motorboating sound. Replace the suspected transistor if the sound disappears. Signal-trace the audio output transistor with the external amp to determine if the noise originates in these transistors. Replace the output transistor if in doubt.

The motorboating noise can be caused by a defective power IC component (Fig. 12-26). Lower the volume control to determine if the noise is still present in the audio output circuits. Momentarily shunt the IC input terminal to ground through a 100-μF electrolytic capacitor. Suspect a noisy component ahead of the power IC if the noise stops. Replace the motorboating IC if the noise is still present.

Audio circuits in large IC

■ **12-26** *Motorboating can be caused by bad transistors and audio ICs.*

Sound and picture do not track

After tuning in a good picture with the best sound possible and the sound has some distortion or poor tracking, suspect a defective component in the sound detector stages or poor sound alignment. Check the sound stages in the TV receiver with remote-control or push-button tuning. Readjust the fine tuning control in older-type tuners for best sound and picture.

When the above adjustments fail to bring the best picture and sound together, check for correct sound alignment of the discriminator or sound coils. Just touching up the discriminator or sound coil adjustment can take care of the sound tracking problem. Follow the manufacturer's sound alignment procedures when the sound and picture cannot be brought together. Look for open coils and defective bypass capacitors in the sound input stages when correct sound alignment cannot be obtained.

Servicing RCA CTC109 sound circuits

One IC contains the entire sound circuit of an RCA CTC109 chassis (Fig. 12-27). You might find this same sound circuit in many of the later RCA solid-state TV receivers. The input signal at pin 15 and the output signal at pin 2 can be signal-traced with a scope or external audio amp. A defective U201 can produce motorboating or weak, distorted, or absent sound.

■ **12-27** *The complete IC sound section in a RCA CTC109 chassis.*

Suspect a defective U201, C201, or improper voltage source (26 V) with weak sound (Fig. 12-28). Check the 32 Ω speaker, U201, and the adjustment of L201 and C209 (7.5 pF) for distorted sound. A touch-up of L201 might cure the distorted and weak sound symptom. Replace C209 if the sound drifts. Intermittent sound can be caused by U201, the 32 Ω speaker, C201 (10 μF), or a dirty volume control (R4201).

For distorted-sound or no-sound symptoms, clip a new speaker across the original. Any 8-Ω or 10-Ω speaker will do as a test speaker. If the sound is still distorted, touch up the sound coil (L201). Check the input and output with the external speaker to determine if U201 is defective. Take voltage measurements on all IC terminals before removing IC201. Be sure to replace the copper heat sink on top of the audio IC component.

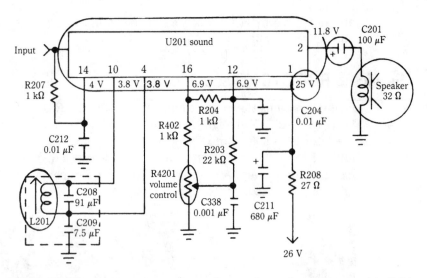

■ **12-28** *Check these components for weak or distorted sound in RCA CTC108 chassis.*

Five actual sound case histories

Here are five different sound problems in different TV chassis, showing the methods used in locating the sound problem.

No sound—normal picture

The color picture was normal, but with no sound in an RCA CTC146B chassis. Audio was signal-traced and was normal at pin 28 of IF/SIF (U1001). No sound was noted on the collector pin of audio amp Q1201. No voltage was noted on Q1201, Q1202, or Q1203. Checking revealed that the 18.5-V supply source in the power supply was higher in voltage than normal. R1211 (5.6 ohm) resistor was found open between the supply source and the collector terminal of Q1202 (Fig. 12-29). Both R1211 and Q1202 were replaced.

Intermittent audio and hum noise

When the audio was intermittent, hum could be heard in the speaker of a Sylvania E32-22 chassis. The audio was monitored at input pin 1 and 11 (Fig. 12-30). No sound was heard from a speaker on pin 11 when sound was intermittent, and audio was heard at pin 1 with an external amp. Voltages seemed to increase on pins 11, 14, and 15 when the sound was in the intermittent state. All components were checked that tied to all pin terminals. Intermittent IC102 was replaced with a universal ECG1231A IC replacement.

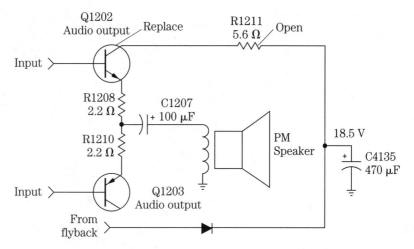

■ **12-29** *R1211 was found open; it and Q1202 were replaced in an RCA CTC146B chassis.*

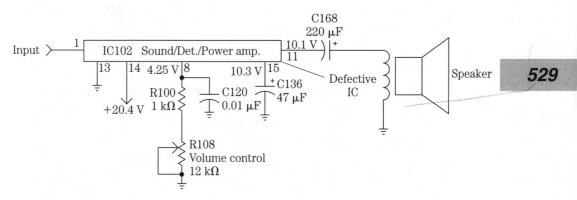

■ **12-30** *A defective IC102 caused intermittent audio with hum noise in a Sylvania E32-22 chassis.*

Low hum in sound

Sometimes the audio played normal, and then would drop down with hum in the sound in a Panasonic CT1320V model. IC201 was monitored at output pin 8, and when volume went down indicated that the output IC, varying supply voltage, or electrolytic capacitors were defective (Fig. 12-31). The supply voltage 7 was monitored with the DMM and was normal throughout the change in sound level. C211 was shunted with another 1000 µF capacitor, which helped some in reducing the hum level. Defective sound IC AN5255 was replaced with a universal SK9324 replacement IC.

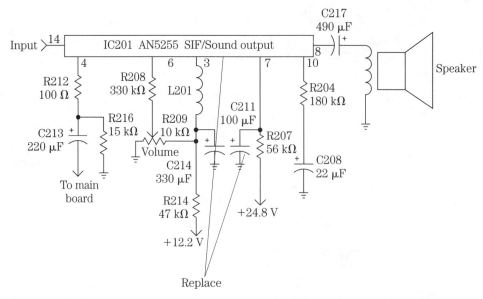

■ 12-31 *Audio with hum was caused by a faulty C24 and IC201 in the Panasonic CT1320V portable.*

Left channel dead—right normal

This audio symptom indicated problems in the stereo audio output circuits of an Emerson MS2250RA. The right and left channels were checked at the output IC feeding each stereo channel of IC1203 (Fig. 12-32). No signal was found at the right channel (pin 3), but operation was normal in the left channel (pin 10). The balance control had no effect on the right channel. Supply voltage at pin 2 was good, with a 10.3-V measurement. The defective AN5836IC was replaced with an NTE1280 universal replacement.

Intermittent, weak sound

When the sound appears weak or intermittent, check the electrolytic output coupling capacitor. In an RCA CTC157, the sound was normal at the audio preamp but weak at the speaker (Fig. 12-33). Clip another electrolytic capacitor (100 µF) across C1914 to see if the sound is normal. In this case, replacing the audio coupling capacitor solved the weak and intermittent sound.

Practical sound hints

The digital multimeter (DMM) and external audio amp are two important test instruments for locating sound problems. Each tran-

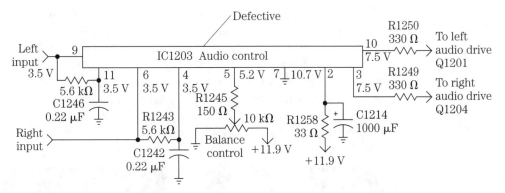

■ **12-32** *A bad IC1203 caused a dead left stereo channel in an Emerson MS2250RA.*

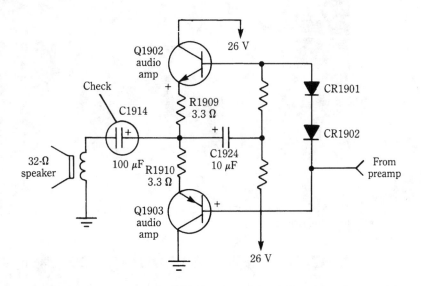

■ **12-33** *A bad capacitor C1914 caused intermittent weak sound in the RCA CTC157 TV chassis.*

sistor and diode can be checked in the circuit for open or leaky conditions. The external audio amp can check the sound signal from the input to the speaker for intermittent, distorted, or no sound, or no weak sound components.

Tack in small electrolytic capacitors across the suspected ones when a capacitor tester is not readily available. Clip a speaker across the original one to determine if the speaker is noisy or open. Use a couple of clip leads, 100 μF electrolytic capacitors, and a speaker to check the audio stages without an external amp.

Check the input and output signal of a suspected IC for correct voltage and resistance measurements before replacing the IC (Fig. 12-34). Before removing the defective IC, make sure you know where pin 1 is located on the PC board. Use a small soldering iron with very small solder to solder the IC terminals (Fig. 12-35). Clean out between each terminal with the back of a pocket knife to remove excess solder and flux. Universal transistors and IC components can be used successfully in the sound circuits as replacements.

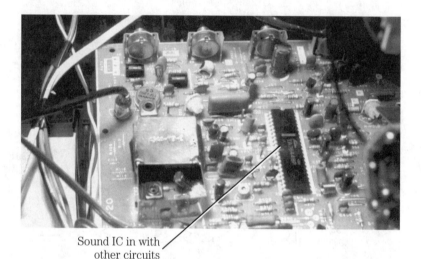

Sound IC in with
other circuits

■ **12-34** *Check the input and output terminals of suspected output IC with a signal tracer.*

■ **12-35** *Solder the IC terminals with a low-wattage soldering iron.*

Troubleshooting the sound circuits

Apply silicone grease on each side of the piece of insulation between the heat sink and transistor of the power output transistors.
Replace detachable heat sinks on all transistors and IC components. Look for burned or overheated base and bias resistors in
transistorized audio circuits. Always keep the volume as low as
possible and the audio circuit loaded down with the correct
speaker after servicing the audio circuits. For more information on
troubleshooting sound circuits, see Table 12-1.

■ **Table 12-1 A troubleshooting chart for the sound circuits.**

What to check	How to check it
No audio.	Sub another speaker.
Signal trace audio circuits with external amp.	Signal trace from volume control to speaker.
Suspect audio output transistor or IC.	Check for correct supply voltages. Test output transistors.
No sound at volume control.	Check sound input circuits.
Sound normal at volume control.	Replace transistors or output IC with sound input.
Signal trace sound input circuits with signal and scope.	Make sure picture is normal.
Distorted sound.	Audio output transistors and IC. Check voltage source.

533

Troubleshooting picture tube circuits

13

MOST PICTURE TUBE SYMPTOMS ARE FIRST SEEN ON THE TV screen, because the CRT displays the picture. Although the picture tube does cause some service problems, very few are replaced compared to other components within the TV chassis. Defective picture tube circuits and circuits connected to the CRT can cause the picture to malfunction (Fig. 13-1). Check all connecting circuits before attempting to replace the picture tube.

The picture tube components consists of a glass envelope, electron gun assembly, focusing system, deflection circuits, and the phosphor screen. The black-and-white and color picture tube comes in many different sizes and shapes. Besides supplying the voltage applied to the anode terminal and electron gun assembly,

■ **13-1** *Video and CRT circuits with high voltage measurements can indicate picture tube problems.*

a deflection yoke provides for the vertical and horizontal sweep of the raster. Static magnets and dynamic convergence circuits with three color gun assemblies are used to provide color pictures on the screen. The latest color picture tube might have a beam bender magnet instead of dynamic coil convergence circuits. We now see flat screens as well as the traditional glass picture tube.

The defective picture tube might have no picture, no raster, a weak or negative picture, low or no control of brightness, poor focus, an intermittent raster, and excessive arcing at the high-voltage anode terminal. The same symptoms can be caused by the picture tube circuits or outside circuits connected to the CRT. A defective picture tube can be detected by visual inspection, testing the tube, and taking voltage measurements.

Besides visual symptoms, low (or no) high voltage applied to the CRT will cause a no-raster symptom. Often high voltage can be detected at the picture tube by holding your forearm near the TV screen. The hair on your arm will stand up. Another method is to hold a small piece of paper next to the screen and watch it pull inward. High voltage can be detected when you hear the deflection yoke expand when the receiver is turned on and collapse when the receiver is turned off. High voltage should be measured at the anode terminal with a high-voltage meter.

For successful picture tube and corresponding circuit tests, the following test instruments are required:

☐ A CRT tester. The picture tube tester checks the performance of each individual gun for proper emission. This tester can operate in conjunction with the tube-charging tester, or it can operate separately. The CRT tester is a must instrument for the complete service shop.

☐ A VOM, DMM, and VTVM.

☐ A VTVM with high-voltage probe.

☐ A separate high-voltage probe (42 kV).

Larger picture tubes

Picture tubes are getting bigger each year. A few years ago the 24-inch tube (with a 25-inch diagonal design around the corner) was the largest size. Today, you will find 26-inch, 27-inch, and 31-inch screens are being produced (Fig. 13-2). You will find the 27-inch screens are quite popular and used in table models and consoles. The 27-inch and 31-inch picture tubes have a flat surface with

■ **13-2** *The 27-inch RCA screen has a flat surface with square corners.*

square corners. Most picture tubes are measured diagonally from corner to corner (Fig. 13-3). Although the larger screens have the same number of scanning lines as the 10-inch screen, the 27-inch screen still produces an excellent picture. Very little distortion is

■ **13-3** *This 27-inch screen is square and has a flat surface.*

found in the larger glass picture tubes. Of course, the larger the screen, the heavier the TV. It takes two people to lift, remove, and replace the larger picture tubes.

Normal high voltage for the 27-inch picture tube is around 27.5 kV, and should not exceed 32 kV at maximum beam current. The 26-inch picture tube should have 26 kV applied and not exceed 29 kV. Check the schematic for the values of high voltage applied to larger direct-view screen picture tubes. Measure the HV with a VTVM (with high voltage probe) or a high voltage probe (meter).

RCA CTC130C kine drive board

The CRT board assembly consists of the RGB driver transistors, CRT socket, and corresponding resistors and capacitors. The RGB bias color signal is applied to the emitter terminals of each color output transistor (Fig. 13-4). Notice that each output collector terminal has a higher voltage compared to other circuits in the TV chassis (160 V to 170 V). This voltage is directly coupled through an isolation resistor in each color output to the color cathode gun assembly.

The grid voltage is taken from a separate voltage source (185 V) that also supplies collector voltage to the color output transistors. The screen and focus voltage is taken from a separate control assembly and is applied to the screen and focus grid terminals. The high voltage (25.8 kV) is fed from the flyback transformer to the

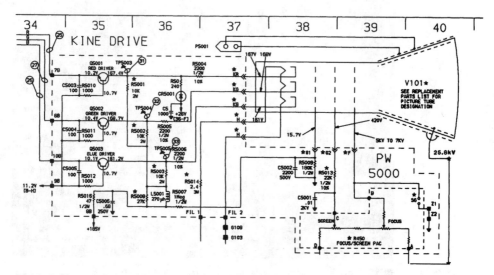

■ **13-4** *The picture tube circuits in the RCA CTC130C chassis.* Thomson Consumer Electronics

538

anode connection on the picture tube. P5001 grounds the CRT aqueduct to the chassis ground.

When the screen is all one color, suspect a collector-emitter short of the color output transistor or CRT gun assembly. Do not overlook a shorted screen gap assembly. If this occurs, the predominate color gun will be fully on, causing an excess of color. Check for a shorted driver transistor by setting the brightness, contrast, and color controls to minimum and measuring the collector voltage. The collector voltage should be about 165 V. If it is lower than 140 V, suspect a shorted color output transistor.

Check the G-Y, R-Y, and B-Y input signals with the scope. Check the output signal with the scope at each output color collector terminal. Measure the dc voltage source (185 V) fed to the CRT board. If normal, check voltages on the CRT and socket. Test the CRT for defects.

Sylvania C9 CRT board assembly

Usually the CRT board contains the RGB driver circuits and the black bias circuits (Fig. 13-5). The CRT cutoff and driver controls are located on the board assembly. The black level from IC640 of the main chassis is approximately 3.5 V. The black bias circuit pro-

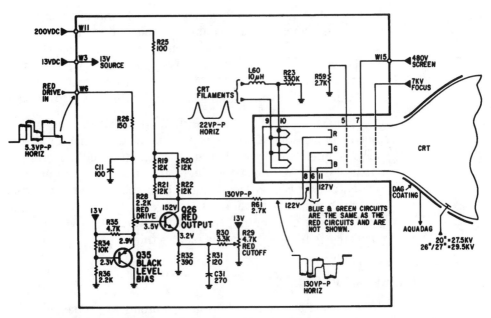

■ **13-5** *Sylvania C9 CRT board circuitry with waveforms.* Phillips Consumer Electronics

vides approximately the same voltage level on the opposite side of the drive controls to minimize a dc difference across the controls.

The black level bias circuit prevents changing the white balance levels of the CRT when the drive controls are adjusted during setup. The black level bias transistor (Q35) acts as an electronic filter to remove the RGB signals from the luminance circuits.

Troubleshoot the CRT board with a scope waveform at the input of each RGB transistor. This signal comes from IC640 and is applied to the base of each color output transistor. Use the scope to check for a loss of color at the base and collector terminals. Check for a 13-V and 200-V source feeding the CRT circuits. A voltage check at the collector terminal (metal) of each output transistor can locate the defective color stage.

CRT bias and driver board circuits

You may find green, red, and blue bias transistor stages ahead of the color output transistors. Each color bias transistor is directly coupled to the same color output transistor. The G-Y, B-Y and R-Y color difference signals from the IC processor are fed to the base circuits of each bias transistor. The color signal is applied to the base circuits, while the luminance or brightness signal is connected to each emitter terminal of the color bias transistor (Fig. 13-6).

The color bias signal is fed to each emitter terminal of color output transistors. Higher collector voltage is applied to each output transistor with a fixed lower voltage upon the base terminal. The output of each color output transistor is coupled through an isolation resistor to cathode pin of CRT. Each color output transistor is coupled to the respective color gun assembly within the picture tube.

Scope the color signal at the G-Y, B-Y, and R-Y terminals on the color IC processor. Next, check the luminance signal applied to the luma buffer transistor. Measure the voltage at the collector terminals of each bias transistor. Check the color signal at output of each output transistor. Measure the high voltage on each output collector terminal. A missing color can result from improper input signal, leaky or open output transistors, low collector voltage, and a defective gun assembly inside picture tube.

Visual symptoms

The following are common visual symptoms of problems in the picture tube and associated circuitry.

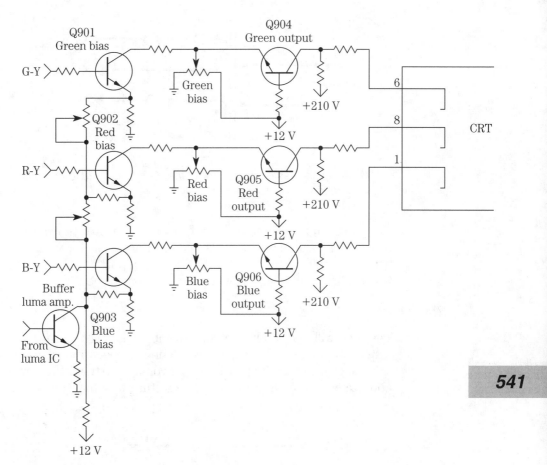

■ 13-6 *The CRT green, red, and blue output with bias transistors are connected to cathodes in the CRT.*

No raster

The front of the picture tube is entirely black with no scanning lines. Improper high or low voltage applied to the picture tube elements can produce a dead raster. The electronic gun assembly will not produce a raster without proper heater or filament voltage. Improper focus voltage can prevent the tube from lighting up. Poor signal voltage at the cathode terminals can prevent a raster from operating, even with a normal picture tube and correct applied voltages.

In an Admiral 3M20 chassis, the high voltage was normal, but no heater light could be seen from the end of the picture tube. In this model the CRT heaters are fused with F901 (Fig. 13-7). The fuse

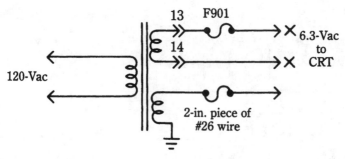

120-Vac

13 F901 →→ ⟩⟩ →X 6.3-Vac

14 →→ ⟩⟩ →X to CRT

2-in. piece of
#26 wire

■ **13-7** *A no-raster symptom in the Admiral 3M20 chassis was caused by a poor heater plug.*

was good. An ac voltage measurement at the heater transformer terminals indicated a poor socket connection at pin 14. When the heater plug was moved on the M900 power supply board, the tube would come to life. Improving the connection solved the problem.

No picture

The tube will light up in the gun assembly with a no-picture, no-raster symptom. If a normal raster is found, suspect problems outside the picture tube circuits. Improper voltages on the cathode and grid circuits of the CRT can cause a no-picture symptom (Fig. 13-8).

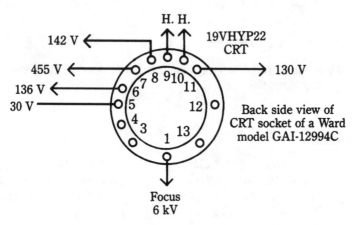

H. H.

142 V ← 19VHYP22
CRT

455 V ← → 130 V

136 V ←

30 V →

Back side view of
CRT socket of a Ward
model GAI-12994C

Focus
6 kV

■ **13-8** *Measure the voltage on the picture tube to troubleshoot a no-picture symptom with a normal raster.*

Weak picture

A poor or weak gun assembly can produce a weak picture. Suspect a defective picture tube when all three color guns test weak with correct element voltage and not enough brightness. Improper high voltage or voltage applied to the picture tube elements can cause a weak picture. Usually with improper high voltage, the raster will pull in at the sides. Check the luminance and output circuits when the picture tube and corresponding circuits are normal.

No brightness

High voltage might be present with a no brightness symptom. Measure the high voltage with a high-voltage probe (Fig. 13-9). Of course, no high voltage will result in a no-brightness, no-raster symptom. An open heater element or no heater voltage produces a no-brightness, no-raster symptom. Improper voltages applied to the picture tube circuits can also cause a no-brightness symptom.

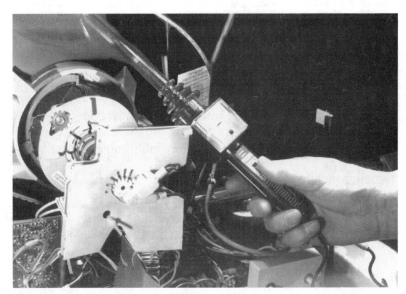

■ **13-9** *Check the high voltage at the anode button on the CRT to determine if high-voltage circuits are normal.*

No control of brightness

Improper brightness control can be caused by a shorted picture tube, improper element voltage, or improper control-signal voltage. Determine if the picture tube or circuits are defective with a tube tester and voltage measurements. Suspect a defective luminance, video, or output circuit with a normal picture tube and voltages applied to the CRT. High brightness can be caused by a defective service switch on some models.

The picture was very bright with no control of brightness in a Goldstar CMT2540. Replace resistor R217 (100 kΩ) to restore the brightness control (Fig. 13-10).

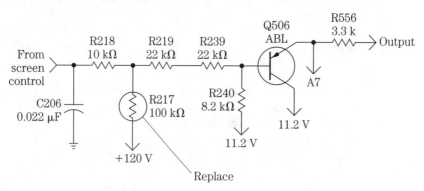

■ **13-10** *Resistor R217 (100 kΩ) was found burned and open in a Goldstar CMT2612 with no control of brightness.*

Washed-out picture

A weak picture tube or improper focus voltage can appear as a washed-out picture. Suspect a defective video or luminance circuit with a normal raster. Often, a washed-out picture can be a combination of a weak gun assembly and improper video signal.

Negative picture

A negative picture can be caused by a shorted picture tube. Sometimes just tapping the end or neck of the tube can cause a negative picture to come and go. Improper element and signal voltage can produce a negative picture.

Blotchy or shiny picture

A weak picture tube can cause a shiny or blotchy picture. Notice if a close up of a person's face is extremely shiny when the brightness and contrast controls are turned up. Test the CRT for low emission. Improper screen or boost voltage can produce a blotchy picture. A defective degaussing coil assembly or poor purity can cause a blotchy colored area on a section of the picture tube.

Poor focus

A weak picture tube can cause poor focus. Improper focus voltage can produce a poor-focus symptom. Intermittent focus can be caused by a defective picture tube socket. Suspect a corroded focus pin at the CRT socket when the focus changes every few seconds. Poor focus can result from a cracked or broken focus control.

Check the focus voltage at the CRT socket with the high-voltage probe. The focus voltage should vary between 3.5 kV and 5.5 kV. You might notice only a slight change on the high-voltage probe because the high-voltage readings are quite close together. A VTVM with a high-voltage probe will give a greater voltage swing with correct focus adjustment.

Suspect a defective focus control or leaky spark-gap assembly at the picture tube socket when low focus voltage is measured at the socket. Only 1.5 kV was measured at pin 1 of the CRT in a Sharp 19C79A (Fig. 13-11). Remove the picture tube socket to measure the focus voltage when no focus voltage points are found outside the socket area. Stick the end of a pigtail fuse down inside the focus socket. Be careful when taking focus voltage measurements. In this case, spark-gap SG855 had arced over several times, providing a low-resistance path with lower focus voltage applied to the second anode and resulting in poor raster focus (Fig. 13-12).

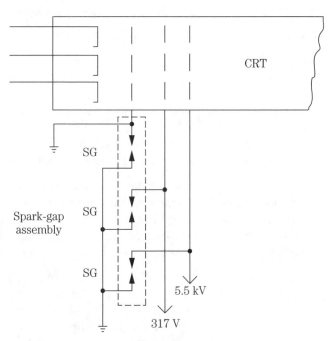

■ **13-11** *A leaky spark gap (SG855) in a Sharp 19C79A reduced the focus voltage (1.5 kV).*

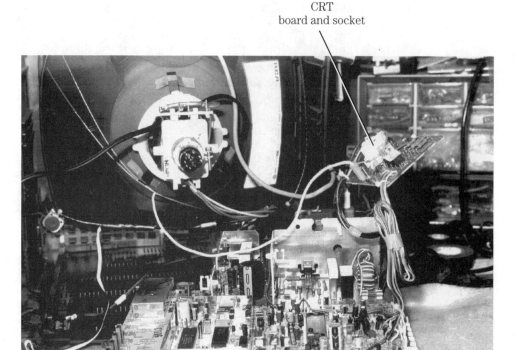

CRT
board and socket

■ **13-12** *Arcing inside the spark gaps can cause a color missing or chassis shutdown.*

Intermittent raster

The raster can become intermittent with poor heater voltage, improper screen or anode voltage, or a defective picture tube. Sometimes the receiver will run for hours or days before going out. Careful voltage tests can help solve the intermittent raster symptom.

Improper heater voltage at the picture tube socket produces most intermittent raster problems. The heater voltage can develop from a power transformer or, in later models, from the high-voltage circuits or flyback winding. Poor plug fittings or a bad picture tube socket can cause the intermittent heater voltage.

Replace the entire socket and harness or repair the heater leads. In the K-Mart SK1310A chassis, a twisted pair of heater wires go to the heater socket terminals and a plug on the chassis (Fig. 13-13). The socket commonly goes bad. Attempting to repair the heater plug-in on the chassis is fruitless. In one particular case, a new pair of twisted wires were soldered directly from pins 8 and 10 of T602

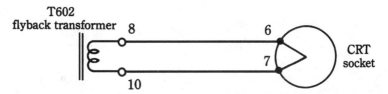

13-13 *Replace the entire socket assembly if it has heater pin terminals in poor condition.*

to pins 6 and 7 of the CRT socket, bypassing the defective socket and fixing the problem.

Besides intermittent voltage applied to the picture tube, suspect a shorted gun or intermittent internal heater assembly. Take an ohmmeter continuity test of the two heater terminals after the light goes out. Remove the TV chassis and connect it to the bench tube to determine if the picture tube is defective on extreme and very difficult raster symptoms. Suspect the picture tube when the chassis operates with the bench test tube for several hours. You might find that the CRT tester will not indicate an intermittent gun assembly.

Retrace lines

Retrace lines can be caused with a shorted picture tube or defective luminance and color output circuits. Improper adjustment of screen and bias controls can produce retrace lines over the whole screen. A defective picture tube can have an extremely bright raster with retrace lines. High brightness and retrace lines with a defective CRT can cause high voltage and chassis shutdown in the latest TV chassis.

Improper boost or screen voltage sources can produce retrace lines in the picture (you may think it is an AGC or video problem). In a Goldstar CMT2612 portable (Fig. 13-14), scanning lines with no control in brightness were caused by very low boost voltage applied to the color output transistors. D502 was burned and open in the 210-V boost source.

All-red raster

Suspect a shorted picture tube, bad color output transistor, improper boost, or low voltage applied to the picture tube circuits when the raster is only one color. The video and sound might be normal, but the entire screen is of one bright color. Check all voltages on the picture tube. Test each gun of the picture tube with the CRT tester.

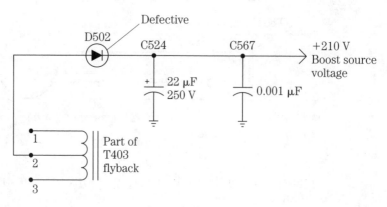

■ 13-14 *Lack of brightness control and retrace lines in a Goldstar CMT-2612 portable were caused by a defective diode D502.*

Improper adjustments of the screen and bias controls can cause one color to stand out. Try to complete the black-and-white adjustment. When one color will not adjust to a horizontal color line with the screen control when the service switch is in service position, suspect a defective gun assembly of the CRT. Make sure the master screen, bias, or individual screen control is turned up. Measure the screen voltage of the missing gun color at the picture tube socket.

No color lines were visible with the service switch in the service position and all screen controls advanced in a General Electric KD chassis. Low voltage was measured at the screen grid terminal of the picture tube socket (Fig. 13-15). R145 was found to be open in the boost voltage circuit supplying screen grid voltage to the CRT.

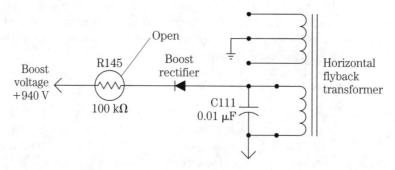

■ 13-15 *Improper boost voltage in a General Electric TV was the result of an open resistor R145 (100 kΩ).*

In an RCA FFR-498WK chassis, the screen was all green, became extremely bright, then shut down. With a very bright colored raster, the spark-gap assembly must be checked for leakage (Fig. 13-16). Just 32 Ω were measured across the green screen spark-gap assembly. In these latest RCA chassis, replace the whole spark-gap assembly with RCA part number 148169 .

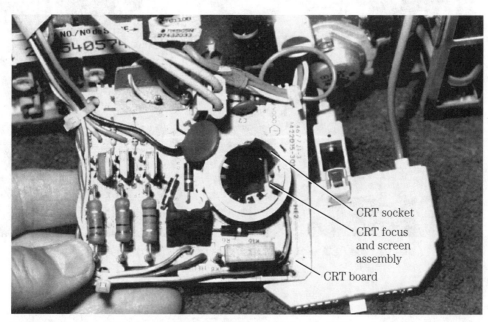

CRT socket

CRT focus and screen assembly

CRT board

549

■ **13-16** *Suspect a defective spark-gap assembly in the CRT socket if the symptom is a bright raster and then shutdown.*

One missing color

Raise the screen control on all three colors to determine which color is missing. Suspect a defective gun assembly or CRT circuits when one color is missing with the service switch in service position. Test all three guns with the CRT tube tester.

A leaky color output transistor can cause one color to be missing from the raster. Some of these color output transistors are located on the picture tube socket assembly (Fig. 13-17). A voltage test on the metal collector terminal can indicate a leaky transistor. Suspect a leaky transistor with very low collector voltage. Compare this voltage with the other two output transistors. Low-voltage on all three output transistors can indicate improper supply voltage.

The color green was missing from the raster in a Goldstar NF9X chassis (Fig. 13-18). A quick test of the video color output tran-

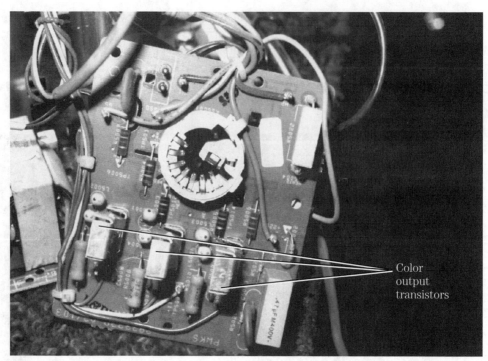

■ **13-17** *Inspect the RCA picture tube socket board for color output transistors.*

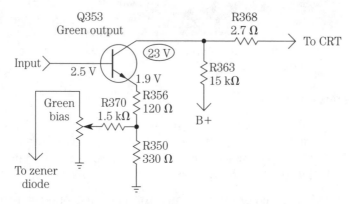

■ **13-18** *Lack of green in the raster in a Goldstar NF9X TV chassis was caused by a faulty green output transistor Q353.*

sistor found only 23 V on the collector terminal of Q353. The transistor was removed and a short was found between the collector and emitter terminals. R363 was replaced because it was running very warm. Replacing both Q353 and R363 solved the no-green problem.

Intermittent color line

Suspect an intermittent gun assembly when the raster operates normally for several hours or days, and then a green line appears with a negative picture. Often, the brightness dims down with no picture. With very difficult intermittent picture-tube problems, the chassis should be connected to the bench test tube to determine if the picture tube or chassis is intermittent. Cover the chassis with a rug or blanket, because most intermittent problems occur after several hours of operation. Replace the intermittent picture tube if the chassis operates normally for several days with the test tube connected.

Chassis shutdown

A shorted picture tube can cause the chassis to shut down. The raster will appear extremely bright and then shut down. Disconnect the anode high-voltage lead, insulate the lead away from picture tube or chassis, and plug in the power cord. If the sound comes up with high voltage at the anode terminal, suspect a shorted picture tube.

Chassis shutdown with only a bright red, green, or blue raster might be a leaky or shorted spark gap. Measure the voltage at the cathode terminal of the picture tube before the chassis shuts down. In an RCA CTC111A chassis, the chassis would shut down after a few seconds with a bright blue raster (Fig. 13-19). Only 29 V was measured at the collector terminal of the blue driver transistor (Q5003). A resistance of 2.2 kΩ was found between the collector and emitter terminals. Q5003 was replaced, but this did not correct the chassis shutdown problem. A dead short was found from pin 11 to ground of the CRT because of a leaky spark-gap assembly.

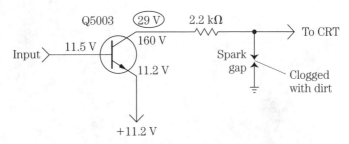

■ **13-19** *A bright CRT followed by chassis shutdown in the RCA CTC111A chassis was caused by a leaky spark gap.*

Yoke problems

The horizontal and vertical deflection yoke is mounted near the bell and on the neck of the picture tube. The defective yoke can contain shorted turns, leakage between the horizontal and vertical windings, and excessive arcing between the windings. The shorted yoke can show a trapezoid pattern or picture on the screen. Arcing between the coil layers can show up as heavy horizontal firing lines in the raster. When excessive arcing is found, puffs of smoke might curl up from the yoke assembly (Fig. 13-20).

The defective yoke can be located with ohmmeter tests, with the flyback-yoke tester, and by feeling for warm spots inside the yoke windings. In the black-and-white TV chassis, the old-timer would disconnect the yoke after 30 minutes of operation and feel for warm areas inside the yoke (indicating shorted turns). A continuity check with the ohmmeter indicates if a winding is open or if the two windings are shorted together.

■ **13-20** *Suspect a shorted yoke assembly when smoke curls up from it.*

High-voltage arcing

Excessive high voltage can produce a loud crack or arcing sounds around the anode connection. Dust and dirt on the CRT anode terminal can cause some corona activity. Discharge the picture tube. Clean off the anode lead and tube area with cleaning fluid or detergent. Inspect the anode lead rubber for excessive firing areas. Replace the entire high-voltage cable and plug if evidence of arcing is found. Check the high-voltage anode for excessive high voltage. High-voltage arc-over at the CRT anode terminally can result from a defective holddown circuit in the TV chassis.

High-voltage arcing can occur with a defective picture tube. With an open heater assembly, a blue area appears at the gun assembly when arcing begins. Check the glass neck for a cracked area. It might be inside the yoke assembly. A shorted picture tube can cause high-voltage arcing around the ground strap and TV chassis.

Firing lines in the picture

Sharp jagged or dotted lines across the picture can indicate outside interference or arcing in the chassis. Remove the outside antenna and notice if the interference lines disappear. Usually, man-made noise picked up from the antenna will appear on channels 2 through 6 (Fig. 13-21).

■ **13-21** *A defective focus-screen assembly in the TCA CTC107 chassis can cause firing lines in the picture.*

Thin firing lines can be caused by improper grounding of the dag (outside area) of the picture tube. Clip a lead from the chassis to the black, rounded area of the picture tube. Sometimes these small springs around the bell assembly become dirty and corroded, leaving a poor ground. Check to see if the ground strap from the chassis to the picture tube mounting bracket is in place. If broken, run a ground wire from the picture tube mounting to the metal chassis.

Arcing lines can be caused by a defective flyback transformer or focus control assembly. Check for corona arcs around the outside plastic body of the horizontal output transformer. Rotate the focus control to see if the lines disappear (Fig. 13-22). Replace the defective focus control if there are poor focus and firing lines in the picture. These printed high-resistance focus controls have a tendency to break and arc over (Fig. 13-23).

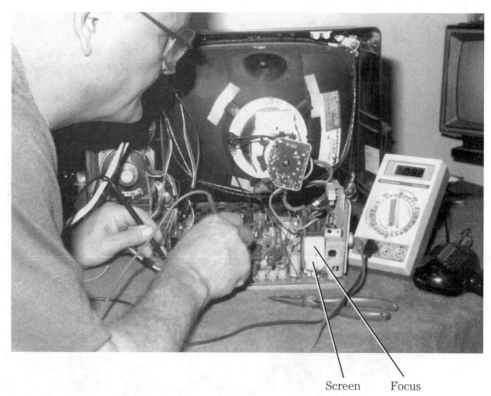

Screen   Focus

■ **13-22** *Rotate the focus control when firing lines are found in the picture.*

CRT anode   HV cable

Focus control   Focus cable   Yoke   Flyback

■ **13-23** *Check the various components that could cause arcing and firing lines in the picture.*

Defective CRT harness

A defective picture tube socket can cause a blurry picture with firing lines. Poor focus can result from a corroded focus pin of the CRT socket (Fig. 13-24). Picture tube spark gaps inside the CRT socket can cause arc-over. Sometimes you can hear the arcing by placing your ear near the tube socket. Besides arc-over, suspect poor heater terminals when an intermittent raster is found. If the heater terminals have changed color forming a high-resistance connection, install a new CRT socket and harness.

A faint spitting noise was heard when the J.C. Penney CTC90JL chassis was first turned on. Intermittent firing lines would appear

Check focus pin

■ **13-24** *Check for a corroded focus pin in the CRT socket if the set exhibits poor focus.*

across the screen. (Under these conditions, suspect either a defective flyback transformer or focus control.) The arcing could be heard at the CRT socket. The arcing noise stopped when the picture tube socket was removed. Installing a new CRT socket and harness solved the intermittent firing in the raster.

Testing the CRT

The defective picture tube can be located with a CRT tube tester and critical voltage measurements. Check for correct high voltage at the anode terminal. Visually inspect the neck of the tube for light in the tube filament. Now test the picture tube with a good CRT tester.

The picture tube can be checked in the home, in the carton, or on the bench with the CRT tester. Intermittent conditions might not be located with the picture tube tester. Visual inspection and high- and low-voltage measurements on the tube elements can help locate the intermittent picture tube (Fig. 13-25). Sometimes tapping the end of the tube can turn up a shorted gun assembly. For very difficult picture tube problems, connect the chassis to the bench tube for observation.

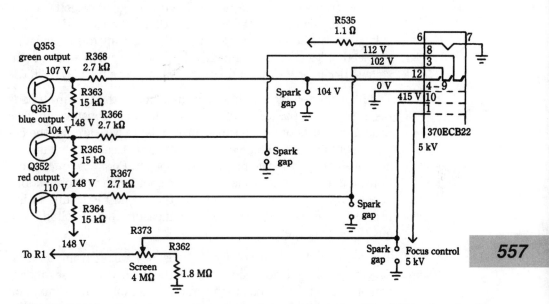

■ **13-25** *Monitor the voltages at the picture tube to determine if the CRT or the circuits are intermittent.*

CRT repairs

Operation of the weak picture tube can be extended by applying a tube brightener or charging the tube with a picture tube charging instrument. The tube brightener plugs on the end of the picture tube with the CRT socket plugged into the end of the brightness assembly. The brightener raises the heater voltage, increasing brightness. Although increasing the tube brightness is not a permanent repair, the life of the TV can be extended for several months. Charging the picture tube might prolong the life of a picture tube for several years. The test instrument renews each gun assembly by removing bombarded ions from the cathode area of the heater gun assembly. Charging the picture tube can help sell a used color TV. Use the instrument to sharpen and brighten the picture after TV repairs. A good picture tube charging and testing instrument can cost over $500.

IHVT-derived voltages for the CRT

Besides supplying low dc voltages for the various circuits of the TV chassis, the integrated flyback transformer (IHVT) can produce high voltage, focus voltage, screen voltage, and filament voltage for the picture tube. The high-voltage winding of the IHVT horizontal output transformer contains high-voltage diodes molded inside the transformer. The high-voltage lead goes directly to the anode button socket on the bell of the CRT. When any of the high-voltage diodes break down, the entire flyback transformer must be replaced (Fig. 13-26). The high voltage is measured with a high-voltage probe or VTVM on the high-voltage terminal of the CRT.

The filament voltage for the three guns inside the picture tube is taken from a separate winding on the flyback transformer. This winding consists of one to three turns of large wire around the ceramic core and metal flange. The two filament cables plug into or are wired directly to the picture tube socket. Poor socket connections, a defective CRT tube socket, or poor soldered connections at the CRT board can cause intermittent or no raster due to improper filament voltage. Do not try to measure this ac voltage with any type of voltmeter.

The focus and screen voltages are usually taken from a focus-screen variable divider network of the high-voltage secondary winding. The focus and screen variable assembly can be found on the end of the CRT board or on a separate mount at the rear of the chassis, A corroded or dirty focus control and CRT pin can produce an out of focus or blurry picture. In some TV chassis, the screen voltage can be taken from another secondary winding with a separate diode rectifier and filter network.

Spark gaps

You will find spark gaps at each grid, at the screen and focus grid terminals, and at all three cathode terminals of the picture tube. These gaps will arc over and protect the picture tube gun assembly if extreme high voltage is present at any picture tube element. If one of these gaps continually arcs over, it can cause an extremely blurry picture, loss of color, a raster of one color, and/or chassis shutdown.

Sometimes when the picture tube sockets become very dirty, excess dust collects in the gap area and causes the entire chassis to shut down (Fig. 13-27). When the raster comes on and goes into a blurry, distorted picture, check the spark gaps in the CRT socket. Usually the spark gaps are molded into the CRT socket or are placed in separate components on the CRT board. Replace the spark gap or CRT socket when excessive arc-over continues or the gap areas cannot be cleaned.

CRT voltages

Critical voltage measurements on the picture tube socket can help determine if the CRT is defective. A no-raster symptom can indicate improper high voltage or a defective gun assembly. No control of brightness can be caused by a shorted picture tube, improper screen voltages, or a poor video signal from the luminance circuits. The all-red screen or a missing color can be caused by a leaky color output transistor lowering the cathode voltage of the picture tube. Poor focus can be caused by improper voltage at the focus pin terminal.

High voltage and focus voltage should be measured with a high-voltage probe or VTVM with a high-voltage probe. Critical screen and boost voltages can be measured with the 1,500-V scale of a VTVM, VOM, or DMM. Correct cathode and grid voltage should be checked with a VTVM or DMM (Fig. 13-28).

Picture tube protection

Remember, all picture tubes employ an integral vacuum, so handle them with care. Do not remove or install the picture tube unless you are wearing goggles to protect your eyes from possible implosion and flying glass. Replace the picture tube with the correct tube replacement. Do not handle the tube by the neck. Grab the picture tube with both hands on the heavy glass front of the tube. Do not force any components off the neck of the picture tube. Remember, some in-line picture tubes have deflection yokes that are permanently attached.

Discharge that picture tube

Always discharge the high voltage at the picture tube when replacing or working around the high-voltage circuits. Remember, the picture tube aqueduct glass around the high-voltage socket acts like a high-voltage capacitor, and can hold a charge for months. Stay alert when discharging or working around the high-voltage circuits to prevent shocks or serious injuries.

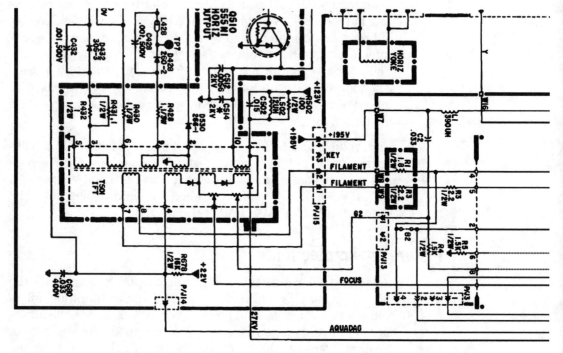

■ 13-26 *The horizontal flyback provides screen, focus, high voltage, and filament voltage for the picture tube.* Phillips Consumer Electronics

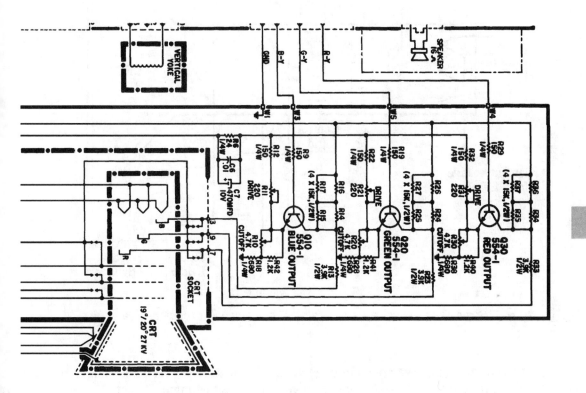

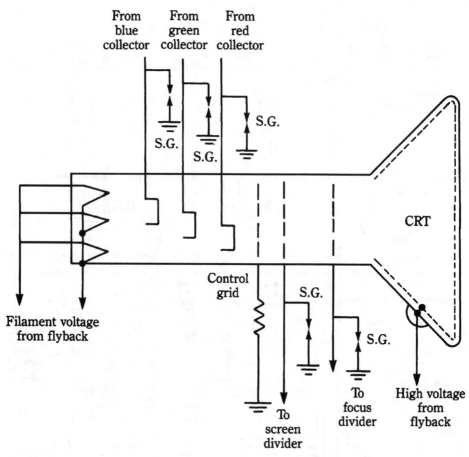

■ 13-27 *Some TVs have spark gaps on all CRT pin terminals except the heater pins.*

In the old black-and-white tube chassis, the picture tube was discharged from the anode connection to the chassis ground. Do not attempt to discharge the CRT in this manner. You can damage transistors and IC components in the solid-state chassis by discharging high voltage to the chassis. Always discharge the high voltage from the button area to the black aqueduct on the outside of the picture tube. Use a long-bladed, well-insulated screwdriver to get under the rubber socket connection, and ground a similar screwdriver blade to the outside ground of the CRT. Hold the metal shanks of the screwdrivers together to discharge the picture tube.

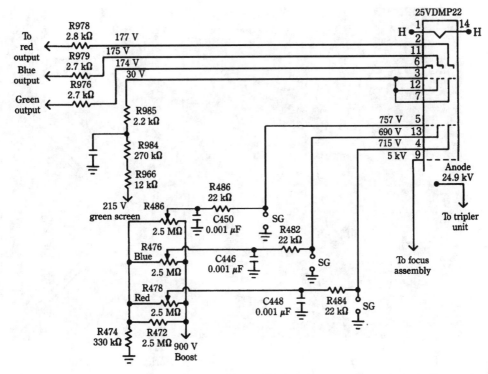

■ 13-28 *Check for critical cathode, grid, and high voltage to determine whether a picture tube is defective.*

Picture tube replacement

After determining the picture tube is defective, take a visual inspection of those components surrounding the CRT (Fig. 13-29). Check the position of the purity and beam magnets on the neck of the tube. Measure the distance between the purity magnet and end of the tube for future reference when installing the new one (Fig. 13-30).

Carefully remove the back cover. Loosen the top screws and then the bottom. Hold the back so it will not fall against the CRT and possibly break or crack the neck of the tube. Remove the TV chassis. Place the front of the tube or cabinet face down on a rubber mat or rug. Notice the mounting of each component from the neck of the CRT and around it. Remove the picture tube and lay it on its

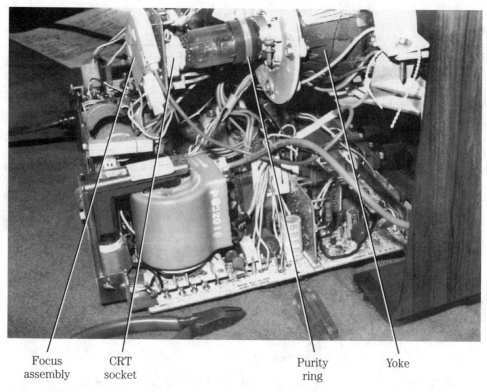

Focus assembly   CRT socket   Purity ring   Yoke

■ **13-29** *Visually inspect all components mounted on a CRT before removing it from the chassis.*

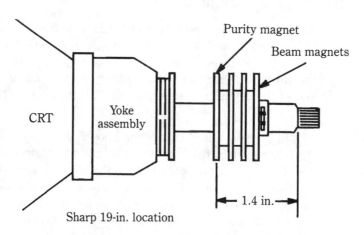

■ **13-30** *Measure the distance of all components on the neck of the CRT for correct location at reassembly.*

Troubleshooting picture tube circuits

face on a rug or carpet so as not to scratch the screen area. Reverse the procedure to install a new tube. Degauss the picture tube before black-and-white setup.

Picture tube removal

Before attempting to remove the picture tube, remove the TV chassis and tuner from the cabinet. Place the cabinet face down on a blanket or soft surface to protect the front screen of the CRT. Remove the metal screw or bolt in each corner to free the bracket or shell holding the degaussing coil. Sometimes the top or bottom degaussing coil is held in place with plastic ties and metal screws (Fig. 13-31).

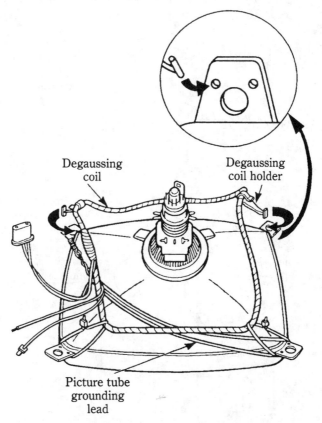

Degaussing coil

Degaussing coil holder

Picture tube grounding lead

■ **13-31** *A picture tube removal schematic for the Realistic 16-281 portable TV.* Radio Shack

Remove the screws or bolts on each corner of the picture tube so the tube can be lifted out of the cabinet. It is best for two people to lift the CRT. Place it on a blanket or soft cloth to prevent scratches. Reverse the procedure when installing a new picture tube. Measure the distance from end of the CRT to the mount tab magnets of the old tube to position on the new replacement. Always wear safety glasses or goggles when removing and installing a new CRT.

RCA CTC157 on-screen display

The signal information for the on-screen display (OSD) is provided by the AIU (U3300), and the system control IC (U3100) determines when to output the OSD signals. During normal operation the luminance information is applied to the base of Q2901. This output is mixed with emitter of Q5002, with chroma information applied to the base of Q5002. The collector of Q5002 drives the green gun assembly of the picture tube (Fig. 13-32).

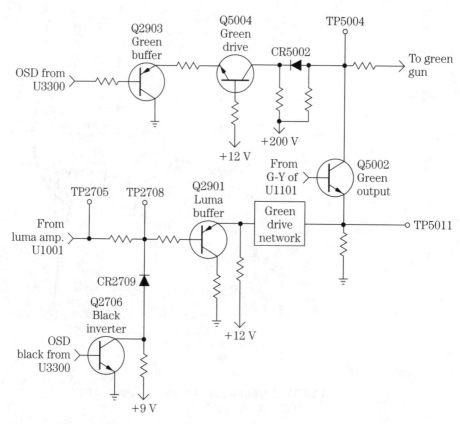

■ **13-32** *RCA CTC157 on-screen display (OSD) circuits.*

When the OSD signal of U3300 is applied to the base of Q2903 and Q5004, both transistors are turned on, which causes CR5002 to become forward-biased. Now Q5002 is pulled low, resulting in a current path to Q5004 instead of Q5002.

Check the OSD circuits by scoping the input terminals of the OSD, luma, and OSD black signals. Scope waveforms at the collector of Q2706 and Q5004. If Q2903 becomes shorted, the picture will be all green. A leaky CR2709 or Q2706 will eliminate black around characters, and when open, makes the picture black all the time. For a green OSD character, check the OSD output of U3300, Q2903, Q5004, and CR5002. If there is no black edge, check the output from U3300 and check Q2706 and CR2709. With no luminance but okay chroma, check the output of U3300 and look for a leaky Q2709.

Five actual picture tube problems

The following five picture-tube-related problems actually happened.

High brightness—retrace lines

The brightness was extremely high and there were retrace lines in an RCA GJR2038P model. Sometimes the chassis would shut down after operating for a few hours. All CRT voltages were checked, with normal boost voltage to the output transistors. Improper voltages indicated that the brightness reference transistor (Q703) was shorted (Fig. 13-33). There have been several of these same transistors replaced in other RCA chassis.

Intermittent brightness

In an Emerson MS250RA portable, the screen brightness would operate for hours or just a few minutes, and come off and on. The brightness and screen controls had little effect on the picture. Both the 11.5-V and 18.8-V voltages were monitored in the video and contrast circuits, and were fairly normal when the brightness went out (Fig. 13-34). Sometimes the brightness would only dim down. No scope waveform out of pin 42 or 22 occurred with the no-brightness problems. Replacing IC201 solved the brightness problem.

Bright screen with shading

The TV screen was very bright with some dark shading in a Sanyo 91C510 TV. Transistors and voltages were checked in the video circuits and were normal. The boost voltage of 196 volts was quite low

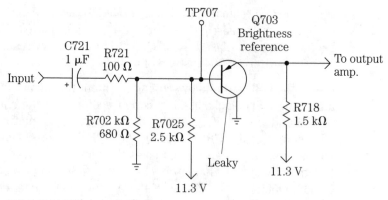

13-33 *High brightness with retrace lines resulted from a leaky transistor Q703 in an RCA GJR2038P.*

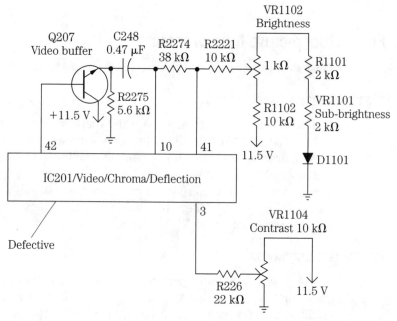

13-34 *An intermittent chip (IC201) caused an intermittent brightness symptom in this Emerson MS250RA portable.*

at the picture tube cathode terminals. D311 and R342 were found to be good in the boost voltage source (Fig. 13-35). C358 (4.7 µF, 250 V) was shunted, which solved the brightness problem.

Bright retrace lines

The picture in an RCA CTC118A chassis looked like there was AGC trouble, with bright retrace lines. In this model several ICs

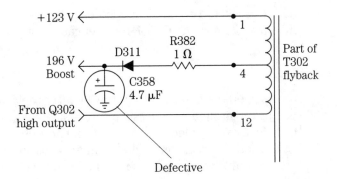

+123 V

196 V
Boost

D311

R382
1 Ω

C358
4.7 µF

From Q302
high output

1

4

12

Part of
T302
flyback

Defective

■ **13-35** *A bright screen with some shading was caused by the failure of capacitor C358 (4.7 µF) in a Sanyo 91C510 TV.*

(U701) were replaced for the same video problems. In this case R114 was open in the color output transistor circuit (Fig. 13-36).

RCA CTC108 warm up, then shut down

In an RCA CTC108 chassis, the raster would come on, begin to fade, go out of focus, and then shut down. High voltage was measured when the TV was turned on and cut off at once in shutdown. With the high-voltage cable unplugged from the CRT, the chassis remained operating. This indicated either a shorted CRT or defective circuits.

The picture tube checked normal with the CRT tester. When the chassis was fired up, some arcing sound appeared at the CRT socket. With the CRT socket pulled off, the high voltage and sound were normal. After blowing excessive dust from the top and inside of

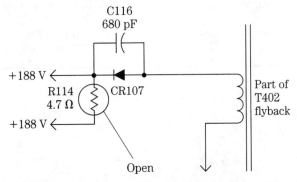

C116
680 pF

+188 V

R114
4.7 Ω

CR107

+188 V

Part of
T402
flyback

Open

■ **13-36** *Bright retrace lines in an RCA CTC118A chassis were caused by an open R114 (4.7 Ω).*

the CRT socket, the chassis was normal. A collection of dust in the CRT spark gaps inside the CRT socket caused chassis shutdown.

Conclusion

Always keep the back cover on the TV after replacing the picture tube or repairing a TV chassis to prevent breakage. Carefully lay the defective tube face down on a rug or carpet to prevent scratching the screen area. Remember, you are out the turn-in value of the tube until the bad tube is returned and accepted.

Installing a rebuilt picture tube can place the TV back in working order for a lower price. Some picture-tube rebuilders will install a new gun assembly in those 19-inch "in-line" picture tubes with the yoke glued to the bell of the tube. This can be done at a much lower cost than purchasing a new tube. Charging the old picture tube can brighten up the picture after repairs. Remember to clean off the TV screen with window cleaner and polish up the TV cabinet before returning it to the customer. More information on servicing the picture tube circuits can be found in Table 13-1.

■ Table 13-1 Servicing the picture tube circuits.

What to check	How to check it
No raster.	Check voltage at HV anode on CRT.
White raster no picture.	Check color output and video circuits.
Black raster with normal high voltage.	Check for no focus voltage.
Poor focus.	Check CRT and focus circuits. Check CRT socket.
Real bright raster, no control.	Check boost voltage. Check all voltages on CRT. Suspect brightness circuits.
Weak picture.	Test CRT; all three color gun assemblies.
Intermittent raster.	Check CRT socket and look for loose particles in grid elements of CRT.

Servicing the
black-and-white and
small-screen chassis

14

THERE ARE MANY DIFFERENT BLACK-AND-WHITE TV SETS on the market. Practically every manufacturer has three or four models in their new TV line each year. They come from Japan, Hong Kong, Taiwan, and many other foreign countries. The screen sizes vary from 1 inch up to the 19-inch portable.

Although the black-and-white portable is fairly easy to service, there are a few new circuits or a combination of components changed each year. One of the big differences between the color and the black-and-white chassis is, of course, the color circuits. A smaller picture tube with a single gun assembly is found in the black-and-white chassis. Very few black-and-white receivers contain separate modules. Of course, the high voltage can be developed in the same way as in the color TV, but the voltage is much lower, eliminating shutdown and holddown circuits. Very seldom do you find expensive tuning mechanisms or remote-control circuits in the black-and-white chassis.

Today you will find that the tiny pocket TV contains either black-and-white or color circuits. These new TVs can contain surface-mounted components and many ICs. The TV screen might be of an LCD instead of a regular picture tube. This makes a very flat and small TV set. A service manual is a "must" item when servicing these small black-and-white or color TVs. The Realistic TV described in the following paragraphs is a tiny color TV. The circuits, surface-mounted components, and LCD picture provide valuable service information about tiny black-and-white or color TV receivers.

Realistic Pocketvision® LCD TV

The Realistic 2-inch color TV with direct-view LCD screen is only $5.125 \times 3.25 \times 1.25$ inches in size. This pocket color TV fits comfortably in the palm of your hand. The TV has digital synthesized tuning with an up/down autosearch system. Also, a backlight hood can be adjusted for clear viewing in dim light.

This small color TV consists mostly of ICs, with a few transistors (Fig. 14-1). The block diagram consists of a color tuner (TU101, part number ENV-76261F1) that selects the desired channel and changes the signal to the IF video signal. Q102 amplifies the IF signal and feeds it to the IF video amp (IC101), which amplifies the signal approximately 10 times. The video, sound FM, AFT detection, and AGC IC eliminate the carrier wave in the IF video signal and pick up the video signal and IF sound signal. Also, the audio is picked up from the IF sound signal by FM detection. IC204 provides audio amplification to a small pin speaker (Fig. 14-2).

The chroma circuit consists of IC103 (M51405FP) which generates the tricolor of red, green, and blue from the video signal. IC202 is an oscillator A/D converter that converts the color signal into the display signal. It also generates the clock pulse for the display and controls the display.

The tuning voltage generator (IC102) generates the tuning voltage from the tuning pulse (TU) output at pin 6. The display voltage generator (IC201) generates the display voltage (V0-V4) from VEE1/2, and VCC6/7 is output from the power supply.

■ **14-1** *The small hand-held Casio TV-470 has a small LCD screen.*

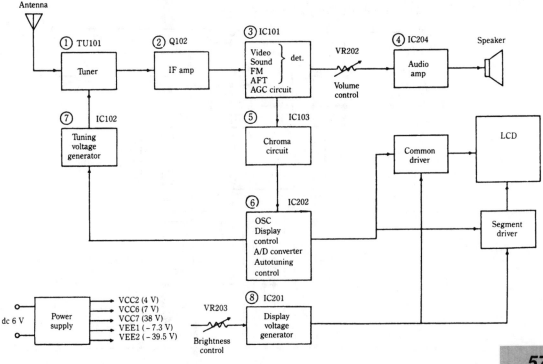

■ **14-2** *The block diagram of a Realistic Pocketvision® LCD TV receiver.* Radio Shack

The dc power supply operates from four AA batteries and builds up the voltage from 6 V to VCC2 (4 V), VCC6 (7 V), VCC7 (38 V), VEE1 (–7.3 V), and VEE2 (–39.5 V).

Tuning voltage generator

The tuning voltage circuit is built around IC102 (MSC1169MSK) as the main component (Fig. 14-3). This circuit generates the tuning voltage (BT). VR103 and VR105 and the soldering pads E through H determine the characteristic of the tuning voltage (BT). If they are not correctly set, the position of the tuning indicator will not match with that of the indicator panel (Table 14-1).

The dc-dc converter power supply

The dc-dc converter power supply operates from a 6-Vdc battery source and converts this voltage for the rest of the color TV circuitry. Q101, Q104, and Q105 provide dc oscillation and feedback circuits in the primary winding of T101 (Fig. 14-4). The output voltage to the various TV circuits is taken from the secondary windings of T101.

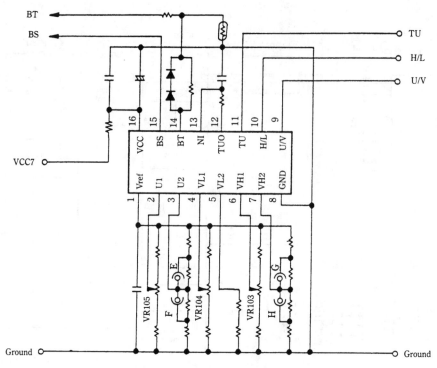

■ 14-3 *The tuning voltage generator (IC102) circuit.* Radio Shack

**■ Table 14-1 The different adjustments
for indicator position in the Realistic 16-159 TV set.**

VR104	VHF-L indicator position adjustment
VR103, G, H	VHF-H indicator position adjustment
VR105, E, F	UHF indicator position adjustment

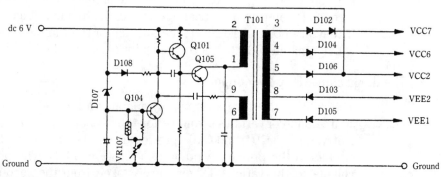

■ 14-4 *The dc-dc converter with five different output voltages.* Radio Shack

Servicing the black-and-white and small-screen chassis

D102 provides 38 V (VCC7) to the display and tuning voltage. D104 rectifies the VCC6 voltage (7 V) for the display voltage. VCC2 voltage comes from D106 and supplies 4 V for the main operating voltage. D103 rectifies the VEE2 voltage (–38.5 V) for the display LCD. VEE1 voltage (–7.3 V) also is applied to the display LCD from diode rectifier D105 (Table 14-2).

■ **Table 14-2 The chart of different voltages and functions of the dc-dc converter.**

Name	Voltage	Function
VCC2	4.0 V	Main voltage
VCC6	7.0 V	Display voltage
VCC7	38.0 V	Display tuning voltage
VEE1	–7.3 V	Display voltage
VEE2	–39.5 V	Display voltage

Radio Shack

Pocketvision®-notes

There are seven main IC components used in this small color TV. IC202, the main tuning IC, contains 56 terminals (Fig. 14-5). All components are surface-mounted parts. Two different types of transistor terminal mounts are found on the chassis. The top view of the PC board is found in Fig. 14-6, while the bottom view is found in Fig. 14-7. The troubleshooting chart for the tiny TV is found in Table 14-3.

SMD components in small TVs

Miniature SMD components found in LCD and small black-and-white portables are the same as in large color TV chassis. Use a pencil-type soldering iron of 30 watts or less in part replacement. Choose a Eutectic solder that consists of 63% tin and 37% lead. Do not apply heat for more than 4 seconds upon replacement terminals. It is best to preheat leadless capacitors for about two minutes before installation; preheat them to 266 to 300 degrees Fahrenheit. Do not reuse SMD components removed from the PC board.

Removing and replacing IC SMD components

A hot-air flat-pack IC desoldering machine can remove flat IC chips within 5 or 6 seconds. Remove the flat pack chip with tweezers while applying hot air. Do not apply hot air on the replacement

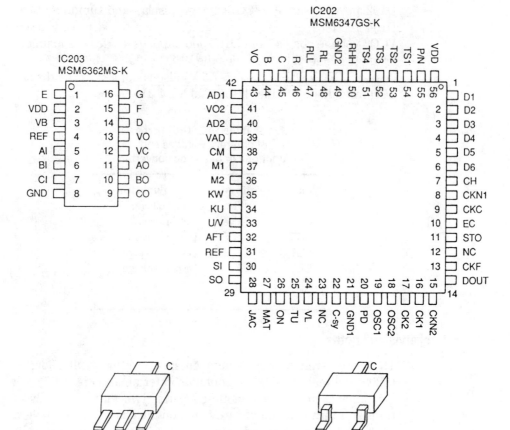

■ **14-5** *The IC202 display control and A/D converter with auto-tuning control has 56 different terminals.*

chip for over 6 seconds. When all electrode joints have melted, remove the component with a twisting motion.

The cheapest method is to use desoldering braid (solder wick) and soldering iron to remove flat IC chips. Apply solder flux to all pins and remove pin or electrode. Lift each lead of flat pack IC up with sharp instrument under each terminal (Fig. 14-8). When heating each pin, use a fine-tip soldering iron.

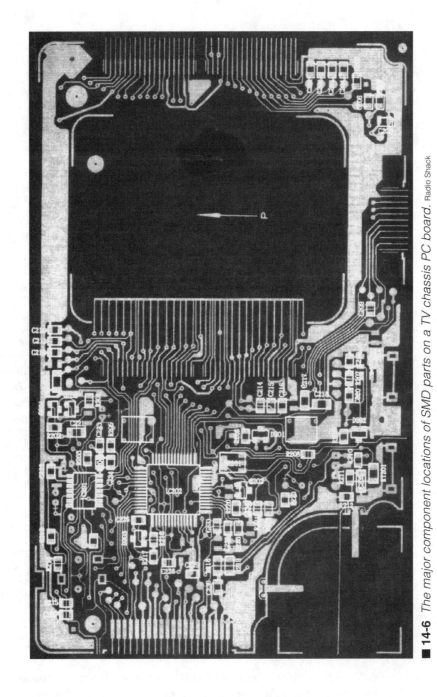

■ **14-6** *The major component locations of SMD parts on a TV chassis PC board.* Radio Shack

Removing and replacing IC SMD components

■ **14-7** *The inside view of the Casio TV-470 small LCD TV receiver.*

■ **Table 14-3 A troubleshooting chart
for the Realistic Pocketvision® 16-159 TV set.**

Symptom	Cause	Solution
No power supply	Defective power switch (SW101)	Replace
No picture	Defective dc-dc converter (T101)	Replace
No sound	Defective tuner (TU101)	Replace
	Defective Q102	Replace
	Defective IC101	Replace
	Defective IC102	Replace
	Defective IC103	Replace
	Defective IC201	Replace
	Defective IC202	Replace
Picture OK but no sound	Defective F104	Replace
	Defective IC101	Replace
	Defective IC204	Replace
	Defective IC202	Replace
	Defective phone jack (JACK201)	Replace
No reception of VHF or UHF	Defective (TU101)	Replace
	Defective IC102	Replace
	Defective power switch (SW101)	Replace

Radio Shack

Before installation of a new SMD IC, remove solder from the foil of each pin, so the new replacement can lay flat. Check the dot or pin number 1 before placing the chip over the PC wiring terminals. Solder all four corners of the flat-pack IC to hold into position (Fig. 14-9). Now solder up all pins of the IC chip, making sure that none of the pins have solder bridges.

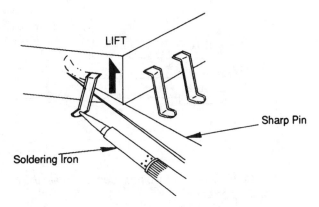

■ 14-8 *Lift up the gullwing terminals with a sharp, pointed tool while applying the soldering iron.*

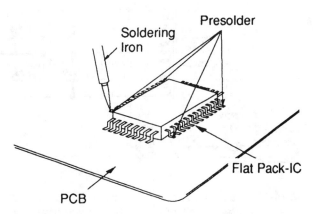

■ 14-9 *Tack each corner of a flat-pack IC to hold it in position, and then solder all the remaining terminals.*

Additional black-and-white circuits

The black-and-white circuits can be found in an AM-FM MPX stereo tape player (Fig. 14-10). Many of the 3- to 5-inch TV chassis are part of an AM-FM radio-combo entertainment center. Usually the small-screen TV operates from batteries or a combination of ac and dc circuits. Some black-and-white TVs operate directly from the 12-Vdc source of a car dashboard.

The 12-V source is fed from a cigarette lighter plug with a separate plug-in power cord to the black-and-white receiver. When the plug is inserted into the portable TV, the ac voltage circuits are switched out of the battery circuits (Fig. 14-11). Most black-and-white TVs operate from a 12- to 15-Vdc source. Often the dc

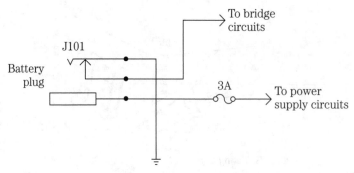

■ 14-10 *The battery circuit connection to the ac-dc black-and-white portable TV.*

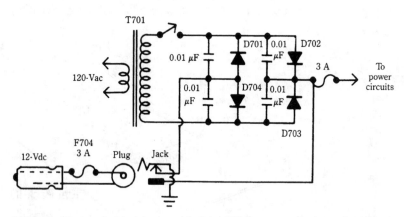

■ 14-11 *The portable black-and-white TV can operate from the power line, the battery pack, or the 12-volt source of an automobile.*

power cord is fused internally, or a dc fuse is located inside the TV chassis.

A three-way black-and-white chassis can operate from the ac power line, from a car, or from self-contained batteries. The 12-V power cord or batteries can be switched into the circuit when the power cord is removed (Fig. 14-12). F2 protects the power supply circuits from overload or wrong battery polarity. Notice that in this circuit, the center terminal (+) of the cigarette plug is to the ground side of the small dc plug, while the negative voltage is fused and goes directly to the TV chassis ground through diode D505. If the wires of the 12-V cord or the polarity are reversed, the diode prevents the TV from operating.

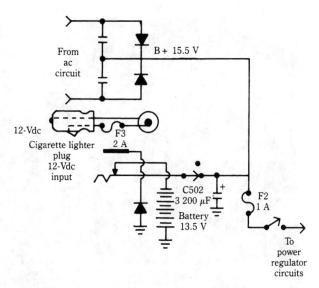

From
ac
circuit

B+ 15.5 V

12-Vdc
Cigarette lighter
plug
12-Vdc
input

F3
2 A

C502
3 200 μF

Battery
13.5 V

F2
1 A

To
power
regulator
circuits

■ **14-12** *When the cigarette lighter power cord plug is pulled from the chassis socket, the small TV can operate from the 110-Vac power line.*

Does not operate on batteries

The black-and-white TV that operates from the power line but not on battery is one of the biggest complaints in a three-way chassis. Check the fuse protecting the battery circuits. If the TV will not work on a 12-Vdc cord, check the fuse inside the male cigarette plug. Remove the metal band that holds the plug together to get at the fuse, or the removable fuse end that can be used as the positive terminal. Simply push in on the fuse and give it a twist to release.

A separate battery fuse might be located inside the TV cabinet. Most of these small fuses are of the pigtail variety, and are soldered into the circuit. Check the continuity of each wire from the male plug to the chassis and fuse to locate a break in the cord or discover a broken dc input switch assembly (Fig 14-13).

The dc fuse can blow with an overload in the TV chassis or the wrong dc polarity of the 12-V cord. In many cases, the operator transfers the TV to a trailer or camper and the TV refuses to operate. The trailer or camper battery might be wired with the wrong polarity, blowing the fuse. Also, the battery in the car or camper might have been charged and reinstalled backwards, blowing the small dc protection fuse. Check for correct battery polarity.

Line fuse   Flyback

■ **14-13** *A small 0.5- or 1-amp line fuse is found in the ac power line circuit.*

Erratic operation

Sometimes when the ac cord is inserted into the plug, the TV will come on and quickly stop. Check the ac-dc plug contacts. Notice if the switch and plug assembly is broken. These switch-type plugs have a tendency to crack, yielding poor switching action. Either repair the chassis jack with epoxy or order a new one. Repairing the jack is much quicker, because special plugs and jacks take awhile to obtain. Spray contact cleaner into the switch area to clean the contacts.

Dead, no picture

To determine where the defective part is located, check the voltage at the dc fuse and body (collector) of the horizontal output transistor, in that order (Fig. 14-14). Correct voltage at the fuse indicates everything is normal through the low-voltage diodes (Fig. 14-15). Voltage found on the collector terminal of the horizontal output transistor can indicate problems within the horizontal and high-voltage circuits. No voltage at the horizontal output transis-

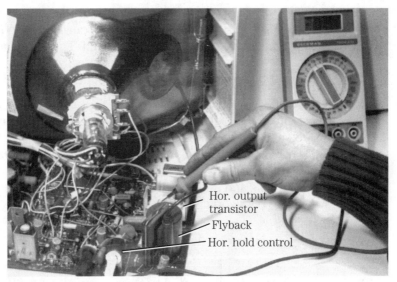

Hor. output
transistor
Flyback
Hor. hold control

■ **14-14** *The leaky horizontal output transistor and flyback can blow the secondary 2-amp and 1-amp line fuses.*

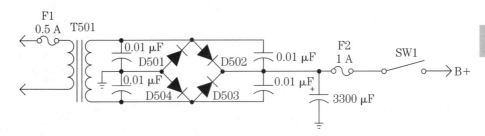

■ **14-15** *Measure the voltage at F2 to see if the power supply is normal.*

tor and voltage at the fuse can indicate dc regulation problems within the low-voltage power supply.

Check the low-voltage diodes, power transformer, and line fuse without any dc voltage at the operating fuse terminal. Because the secondary winding of the power transformer seldom opens, check the primary winding for open conditions with the low-ohm scale of the VOM. Check each diode with one terminal removed from the circuit with the diode test of a DMM. Lightning damage on the power line pops most line fuses and diodes. Replace each defective diode with 2.5-A types.

With dc voltage at the fuse and no voltage at the horizontal output transistor, suspect a defective component within the B+ regulation circuits. No dc voltage at the collector terminal of the regula-

Additional black-and-white circuits

tor can be caused by a broken wire, poor board connection, or open on/off switch. Test the B+ regulator transistor within the circuit with the transistor diode test of a DMM. These regulator transistors can open under load. Automatically replace the B+ regulator transistor when no other component is found defective.

Very low B+ voltage was measured at the collector terminal in a Goldstar VR230 black-and-white TV. The collector terminal wire was removed from the horizontal output transistor with only a slight increase in dc voltage (Fig. 14-16). TR901 was found to be open in the B+ regulator circuit. With TR901 out of the circuit, TR902, TR903, and R901 were tested. Replacing TR901 solved the low-voltage regulation problem.

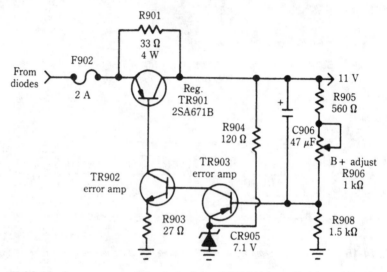

■ **14-16** *An open regulator transistor (TR901) in a Goldstar VR230 TV produced a very low power-supply voltage.*

Keeps blowing fuses

If the main line fuse opens after replacement, check each diode rectifier for leakage. Take out the B+ fuse to the horizontal circuit. Remove the collector lead or horizontal output transistor from the circuit to determine if the horizontal circuits are blowing the fuse. You might find a leaky bypass capacitor across the low-voltage diode that is blowing the fuse. Check all diodes and capacitors for damage when lightning has entered the TV chassis.

Suspect a leaky regulator or horizontal output transistor when the B+ fuse will not hold. Remove the collector terminal of the horizontal output transistor to determine if the problem is within the regulation circuits. Often a leaky output or regulation transistor can lower the dc voltage source and not open the B+ fuse.

Hum bars

Hum in the speaker with 120-Hz or 60-Hz dark bars in the raster can indicate a defective filter capacitor or poor B+ regulation. Shunt a large filter capacitor across the suspected one to isolate a defective filter capacitor (Fig. 14-17). Try to raise or lower the voltage with the AVR or B+ control. Improper adjustment of the B+ control can cause small dark bars in the picture.

Filter capacitor

■ **14-17** *Hum bars and a narrow picture can result from a defective main filter capacitor.*

If the B+ voltage adjustment does not lower or raise the voltage, suspect a leaky voltage regulation transistor. Several dark bars were found in the raster of a Quasar XP3222 (Fig. 14-18). Adjustment of VR71 had little effect on the low output voltage. TR73 was found to be leaky in the circuit with the transistor diode test of the DMM. Always readjust the B+ voltage after replacing defective components within the power supply.

Poor width

Suspect low power supply voltage or a leaky horizontal output transistor if the set exhibits poor width. You might find the width

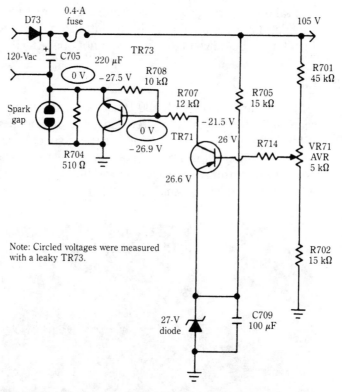

Note: Circled voltages were measured with a leaky TR73.

■ **14-18** *A leaky transistor (TR73) caused hum bars in a Quasar XP3222 TV chassis.*

is pulled in so much that the raster looks like the outline of a barrel. Check the voltage of the horizontal output transistor and the input voltage at the B+ regulator transistor. Often the B+ control has no effect on the raster.

Check the B+ regulator transistor for open and leakage conditions. Test all zener diodes for leakage. Inspect critical regulation resistors for burned marks and for correct resistance. No sound with poor width was found in a Goldstar KMB1230G model. The collector voltage at the horizontal output transistor was only 5.4 Vdc and should be 15 V (Fig. 14-19). The B+ regulator (TR901) was found to be open in the low-voltage power supply. TR901 was replaced with an ECG152 universal replacement.

Servicing the horizontal circuits

Besides the low-voltage power supply, the horizontal circuits cause more problems than any other circuit in the black-and-white

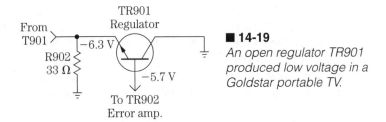

From T901
R902 33 Ω
−6.3 V
TR901 Regulator
−5.7 V
To TR902 Error amp.

■ 14-19
An open regulator TR901 produced low voltage in a Goldstar portable TV.

chassis. No sound, no raster, no picture, picture pulling, and insufficient width are common symptoms caused by problems in the horizontal circuits. The horizontal circuits can cause excessive arcing and firing within the black-and-white raster (Fig. 14-20). Check the flyback transformer if the problem is a high-pitched squeal with normal picture and sound.

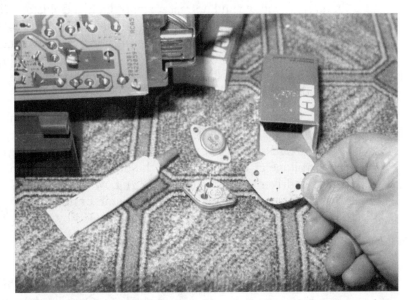

■ 14-20 *Check the horizontal hold coil, horizontal output transistor, and flyback for most horizontal problems. Place silicone grease in the output transistor before replacement.*

No sound and no raster

A voltage and waveform check at the horizontal output transistor can indicate what stage is defective in the horizontal circuit. Higher-than-normal voltage at the collector terminal with no waveform at the base of the horizontal output transistor can indicate a defective horizontal oscillator or driver stage. A high collector voltage at the horizontal output transistor can indicate that the transistor is open. Adequate drive voltage with no output waveform indicates an open horizontal output transistor or emitter resistor. Low voltage on the collector terminal of the horizontal output transistor can be caused by insufficient drive voltage. Very low voltage at the horizontal output transistor can be caused by a shorted flyback transformer, leaky damper diode, or boost rectifiers.

Go directly to the horizontal oscillator and driver transistor when there is no drive waveform on the base terminal of the horizontal output transistor. Proceed to the driver collector terminal. Next check the collector terminal of the horizontal oscillator stage. Take voltage measurements on the collector and emitter terminals of the oscillator transistor. Higher voltage than normal on the collector and no voltage at the emitter terminal can indicate an open transistor or emitter circuit.

A no-high-voltage and no-raster symptom in a Goldstar VR230 TV was caused by a broken horizontal oscillator coil (Fig. 14-21). Often the horizontal hold control is either broken or twisted to one side, pulling the coil loose from the PC board (T701). A low resistance measurement between the emitter terminal and chassis ground will show if the coil is open.

If a waveform is found at the collector terminal of the horizontal oscillator and not at the collector of the driver transistor, suspect a leaky transistor, driver transformer, or voltage-dropping resistor. High voltage at the collector of the driver transistor can indicate an open driver transistor (Fig. 14-22). Low voltage at the collector terminal can indicate a leaky transistor or improper oscillator drive signal. R507 overheating can be caused by a leaky Q402 or poor drive voltage.

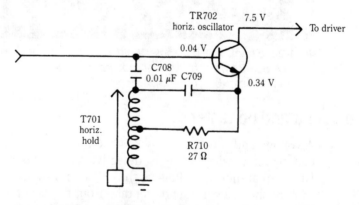

■ 14-21 *Check for a broken horizontal hold control with a no-picture, no-sound system.*

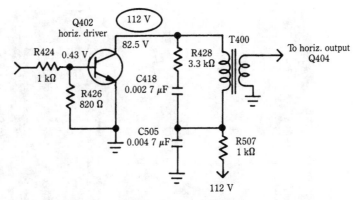

Q402
horiz. driver

112 V

82.5 V

R424   0.43 V
1 kΩ

T400

R428
3.3 kΩ

To horiz. output
Q404

C418
0.002 7 µF

R426
820 Ω

C505
0.004 7 µF

R507
1 kΩ

112 V

Note: Circled voltage was measured with an open Q402.

■ **14-22** *An open transistor (Q402) caused the absence of drive voltage on the base of transistor Q404.*

Horizontal drifting

Horizontal drifting is when the chassis operates for a few minutes and then goes into horizontal lines. The horizontal hold control might straighten up the picture for a few minutes, then the picture starts tearing in another direction. Check for defective bypass and coupling capacitors in the horizontal hold coil circuits. Spray each capacitor with coolant (Fig. 14-23). Several coats of coolant might be needed. Applying heat and coolant to critical components within the horizontal oscillator circuit can locate the defective component. Check the components shown in Fig. 14-24 as a possible cause of horizontal drifting.

589

Red-hot horizontal output transistor

Suspect a leaky horizontal output transistor, improper drive voltage, or a shorted flyback transformer with an overheated horizontal output transistor. Remove the transistor and test it out of the circuit. Now check for adequate drive voltage. Of course, the waveform might not look like it should with the transistor out of the circuit, but a 13-V to 20-V horizontal waveform should be found.

Usually an overheated horizontal output transistor will have a lower collector voltage. Check the damper diode for open conditions. Remove one lead of the horizontal deflection yoke from the circuit to determine if the yoke has shorted turns. Suspect a shorted yoke if the voltage returns. Replace the flyback transformer when there is an overheated horizontal output transistor (after all other components are checked).

■ **14-23** *Spray suspected components with coolant spray to detect the source of drifting and intermittent TV problems.*

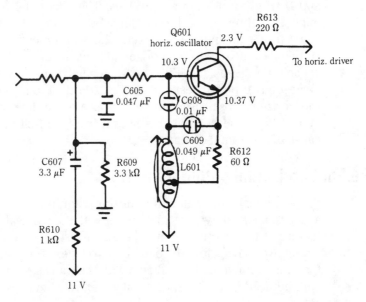

■ **14-24** *Check the circled components for drifting in the early black-and-white TV chassis.*

Checking the yoke assembly

Suspect a shorted yoke assembly when a trapezoid pattern is found on the raster. A yoke with shorted turns can lower or prevent high voltage from appearing at the anode terminal. If the high voltage increases with the red lead of the yoke removed from the flyback transformer circuit, replace the yoke assembly. You might see signs of smoke or firing within the yoke windings.

It's very difficult to locate a few shorted turns in a yoke assembly with the low-ohm scale of any meter. Each section of the yoke can be checked for open windings. Sometimes a yoke wire can break where it enters the terminal connection or at the yoke winding. Some of these connections can be repaired. One easy method to determine if the yoke has shorted turns is to let the chassis operate for 30 minutes, then shut down the chassis. Pull the yoke assembly off the tube and feel the inside of the yoke. A hot spot in the yoke assembly indicates a shorted winding.

Checking the damper and boost diodes

A damper diode might be found in the collector circuit of the horizontal output transistor or the flyback transformer circuit (Fig. 14-25). A leaky damper diode can produce a no-sound, no-picture symptom. The leaky diode can cause the B+ fuse to open. The diode should be replaced with one with a 1,200-kV rating. When the horizontal output transistor runs quite warm, check the damper diode for open conditions. Remove one terminal of the diode for accurate leakage tests, or test it when the horizontal output transistor is out of the circuit.

The boost rectifier in the black-and-white chassis supplies voltage to the picture-tube elements. The sound might be normal with no raster and a leaky boost rectifier. Sometimes only a hum is heard in

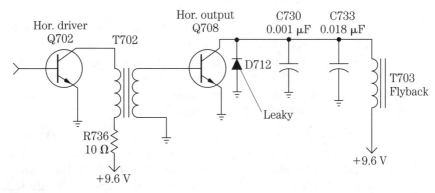

■ **14-25** *A leaky damper diode can open the secondary and main line fuses in a TV chassis.*

the speaker with a high-voltage measurement. Most black-and-white boost rectifiers can be replaced with 2.5-A or 3-A, 1-kV types.

A loud hum with no raster was found in a Sharp 3K-91. The high voltage at the CRT measured 10.5 kV. A voltage measurement on the picture-tube socket indicated no voltage at pins 6 and 7. After checking the schematic, boost diode D607 was found to be leaky (Fig. 14-26). Boost diode D607 connects to a winding on the flyback transformer (T602).

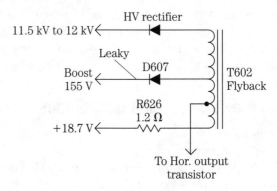

■ **14-26** *A leaky boost diode (D607) in a Sharp 3K-91 TV produced a loud hum and no raster.*

The flyback transformer

A shorted horizontal output transformer can produce an overload to the horizontal output transistor and, in turn, cause the fuse to blow. The leaky flyback transformer can keep destroying the horizontal output transistors (Fig. 14-27). Suspect a defective horizontal output transformer if the output transistor becomes warm after replacement.

Check the flyback transformer when arcing lines are found in the raster. Inspect the plastic or rubber case for arcing marks. Excessive arcing can produce an ozone odor. Attempting to patch arcing paths in a flyback transformer is useless. Suspect a defective transformer with a warm plastic body. Replace the transformer if the cost of the repair is warranted.

The defective flyback can be located with a flyback tester and sweep analyzing methods. One quick method is to remove the red yoke lead and one end of each diode from the flyback circuit. Often the drive voltage and transistors are checked before suspecting the flyback. Replace the horizontal output transistor. If the

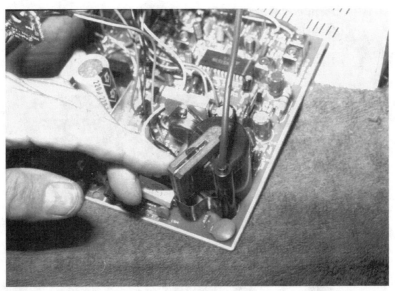

■ **14-27** *Shorted turns within the horizontal output transformer can destroy it.*

high voltage is low when the output transistor is warm, replace the defective transformer.

High-voltage rectifier

The high-voltage rectifier can be found molded inside the flyback transformer or as a "stick rectifier." A stick rectifier is made up of many layers of selenium rectifiers inside a long ceramic or fiber tube. The high-voltage rectifier rectifies the RF voltage of the horizontal output transformer. This rectifier can plug into the flyback transformer or be wired to the transformer and anode lead of the picture tube.

If the rectifier becomes leaky, low or no high voltage is measured at the CRT. You might smell a sweet odor with a shorted rectifier. The stick rectifier can show burn marks and run quite warm. Replace the high-voltage rectifier with one that has the original's part number.

Black-and-white vertical problems

Insufficient vertical and no vertical sweep are two of the most common problems found in the vertical circuits. A horizontal white line indicates no vertical sweep. Vertical rolling can result from insufficient vertical sync from the sync separator stage. Defective output transistors or capacitors in the output stage can cause vertical bouncing. Check electrolytic capacitors in the vertical circuits or the power source if the set exhibits vertical crawling (Fig. 14-28).

■ **14-28** *The vertical output transistors may look like the T-040, T-041, or T-048 transistor outlines.*

No vertical sweep—white line

Almost any component within the vertical circuits can cause a horizontal white line. Check the vertical waveform at the yoke capacitor and collector terminal of the vertical oscillator. A normal waveform at the coupling capacitor indicates a possible open capacitor or yoke winding. No waveform at the vertical oscillator transistor or IC can be caused by a leaky IC component. Each succeeding stage can be signal-traced with the scope if an oscillator waveform is found.

Check the voltage on each vertical output transistor. Lower than normal collector voltage can indicate a leaky transistor. Higher-than-normal collector voltage can indicate an open transistor or improper drive voltage. Low voltage was found at the output transistor in a Zenith 16EB12X chassis (Fig. 14-29). Because Q202 tested normal in the circuit, the vertical and horizontal module (9-100B) was suspected of causing the white horizontal line. Replacing Q202 with a GE-37 replacement solved the no-vertical-sweep problem when Q202 tested okay.

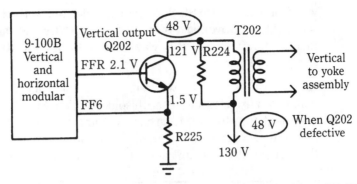

14-29 *A leaky transistor Q202 was replaced with a GE-37 universal replacement in a Zenith 16EB12X chassis.*

Insufficient vertical sweep

Go directly to the vertical output circuit with insufficient vertical sweep and normal voltages. A leaky or open vertical output transistor can produce poor vertical sweep. Replace both vertical output transistors. Check bias and base resistors while the output transistors are out of the circuit (Fig. 14-30).

Vertical foldover

Suspect the vertical output transistors or coupling capacitors for vertical foldover problems. Insufficient vertical sweep with foldover

14-30 *The vertical output circuits have small power output transistors.*

595

was found in a Sharp SK-82A. In this model, one vertical IC includes the oscillator, driver, and output circuits (IC601). Voltage measurement on pins 9 and 10 were way off (Fig. 14-31). Pin 10 had 7.5 V. It should have 13.8 V.

At first, IC601 and C608 were suspected of leakage. Pin 10 was disconnected from the circuit with solder wick, and a resistance measurement from pin 10 to ground eliminated IC601 as a potential problem. One terminal of C608 was disconnected and tested good. Replacing C607 (100 µF) solved the vertical foldover problem.

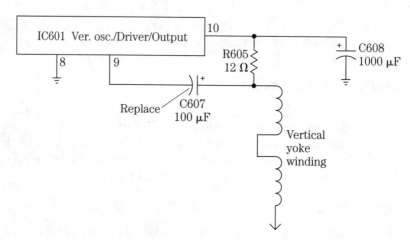

■ **14-31** *A leaky C607 (100 µF) in a Sharp SK-82A TV caused insufficient vertical sweep and foldover.*

Vertical crawling

In some TVs the raster will begin to crawl upward, or sections of the raster will have dark lines with portions of the picture moving. This is called vertical crawling. In the tube days, a leaky vertical output tube would produce vertical crawling. Although a leaky output transistor can cause some vertical crawling or foldovers, most vertical crawling in the solid-state chassis is caused by poor filtering in the power-supply circuits.

Vertical crawling was noticed in the picture of a Westinghouse V-2446-1 (Fig. 14-32). A low negative supply voltage (–8.5 V) was measured on one side of the vertical linearity control. Tracing the negative voltage back to the power supply, the voltage should be –22 V. D1 and R76 were normal. When C2 was shunted, vertical crawling was eliminated. Replacing a dried-up capacitor (C2) solved the vertical crawling problem.

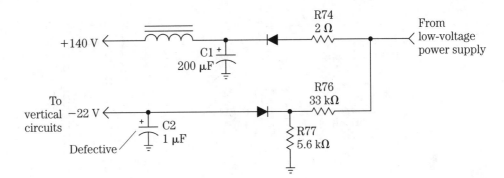

■ 14-32 *An open filter capacitor C2 caused vertical crawling in the small black-and-white TV.*

Video and CRT problems

The most common problems found in the video and picture-tube circuits are very little brightness, no control of brightness, an all-white screen, and intermittent video images. A fairly dark screen with very little brightness can be caused by a weak picture tube or improper boost voltage. Lack of control over the brightness can be caused by a defective picture tube or video output circuit. Check the video circuits if the set has a white raster without a picture. The intermittent video picture can occur in the video stages or the picture-tube circuits.

Very little brightness

First check the voltages on the picture-tube elements for improper voltages. Often the brightness cannot be raised in a fairly dark picture. The picture was fairly dark in a Sharp 3M-45 black-and-white model (Fig. 14-33). Voltage measurement at CRT pins 6 and 7 were very low. Low anode or boost voltage to the CRT circuits can cause low brightness. R726 appeared burned and had increased in resistance producing a dark picture.

Check the picture tube for poor brightness. The picture tube might be weak or have improper heater voltage. In the latest black-and-white TVs, the heater voltage is developed by a separate winding from the horizontal output transformer or directly from the B+ source in the power supply.

Very little brightness was found in a K-Mart KMB1221G. Voltages on the CRT were fairly normal, but the heater was not very bright at the end of the tube. The heater resistance had increased to 55 Ω

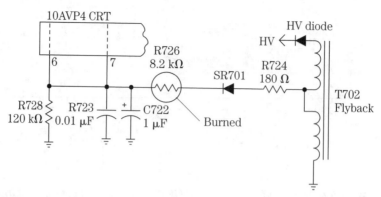

14-33 *A burned resistor (R726) in the boost circuit of a Sharp 3M-45 TV produced a very dark picture.*

in the 310EUB4 CRT (a normal black-and-white heater resistance should be under 20 Ω). The defective picture tube was replaced.

All-white screen

Suspect a defective video stage with the all-white screen. Determine if both sound and picture are missing. Check each video transistor with voltage measurements to locate the defective stage. Also, signal-tracing each video stage with the scope is very effective. Most video problems are caused by leaky transistors and ICs.

No sound and no picture with an all-white raster was noted in a K-Mart KMC0920G black-and-white TV. IC202 was suspected, with no sound at the speaker (Fig. 14-34). Correct voltage measurements indicated zero voltage at pin 4, with higher-than-normal voltage on pins 2 and 3 of IC202. Both picture and sound were restored by replacing the leaky IC202 with an ECG747 universal replacement IC.

Smeary picture

A flashing or smeary picture can result from a defective video output transistor. Take voltage measurements on the transistor. Usually low collector voltage indicates a leaky video output transistor. Suspect a wrong type of transistor replacement when the picture is smeary after replacing the video output transistor. Check the schematic and transistor manual for the correct replacement.

Poor brightness and a smeary picture was found in a Broadmoor 2612 black-and-white TV. The collector voltage was low (68.5 V). Q106 was replaced with a universal replacement. Although the brightness was restored, the picture was still smeary (Fig.

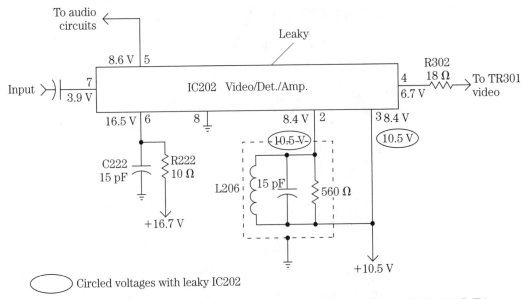

■ 14-34 *Low voltages on pin 4 of IC202 indicated a leaky IC in a K-Mart KMC0920G TV.*

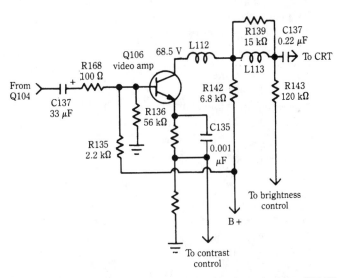

■ 14-35 *A leaky transistor Q106 was replaced with a GE-27 universal replacement because it was producing a smeary picture with poor brightness.*

14-35). Because original transistors are difficult to obtain in these models, universal transistors must be used. The picture returned to normal when Q106 was replaced with another GE-27 transistor.

Video and CRT problems

Genuine replacement part mail-order firms

Audio Video Parts, Inc.
1071 South LaBrea Ave.
Los Angeles, CA 90019
213/933-8141

B-B & W Electronics
2137 S. Euclid Ave.
Berwyn, IL 80402-0710
312/242-1533

C & S Electronics
1924 Silver Star Rd.
Orlando, FL 32804
407/299-4300

E & K Parts
2115 Westwood Blvd.
Los Angeles, CA 90025
312/475-6848

E.A. Ross & Co., Inc.
184 Griffin St.
Fall River, MA 02724
800/828-4901

Eiger Electronics
91 Toledo St.
Farmingdale, NY 11735
800/835-8316

ESP Electronics
2901 E. Washington St.
Indianapolis, IN 46201
800/382-9976

Howard Electronics
4573 S. Archer St.
Chicago, IL 60632
312/254-1777

Joseph Electronics
8830 N. Milwaukee Ave., Dept. K
Niles, IL 60648
800/323-5925

Mill Electronics, Inc.
2026 McDonald Ave.
Brooklyn, NY 11223
800/346-8994

Pacific Coast Parts Distributors, Inc.
15024 Staff Ct.
Gardena, CA 90248
800/421-5080

P.I. Burks Co.
Broadcast Electronics
824 S. 7th St.
Louisville, KY 40203
800/821-4893

Service Electronic Supply Co.
1046 New Circle Rd. NE
Lexington, KY 40505
800/432-0817

Standard Electronics, Inc.
215 John Glenn Dr.
Amherst, NY 14150
800/333/1519

Tee-Vee Supply Co., Inc.
407R Mystic Ave., Unit 14
Medford, MA 02155
800/255-5854

Union Electronic Distributors
16012 S. Cottage Grove
South Holland, IL 60473
312/468-7300

United Electronics
3860 Tenth Ave.
New York, NY 10034

International distributors

Consolidated Electronics
705 Watervliet Ave.
Dayton, OH 45420-2599
800/543-3568

Fox International
752 S. Sherman
Richardson, TX 75081
214/231-1826

Fox International
23600 Aurora Rd.
Bedford Heights, OH 44146

Five actual black-and-white case histories

Here are five different black-and-white case histories that might help you to solve any comparable problem within the various sections of a TV chassis.

Improper height and poor sync

Not enough vertical sweep and poor sync was found in a Sylvania BWE150S black-and-white portable. Critical voltage measurements in the vertical and sync circuit revealed low supply voltages. The 19.9-V sync source was missing, with a low vertical source to the output transistors. D474 was found burned and shorted with overheated R470 resistor (Fig. 14-36). Both D474 and R470 (390 ohm) were replaced to restore normal reception.

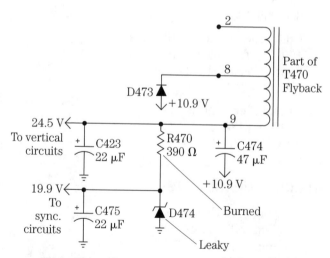

■ **14-36** *Improper height and poor sync was caused by a leaky IC202 in a Sylvania BWE150S portable.*

Keeps blowing 1-A fuse

The 1-A fuse would not hold in an RCA KCS2010 chassis after replacement (Fig. 14-37). Both the 1-A fuse and 13-Ω resistor were replaced when the black-and-white TV came in for repair. Lightning had struck the power line, taking out R101, and D101.

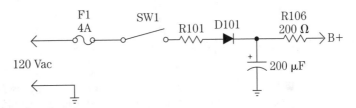

■ **14-37** *Check the line fuse F1, switch SW1, resistor R101, and diode D101 when lightning strikes the black-and-white portable.*

No high voltage, no sound

The scope was used to trace the horizontal waveform in a Goldstar VR230 black-and-white TV. No waveform was found on the base terminal of TR704 or the collector terminal of the horizontal drive transistor (TR703). The horizontal oscillator waveform was fairly normal (Fig. 14-38).

The collector voltage of TR704 was higher than normal, with very little voltage at the collector terminal of TR703 (0.05 V). The collector terminal of TR703 feeds through the driver transformer and is at ground potential. Under normal operation, the collector voltage measures only 0.17 V. A resistance measurement of the primary winding at the driver transformer was only 0.2 Ω and should be 1.98 Ω. Replacing T701 with an original replacement component solved the no-raster and no-sound problem.

Good sound, no raster

With normal sound in a Goldstar black-and-white portable and no raster, the defective component must be located in the output circuits. Often you can determine that the horizontal output stages are normal because the sound is good. A voltage measurement at the CRT indicated poor boost voltage (17 V). CR302 checked out (Fig. 14-39). Shunting another 1-μF capacitor across C302 brought the picture and boost voltage (107 V) back again.

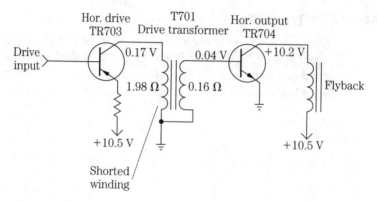

■ 14-38 *A shorted primary winding of T701 caused a no-high-voltage/no-sound problem in a Goldstar VR230 portable TV.*

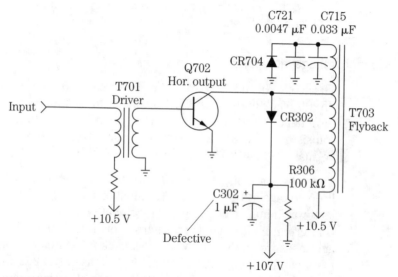

■ 14-39 *A leaky capacitor C302 (1 μF) in the boost circuit caused the no-raster symptom.*

AGC or tuner

In an Emerson B-120A (Fig. 14-40), the tuner-subber was connected to the IF input cable to determine if the tuner was bad or if problems were in the AGC circuits. The picture was very snowy with no picture and very little sound. No video signal was present at pin 3 of IC201 and voltage measurements were low on pins 5, 7, and 12. All components were checked in the AGC circuit and appeared normal. The defective IC201 (UPC1366C) was replaced with an ECG1522 universal replacement.

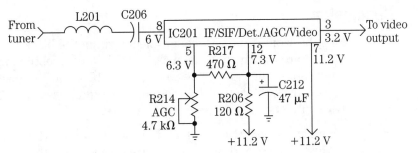

■ 14-40 *A defective IC201 caused a very snowy picture in an Emerson B-120A chassis.*

Unusual black-and-white picture

Only a faint outline of figures could be seen in a Goldstar VR-230. White light could be seen around the outside edge of the picture tube. Otherwise the picture was almost completely dark, even with the brightness and contrast controls wide open.

The high voltage was normal at the anode connection of the CRT. Voltage on pins 6 and 7 of the picture tube was about 5 V. Because the boost voltage is developed in the flyback circuits, CR302 and R308 were checked. Both were normal (Fig. 14-41). Replacing electrolytic capacitor C302 restored the picture and raster.

605

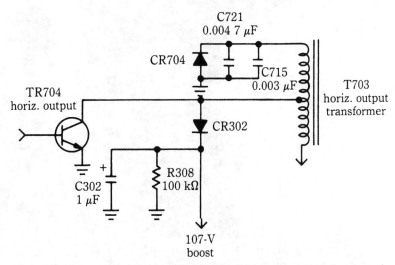

■ 14-41 *An open capacitor C302 caused poor boost voltage and a faint picture in a Goldstar VR-230 chassis.*

Conclusion

Most horizontal and vertical problems within the black-and-white chassis are similar to those in a color TV. Of course, the high voltage is somewhat lower (between 8 and 15 kV in the black-and-white chassis). Picture tube circuits are much easier to service with only a single gun assembly (Fig. 14-42).

There are fewer IF stages found in the black-and-white TV. Many of the IF, video, and AGC circuits are developed with one IC component. Sometimes replacing one IC component can solve many other picture problems. With less components, the black-and-white chassis is relatively small compared to the color set. Most black-and-white TVs still have the old standard wafer-type tuner to select the station.

Remember, today you might find surface-mounted components and LCD TV screens in tiny pocket TV receivers. Testing and replacing the various components is a whole new ball game, so to speak. Always attempt to purchase exact replacement parts from the manufacturer or parts supplier. Small universal surface-mounted resistors and capacitors can be obtained from local or large mail-order electronics parts stores.

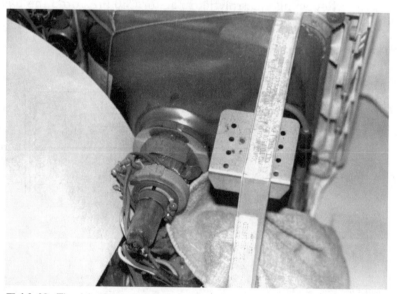

■ **14-42** *The black-and-white picture tube has only one gun assembly and very few parts on the neck of the tube, compared to the tube in a color chassis.*

Index

607

609

610

613

614

615

618

Z

619

About the author

Homer L. Davidson is an electronics technician who owned and operated a successful small appliance repair business for 38 years before retiring to write full-time. He is a regular contributor to Electronic Servicing and Technology magazine and the author of 18 TAB/McGraw-Hill books.